Electron Microscopy in Biology

AF702960

The Practical Approach Series

SERIES EDITORS

D. RICKWOOD
Department of Biology, University of Essex
Wivenhoe Park, Colchester, Essex CO4 3SQ, UK

B. D. HAMES
Department of Biochemistry and Molecular Biology, University of Leeds
Leeds LS2 9JT, UK

Affinity Chromatography
Anaerobic Microbiology
Animal Cell Culture
Animal Virus Pathogenesis
Antibodies I and II
Biochemical Toxicology
Biological Membranes
Biosensors
Carbohydrate Analysis
Cell Growth and Division
Cellular Calcium
Cellular Neurobiology
Centrifugation (2nd edition)
Clinical Immunology
Computers in Microbiology
Crystallization of Proteins and Nucleic Acids
Cytokines
Directed Mutagenesis
DNA Cloning I, II, and III
Drosophila
Electron Microscopy in Biology
Electron Microscopy in Molecular Biology
Essential Molecular Biology I and II
Fermentation
Flow Cytometry
Gel Electrophoresis of Nucleic Acids (2nd edition)
Gel Electrophoresis of Proteins (2nd edition)
Genome Analysis
HPLC of Small Molecules
HPLC of Macromolecules
Human Cytogenetics
Human Genetic Diseases
Immobilised Cells and Enzymes
Iodinated Density Gradient Media
Light Microscopy in Biology

Liposomes
Lymphocytes
Lymphokines and Interferons
Mammalian Cell Biotechnology
Mammalian Development
Medical Bacteriology
Medical Mycology
Microcomputers in Biology
Microcomputers in Physiology
Mitochondria
Molecular Neurobiology
Molecular Plant Pathology I and II
Mutagenicity Testing
Neurochemistry
Nucleic Acid and Protein Sequence Analysis
Nucleic Acids Hybridisation
Nucleic Acids Sequencing
Oligonucleotide Synthesis
PCR
Peptide Hormone Action
Peptide Hormone Secretion
Photosynthesis: Energy Transduction
Plant Cell Culture
Plant Molecular Biology
Plasmids
Post-implantation Mammalian Development
Prostaglandins and Related Substances
Protein Architecture
Protein Function
Protein Purification Applications
Protein Purification Methods
Protein Sequencing
Protein Structure
Proteolytic Enzymes
Radioisotopes in Biology
Receptor Biochemistry
Receptor–Effector Coupling
Ribosomes and Protein Synthesis
Solid Phase Peptide Synthesis
Spectrophotometry and Spectrofluorimetry
Steroid Hormones
Teratocarcinomas and Embryonic Stem Cells
Transcription and Translation
Virology
Yeast

Electron Microscopy in Biology

A Practical Approach

Edited by
ROBIN HARRIS
Institute of Cell and Tumor Biology
German Cancer Research Center
Heidelberg

IRL PRESS
at
OXFORD UNIVERSITY PRESS
Oxford New York Tokyo

Oxford University Press,
Walton Street, Oxford OX2 6DP

Oxford is a trade mark of Oxford University Press

Published in the United States
by Oxford University Press, New York

© Oxford University Press 1991

All rights reserved. No part of this publication may be reproduced, stored in a retrieval system, or transmitted, in any form or by any means, electronic, mechanical, photocopying, recording, or otherwise, without the prior permission of Oxford University Press.

This book is sold subject to the condition that it shall not, by way of trade or otherwise, be lent, re-sold, hired out, or otherwise circulated without the publisher's prior consent in any form of binding or cover other than that in which it is published and without a similar condition including this condition being imposed on the subsequent purchaser.

A catalogue record of this book is available from the
British Library

Library of Congress Cataloging in Publication Data
Electron microscopy in biology : a practical
approach / edited by Robin Harris.
p. cm. — (Practical approach series)
Includes bioliographical references and index.
1. Electron microscopy—Technique. 2. Transmission electron
microscopes. I. Harris, James E. II. Series.
[DNLM: 1. Cytological Techniques. 2. Histological Techniques.
3. Microscopy, Electron. QH212.E4 E3887]
QH212.E4E3983 1991 578'.45—dc20 91–2476

ISBN 0–19–963219–7
ISBN 0–19–963215–4 (pbk.)

Set by Footnote Graphics, Warminster, Wilts
Printed by Information Press Ltd, Oxford, England

Preface

This techniques book *Electron Microscopy in Biology* was conceived following the publication of the more specialist book within the *Practical Approach* series, entitled *Electron Microscopy in Molecular Biology*. An attempt is made here to cover some of the topics omitted or not fully covered in this previous book and to provide useful up-to-date methodological guidelines for those wishing to utilize electron microscopy as one of their research tools.

No attempt is made here to compete in terms of depth of background, experimental details, biological applications, and referencing with the other established books and series dealing with EM techniques. Indeed, several of the chapters within this book have been covered elsewhere as single or more books. Rather, each chapter presents concise accounts of the current methodology. Authors, who are specialists within their field of study, have selected what they consider to be relevant and important techniques for inclusion, most of which are presented in the 'protocol' format, supplemented by technical discussion. This being the case, there is a strong personal emphasis and flavour to almost all the contributions, which I look upon as a point of strength, within the limited framework of this short book. An attempt has been made to seek authors based within universities and research institutes throughout the world. This has been achieved to some considerable extent, although an apology is perhaps due for the predominance of UK-based contributors. The freedom given to authors with regard to chapter content has led to a small amount of technical repetition between chapters. In general this is of value, since slightly varying technical emphasis is likely to prove useful. Thus, in the main, each chapter stands as a fairly complete subject area in its own right, without the need for extensive cross-referencing to other parts of the book.

The book commences with the two fundamentally important topics of tissue fixation, covered by David Hopwood and Gordon Milne, and embedding and thin sectioning, covered by Michael Smith and Simon Croft. These two chapters represent the core strength of classical transmission electron microscopy (TEM), leading to the production of 'thin plastic/resin sections'. Many of the procedures subsequently dealt with in later chapters extend from the extensive groundwork of investigators in these areas over the past 40 years. In the third chapter, low temperature techniques are introduced as a main theme for the first time, although some mention of low temperature embedding has already appeared in the second chapter. Norbert Roos deals here with freeze-substitution and other low temperature embedding methods in an extremely competent manner. Then comes a chapter from Alex D. Hyatt that deals very broadly with immunogold labelling techniques. It is easy to forget that immunolabelling techniques can be applied to freeze-fractured and negatively stained materials (and indeed to metal shadowed preparations) as well as to sectioned specimens. Alex Hyatt appropriately balances the subject and includes his own unique approach making use of negative staining. In the fifth chapter Gérard Morel covers in an extremely thorough manner the topic of electron microscopic autoradiography. Of equally high standard is the chapter on electron microscopical enzyme histochemistry from Cornelis Van Noorden and Caesar Hulstaert, who select for presentation the principal cytochemical techniques from what is a very large field of study. In his chapter on high resolution shadowing, Henry Slayter, places emphasis upon macromolecular studies by darkfield electron microscopy. In reality, the advanced shadowing techniques that Henry Slayter presents are likely to have a much broader usage in the future, particularly in relation to isolated organelles and membranes, and also for the study of the surface of intact cells. The field of freeze-fracture has expanded steadily and this is refected in the large number of technical procedures presented by Seizo Fujikawa. Negative staining, to some extent a neglected area in the existing texts on electron microscopical techniques, is dealt with by Bob Horne and myself. We have not attempted to cover every aspect of this subject, but have tried to give a thorough background upon which we place some of the existing negative staining techniques of tried and tested success, together with a new technique that has exciting cellular possibilities. Advancement in the field of cryoelectron microscopy of unstained frozen-hydrated specimens continues apace. This subject is handled in a brief yet professionally thorough manner by Murray Stewart in his chapter on the TEM of vitrified biological macromolecular assemblies. Emphasis upon cryotechniques continues in the chapter by Alice Warley and Brij Gupta, who present a detailed account of the procedures for performing quantitative biological X-ray microanalysis. In the final chapter of the book, Nigel James handles the difficult topic of three-dimensional electron microscopy of tissues and cells, from a mathematical yet practical point of view. Dr James indicates that the 3D reconstruction techniques available can be performed in any laboratory possessing a suitable microcomputer and prepared to purchase and utilize the necessary software.

In the space made available by the publisher not all electron microscopical techniques could be included, even within the subject areas of the chapter titles selected for inclusion in the book. Nevertheless, it is hoped that the book will prove to be useful for those performing or wishing to perform biological or biomedical experimentation using electron microscopy. All chapters are thoroughly referenced and provide an entry point to the relevant literature.

October 1990 R.H.

Contents

Contributors

SIMON CROFT
Electron Microscope and Histopathology Unit, London School of Hygiene and Tropical Medicine, Keppel Street, London WC1E 7HT, UK

SEIZO FUJIKAWA
Institute of Low Temperature Science, Hokkaido University, Sapporo, 060 Japan

BRIJ L. GUPTA
Department of Zoology, Cambridge University, Downing Street, Cambridge CB2 3EJ, UK

JAMES ROBINSON HARRIS
Institute of Cell and Tumor Biology, German Cancer Research Center, Im Neuenheimer Feld 280, D-6900 Heidelberg, Germany

DAVID HOPWOOD
Department of Pathology, Ninewells Hospital and Medical School, University of Dundee, Dundee DD1 9SY, UK

ROBERT W. HORNE
School of Biological Sciences, University of East Anglia, Norwich, Norfolk NR4 7TJ, UK

CAESAR E. HULSTAERT
Department of Histology and Cell Biology, University of Groningen, Oostersingel 69–1, 9713 EZ Groningen, The Netherlands

ALEX D. HYATT
Australian Animal Health Laboratory, Division of Animal Health, Institute of Animal Production and Processing, Geelong, PO Bag 24, Victoria, 3220, Australia

NIGEL T. JAMES
Department of Biomedical Science, The University of Sheffield, Sheffield S10 2YN, UK

GORDON MILNE
Department of Pathology, Ninewells Hospital and Medical School, University of Dundee, Dundee SS1 9SY, UK

GÉRARD MOREL
CNRS–URA 559, Laboratory of Histology and Embryology, Faculty of Medicine Lyon-Sud, BP 12–69600 Oullins, France

NORBET ROOS
Electron Microscopical Unit for Biological Sciences, Department of Biology, University of Oslo, P.b. 1062 Blindern, N–0316, Oslo 3, Norway

HENRY S. SLAYTER
Laboratories of Electron Microscopy and Structural Molecular Biology, Dana-Farber Cancer Institute, 44 Binney Street, Boston, MA 02115, USA

MICHAEL SMITH
Safety Adviser, London School of Hygiene and Tropical Medicine, Keppel Street, London WC1E 7HT, UK

MURRAY STEWART
MRC Laboratory of Molecular Biology, Hills Road, Cambridge CB2 2QH, UK

CORNELIS J. F. VAN NOORDEN
Laboratory of Cell Biology and Histology, University of Amsterdam, Academic Medical Centre, Meibergdeef 15, 1105 AZ Amsterdam, The Netherlands

ALICE WARLEY
Division of Biochemistry, Block 9, United Medical and Dental School of Guy's and St. Thomas's Hospitals, Lambeth Palace Road, London SE1 7EH, UK

Abbreviations

ADP	adenosine diphosphate
ATP	adenosine triphosphate
BPST	2-(2-benzothiazolyl)-3-(4-phthalhydrazidyl)-5-styryl-tetrazolium
BSA	bovine serum albumin
CBZ	benzyloxycarbonyl
CG	colloidal gold
Ci	Curie
Con A	concanavalin A
CRT	cathode ray tube
CTEM	conventional transmission electron microscope
CTF	contrast transfer function
2D	2-dimensional
3D	3-dimensional
DAB	diaminobenzidine
DMP	dimethoxy-methane
DMSO	dimethyl sulphoxide
d.p.m.	disintegrations per minute
EDS	energy dispersive spectrometer
EDTA	ethylene diaminetetraacetate
EDX	energy dispersive X-ray
EG	ethylene glycerol
EGTA	ethyleneglycobis (β-aminoethyl) ether *N*,*N*,*N′*,*N*' tetra acetic acid
EM	electron microscope/microscopy
EMT	electron microscope tomography
GACH	glutaraldehyde/carbohydrazide
GMA	glyco methacrylate
eV	electronvolt
HD	half distance
HES	hydroxyethyl starch
HPL	*Helix pomatia* lectin
HR	half radius
HPL–CG	*Helix pomatia* lectin-colloidal gold
HVEM	high voltage electron microscopy
Ig–CG	immunoglobulin coated colloidal gold
IEM	immunoelectron microscopy
IGS	immunogold staining
ITP	inosine triphosphate
IVEM	intermediate voltage electron microscopy
K	kelvin
kV	kilovolts
LN_2	liquid nitrogen
L_V	length per unit volume of tissue

MNA	4-methoxy-2-naphthylamine
NADH	nicotinamide adenine dinucleotide (reduced form)
NADP	nicotinamide adenine dinucleotide phosphate
NADPH	nicotinamide adenine dinucleotide phosphate (reduced form)
NCIEM	negative contrast immunoelectron microscopy
nitro BT	nitro blue tetrazolium
N_V	numerical density
Pa	pascal
PA–CG	protein-A-coated colloidal gold
PBS	phosphate buffered saline
PEG	polyethylene glycol
PLT	progressive lowering temperatures
PMSF	phenyl methyl sulphonyl fluoride
PVP	polyvinyl pyrrolidone
RBC	red blood cell
RCA I	*Ricinus communis* I
RCA I–CG	*Ricinus communis* I–colloidal gold
RT	room temperature
SEM	scanning electron microscope
SNR	signal-to-noise ratio
SIEM	scanning immunoelectron microscopy
STEM	scanning transmission electron microscope
S_V	surface density
TCH	thiocarbohydrazide
TEM	transmission electron microscope
UA	uranyl acetate
UDP	uridine diphosphate
UDT	uridine triphosphate
VDU	visual display unit
V_V	volume fraction
WDS	wavelength dispersive spectrometer
WGA	wheat germ agglutinin
ZAF	atomic number (Z), absorption of X-rays (A) and secondary fluorescence (F)

1

Fixation

DAVID HOPWOOD and GORDON MILNE

1. Introduction

The techniques chosen to fix tissues should enhance the sort of microscopical information which is eventually produced. Ideally, the tissue should be retained in as near life-life condition as possible, but the better the morphological preservation, the less the biological activity and vice versa. Some compromise has to be reached. Fixation should also prepare tissues to withstand the subsequent rigours of the preparative techniques (Chapter 2) and finally irradiation by the electron beam (1, 2).

Conventionally, tissues are fixed chemically and a number of chemical primary fixatives are in wide use for electron microscopy. Those most commonly used are the dialdehyde glutaraldehyde, (*para*)-formaldehyde, and osmium tetroxide. Others which may have a role include picric acid and acrolein. These fixatives are not necessarily interchangeable. Different results can be obtained in freeze-fracturing by using different fixatives (3). It is important to choose the right buffer for the fixative to be used. Some buffers will react with the fixative, thereby reducing the effective concentration of both reagents, for example, amine-containing buffers and aldehydes. Other buffers will extract material from the tissues. The most commonly used buffers for electron microscopy are sodium cacodylate and phosphate.

The temperature of fixation for electron microscopy is by convention 4°C. This will affect the rate of penetration of the fixative into the tissue, its rate of reaction with the tissues and the rate at which substances are extracted from the tissue. It is possible to use other temperatures successfully.

The rate at which fixatives diffuse into tissues is governed by

$$d = k\sqrt{t}$$

where d is the depth of penetration at positive t and k, the constant of diffusibility, a characteristic of each fixative. The values for the commonly used fixatives are given in the table below, as measured by standard techniques. The slow rate of penetration is the rationale behind small tissue blocks for electron microscopy. Even so, different layers of cells will be fixed at different rates, at times, producing a noticeable artefact.

Fixative	Concentration	*k*	Temperature
Osmium tetroxide	1%	0.3–0.6	20
Formaldehyde	4%	0.8	20
Glutaraldehyde	3%	0.3	4

In histopathology laboratories, one now frequently finds that surgical specimens received in formaldehyde need to be processed for electron microscopy to achieve a diagnosis. The specimens are sub-optimally fixed in regard to the fixative, the agonal period and the size of the specimen. None the less, it is still worthwhile to process material on into glutaraldehyde for electron microscopy, preferably taking blocks from a face of the tissue which has been exposed to the formaldehyde since the specimen was obtained. The appearances may show features of damage or extraction of cytoplasmic matrix. If specific features are being sought, such as melanosomes or neurosecretory granules, then the diagnosis can usually be made. No general fixative which is ideal for the needs of routine histopathology in District General Hospital immunocytochemistry and electron microscopy has yet been devised, but we have found that formaldehyde and phenol formaldehyde work reasonably well (4, 5). For ideal fixation with human material, communication with the clinicians is necessary to develop a mutual understanding of each other's difficulties.

For special purposes, it is possible to stabilize/fix tissue for electron microscopy by non-chemical means using microwaves (6). This technique is discussed in greater detail in Section 3.

Many of the general methods apply to plant material as well as animal tissues. Here there are problems of diffusion related to the thicker cell walls encountered in botanical specimens. The preservation of marine animals may prove difficult due to osmotic problems and great metabolic differences between species. A good starting point is to use a primary fixative in sea-water and to give the specimen prolonged washing between steps.

Tissues are usually fixed by one of two methods. The simpler is immersion. Blocks of tissue about 1 mm^3 are cut using a fresh razor blade, cutting on to a sheet of dental wax. The blocks of tissue are simply immersed on the fixative for whatever time may be required. There are problems of diffusion of the fixative into the tissue and gradients are set up. To try to overcome these difficulties, perfusion fixation has been used. With animal tissue, a cannula can be placed in the left ventricle and a cut made in the inferior vena cava. The vasculature is first washed out with heparinized saline to remove the blood. Formaldehyde or glutaraldehyde is then run in. This gives much more even fixation. It may also be applied to human tissue, if the artery to the tissue of interest can be cannulated.

2. Standard fixative methods

The following are the basic fixatives and buffer solutions used in electron microscopy (Sections 2.1.1–2.1.6). A fume cupboard should be used to prepare fixatives.

2.1 Glutaraldehyde fixation

(a) Fix blocks of tissue not exceeding 1 mm^3 in 3% glutaraldehyde in 0.1 M phosphate or cacodylate buffer pH 7.3 for at least 4 h at 4°C.

(b) Generally, mix the reagents together in the following volumes/100 ml:
- 25% glutaraldehyde (stock) 12 ml
- 0.2 M buffer 50 ml
- distilled water 38 ml

(c) Wash tissue in 0.1 M holding buffer (Section 2.1.6) overnight.

(d) Post-fix tissue in 1% buffered osmium tetroxide for (Section 2.1.3) 1 h at 4 °C.

2.1.1 Preparation of 3% glutaraldehyde fixative in 0.1 M phosphate buffer

- 0.2 M phosphate buffer pH 7.3–25 ml
- 25% glutaraldehyde 6 ml
- distilled water 19 ml

2.1.2 Preparation of 3% glutaraldehyde fixative in 0.1 M sodium cacodylate buffer

- 0.2 M sodium cacodylate 25 ml
- 25% glutaraldehyde 6 ml

(a) Adjust pH to 7.3 with 1 M HCl

(b) Dilute to 50 ml with distilled water.

2.1.3 Preparation of 1% buffered osmium tetroxide

(a) Prepare stock solution of 2% osmium tetroxide by breaking ampoules containing 1 g of osmium tetroxide and dropping into a clean glass brown bottle to which is added 50 ml distilled water.

(b) Allow at least 24 h for the osmium tetroxide to dissolve.

(c) Store in a fume cupboard.

(d) To prepare working solution:

Take 5 ml of the 0.2 M stock buffer solution pH 7.3 used for preparing the primary fixative (see Sections 2.1.4 and 2.1.5) add 5 ml of stock osmium tetroxide solution

2.1.4 Preparation of phosphate buffer

- 0.2 M dibasic sodium phosphate
- Na_2HPO_4 28.39 g
- distilled water 1000 ml
- 0.2 M monobasic sodium phosphate
- NaH_2PO_4 $2H_2O$ 31.21 g
- distilled water 1000 ml

(a) Prepare 0.2 M phosphate buffer by mixing *x* ml of 0.2 M dibasic sodium phosphate with *y* ml of 0.2 M monobasic sodium phosphate.

(b) Store at 4°C.

pH (at 25°C)	*x* ml	*y* ml
7.2	36.0	14.0
7.3	39.0	11.0
7.4	40.5	9.0

2.1.5 Preparation of 0.2 M cacodylate buffer

(a) 4.28 g sodium cacodylate diluted to 100 ml with distilled water

(b) Store at 4°C

2.1.6 Rinsing buffer/holding buffer

(a) 0.2 M sodium cacodylate, 50 ml

(b) Adjust to required pH with 0.1 N HCl.

(c) Add 4.2 g sucrose and dilute to 100 ml with distilled water.

3. Microwaves

Microwave ovens can be used to deliver heat in a controlled manner uniformly throughout the tissue (6). The heating is produced by the rapid movement and relaxation of polar molecules and groups throughout the tissue at a frequency of 2.45 GHz. The effective penetration of microwaves into tissues is limited to about 2 cm from the surface. Specimens should be thin slices, and the container raised from the bottom of the oven. The rate of heating of tissues is controlled by the following:

- power setting of the oven;
- the nature of the tissue;
- the position of the tissue in the oven;
- size of ballast in the oven.

Microwaves contrast with conventional heating, which produces a temperature gradient across the tissue.

Microwaves can be used for:

- tissue stabilization
- speeding up of tissue processing and staining, including washing in immunohistochemistry
- rapidly stopping cellular processes throughout the tissue

For microwaving tissues (*Protocols 1* and *2*, and Section 3.1.1) items are required:

- simple domestic microwave oven
- rapid-reading thermometer
- incubation solutions—phosphate buffered saline or Karnovsky fixative; see Section 4.1.2
- beaker 500 ml
- marker pen
- stop-watch
- Lego pieces to make a table for the bottom of the oven or a Perspex table

---

Protocol 1. Establishing microwave heating curves

Proceed as follows:

1. Place beaker with say 200 ml fluid on the table on the bottom of the oven.
2. Mark their positions with the pen to give repeatable relocation.
3. Set power output on oven to maximum.
4. Heat for 15 seconds.
5. Measure temperature rapidly.
6. Repeat heating process to 65 °C.
7. Draw heating curve.
8. Repeat for some other oven power outputs.

This will allow the heating of the specimens to be known. The most useful temperature range for mammalian tissue is 45–55 °C.

Protocol 2. Microwave fixation/stabilization of tissues

1. Place tissue in labelled cassette if necessary in 200 ml chosen fluid.
2. Heat specimen to chosen temperature.
3. Check temperature with thermometer.

This tissue is in reality stabilized only and not fixed. It may be stored at 4 °C for a period. It must be post-fixed in osmium tetroxide (see Section 2.1.3) before final processing for TEM and SEM.

3.1 Temperature control

The temperature in microwave ovens can also be assessed using Giemsa-stained agar gels. These change colour from purple to blue at 45–55 °C (see Section 3.1.1).

3.1.1 Control of temperature with microwaves: Giemsa-stained suppositories for temperature control of microwaves (after Login and Dvorak, ref. 7)

- 0.9% sodium chloride 100 ml
- Agar 29 ml

(a) Warm to make solution

(b) Add Giemsa stain 0.5 ml

(c) Pour in suppository or similar moulds about 10 × 5 × 5 mm.

(d) Cool to room temperature (RT).

Colour change on heating: purple to blue at 45–55 °C.

4. Additives: mixtures

No one fixative is ideal for all problems. To try to overcome this difficulty, various mixtures of additives and fixatives have been devised in the hope that the shortcomings of one reagent may be overcome by the mixture.

Glutaraldehyde and osmium tetroxide are two of the longest established fixatives for electron microscopy. Mixtures of these two substances have been used successfully to fix some difficult specimens. One feature of this mixture is that the two components react with each other, thereby lessening the active concentration of each other. This mixture has also been recommended for marine invertebrates (8).

Another reagent which has been found to be a useful additive for fixation is tannic acid (9, 10). This will react with lipid and proteins, holding them within tissues. It also can demonstrate damaged cells, showing them to be particularly electron-dense (11).

4.1 Fixatives and additives

4.1.1 Tannic acid–3% glutaraldehyde fixative

- 25% glutaraldehyde 6 ml
- 0.2 sodium cacodylate 25 ml
- tannic acid 0.5 g

(a) Mix the three above reagents together and adjust pH to 7.2 with 2 N HCl.

(b) Make final volume 50 ml with distilled water.

(c) Fix for at least 6 h before post-osmication (see Section 2.1.3).

4.1.2 Paraformaldehyde–glutaraldehyde fixative (Karnovsky)

Formaldehyde penetrates much more rapidly than glutaraldehyde and is thought to stabilize structures which are subsequently fixed more permanently by glutaraldehyde. Primary fixation in this mixture is followed by secondary fixation in osmium tetroxide.

Preparation of fixative:

Dissolve 2 g of paraformaldehyde in 20 ml distilled water by heating to 65 °C in a fume cupboard and by adding one or two drops of 1.0 N sodium hydroxide until the solution clears. When the solution is cool, prepare the fixative as follows:

- 10% paraformaldehyde in water — 20 ml
- 0.2 M phosphate or cacodylate buffer at required pH — 50 ml
- 25% stock glutaraldehyde — 10 ml
- distilled water — 20 ml

4.1.3 Picric acid formaldehyde (12)

This fixative penetrates rapidly and gives good preservation. It is very stable, not light-sensitive and has a shelf-life of more than one year at RT. It has been recommended for the fixation of renal biopsies.

Preparation of fixative:

To 150 ml of saturated picric acid in distilled water add 20 g of paraformaldehyde and heat to 65 °C in a fume cupboard. Add several drops of 1 N sodium hydroxide to clear the solution. Filter and allow to cool. Make up to 1000 ml with phosphate buffer pH 7.3. Primary fixation in this mixture should be followed by secondary fixation in osmium tetroxide.

4.1.4 Periodate–lysine–paraformaldehyde (13)

This fixative was designed specifically for immunoelectronmicroscopy to stabilize antigens *in situ* without destroying their antigenicity, although the mechanism of fixation may not be the one proposed (14).

Preparation of fixative:

(a) Lysine phosphate buffer:
To 0.2 M lysine hydrochloride in distilled water, add 0.2 M dibasic sodium phosphate until the pH is 7.4 then dilute to 0.1 M with 0.1 M phosphate buffer pH 7.4

(b) Paraformaldehyde:
Dissolve 40 g paraformaldehyde in 100 ml distilled water. Heat to 65 °C and add a few drops of 40% sodium hydroxide until the solution becomes clear, then add 5.4 g of glucose.

Just before use add 3 parts of (a) to 1 part of (b) and add solid sodium metaperiodate to bring its concentration to 0.01 M.
Fix small pieces of tissue for three days at 4°C. Subsequently, wash in phosphate buffered sucrose.

4.1.5 Osmium tetroxide–potassium ferricyanide

Karnovsky (15) reported that an osmium tetroxide–potassium ferricyanide fixative greatly increased the contrast of the outer surfaces of membranes. This fixative has also been used (16) for the selective staining of glycogen. It is used as a secondary fixative, after primary fixation in glutaraldehyde.

Preparation of fixative:

To 1% osmium tetroxide in 0.1 M cacodylate or phosphate buffer pH 7.4 add potassium ferricyanide to give a 0.05 M solution.
Secondary fixation in this solution should be carried out for 2 to 24 h at 4°C.

There are problems with demonstrating carbohydrates and mucosubstances at ultrastructural level. Molecules such as glycosaminoglycans and cell surface proteins collapse. A number of additives, such as Cuprolinic Blue, have been found to be useful as agents with varying specificity (17).

Non-specific preservation and demonstration of cell surface glycocalyx can be achieved with Ruthenium Red (18).

Protocol 3. Fixation with Ruthenium Red

1. Fix for 4 h at RT in
   - Ruthenium Red 1 mg/ml
   - glutaraldehyde 3%
   - 0.2 M cacodylate buffer pH 7.3.
2. Wash 3 × 10 min in holding buffer.
3. Secondary fixation in buffered 1% osmium tetroxide (see Section 2.1.3) containing 1 mg/ml of Ruthenium Red for 2 h at RT.

Protocol 4. Cuprolinic Blue (17)

1. Fix fresh tissue in 0.25 M sodium acetate buffer pH 5.6 containing 2.5% Cuprolinic Blue, 0.2% magnesium chloride, and 3% glutaraldehyde. Fix overnight at RT.
2. Wash 3 × 10 min in fixative solution without Cuprolinic Blue.
3. Wash 3 × 10 min in 1% aqueous sodium tungstate.
4. Dehydrate in ascending concentration of ethanol, the 30 and 50% concentrations containing 1% sodium tungstate.
5. Embed as normal.

The glycosaminoglycan heparan sulphate can be identified in tissues fixed in the presence of Cuprolinic Blue. Removal by enzymic digestion should be used as a control (pronase, neuraminidase, chondroitinase, and nitrous acid).

5. Carbohydrates: mucosubstances

Carbohydrates and mucosubstance are materials with which there are difficulties in preservation at both light and electron microscopy. These include loss, false localization, and collapse of molecules such as glycosaminoglycans. Up to 50% of the glycogen is lost from glutaraldehyde fixed tissues and even greater amounts with formaldehyde. The standard methods of fixation do demonstrate glycogen satisfactorily, but in general, the glycocalyx is poorly preserved (9) (see *Protocol 3*). It has been found that various substances added to the fixation fluid can overcome this problem. These include Ruthenium Red and tannic acid (see Section 4). There is evidence that microwave treatment can retain these substances in tissues.

More recently, osmium ferricyanide has been introduced as a primary fixation for carbohydrates. This also demonstrates membranous structures within cells (see Section 4.1.5).

Silver techniques are used to demonstrate carbohydrates at the ultrastructural level, using oxidation of sugars to form aldehydes, which in turn, are revealed by the silver. Clearly, glutaraldehyde should not be used as a fixative for these techniques, but rather (*para*)-formaldehyde.

Microwaves have been found effective in some examples of carbohydrate retention within tissues.

6. Enzyme histochemistry

At the electron microscopic level, a number of enzymes have been successfully demonstrated (see Chapter 6), including the phosphatases and various glucosidases. For both groups of enzymes, similar preparative steps are used.

Protocol 5. Fixation for electron enzyme histochemistry

1. Fix a thin (1–2 μm) slice of tissue in buffered 4% formaldehyde or 2% cacodylate overnight.
2. Wash repeatedly in 0.2 M cacodylate 2–3 h.
3. Cut 40 μm sections on a freezing microtome and collect in 5% sucrose.
4. Take sections to incubation medium.

7. Lipids

In the study of lipids the major problem is the loss of material during dehydration and embedding. To limit this loss when using conventional embedding media, initial fixation is an important factor.

In general, following aldehyde fixation, the majority of lipids are still labile (19). Osmium reacts with unsaturated lipids in tissues, but the type of buffer used as a vehicle is also important. Imidazole buffered osmium reacts much more intensely with unsaturated fatty acids than aqueous osmium solutions and provides an excellent stain for tissue lipids (20).

The addition of potassium ferricyanide to fixatives for the preservation of phospholipids was first suggested by Elbers *et al.* (21). De Bruijn preserved labile lipids by fixing in 1% osmium tetroxide containing 0.05 M calcium chloride after primary glutaraldehyde fixation.

Fruhling *et al.* (22) developed a method for localizing cholesterol based on its formation of an insoluble complex with digitonin.

Protocol 6. Imidazole buffered osmium tetroxide

1. Primary fixation in buffered glutaraldehyde.
2. Rinse in holding buffer.
3. Post-fixation in 0.2 M imidazole pH 7.5 containing 2% osmium tetroxide for 1 h.
4. Dehydrate and embed as normal.

---

8. Nucleic acids

There is evidence that nucleic acids do not react to any extent with formaldehyde or glutaraldehyde, especially DNA (23). Reaction occurs only at temperatures greater than 45°C for RNA and 65°C for DNA, but most processing is at lower temperatures than these. At RT, reaction will occur over prolonged periods (18–40 days). The ability to perform *in situ* hybridization is a further demonstration that little cross-linking has occurred. Probably, nucleic acids are trapped within cells by the fixatives reacting with the associated histones.

9. Immunocytochemistry and cryoultramicrotomy (*see also* Chapters 3 and 4)

Fixation is usually achieved with paraformaldehyde or glutaraldehyde or mixtures of both aldehydes in 0.1 M sodium phosphate buffer.

If the antigen epitope under investigation is glutaraldehyde sensitive, simple formaldehyde fixative can be used (*Protocol 6*). For studying soluble compounds, (2%) glutaraldehyde fixation may be needed to prevent extraction of proteins from the section during cryoultramicrotomy. The immunostaining of substances attached to membranes or cytoskeleton is often enhanced by the extraction of surrounding soluble protein. Therefore, membrane proteins may be better studied in formaldehyde fixed specimens.

Acrolein periodate–lysine–paraformaldehyde and picric acid containing fixatives have also been used for electron immunocytochemistry. Osmium tetroxide masks or denatures many antigens but there are exceptions and there is also evidence to suggest that the action of osmium tetroxide may be reversible, by treatment with a saturated aqueous solution of sodium metaperiodiate (24). Grids should be floated on or immersed in the solution for 10–60 min at RT.

The fixed tissue can be stored:

- in buffer
- in 4% formaldehyde solution in buffer (desirable when no glutaraldehyde is used; see Section 4.1.2) and adjust strength appropriately
- in liquid nitrogen after sucrose infusion and mounting on specimen holders for cryoultramicrotomy

Freeze-substitution and freeze-drying techniques have been used successfully in immunocytochemistry. In the latter procedure, the tissue is snap frozen, dried *in vacuo* for 1–2 days, osmium vapour fixed, and embedded in Araldite resin.

10. Cultured tissues, isolated cells, and organelles

Monolayers of cells grown directly in Petri dishes or other containers are generally fixed by replacing the medium with fixative. Cells growing on glass coverslips or substrates such as Millipore filters can be fixed by using a Columbia staining jar in which they can stand vertically.

Satisfactory fixation is obtained by primary fixation for 15–30 min at RT in 1–3% glutaraldehyde in 0.1 M phosphate or cacodylate buffer pH 7.2–7.4 containing 2–3 mM calcium chloride.

Good fixation is sometimes obtained by fixing the cells in the presence of the growth medium. The medium is not removed and the culture is fixed by adding a volume of warm (37°C) double-strength fixative equal in volume to the growth medium.

It may be necessary to rinse cultures with serum free saline prior to fixation to remove serum proteins and cell secreted materials from cell surfaces. The rinsing solution should be as close to the culture medium in composition, temperature, and pH as possible to avoid spurious changes in cell morphology.

When fixing isolated cells or organelles it is best to centrifuge the cells into a pellet, remove the supernatant, and then add the fixative by slowly runing it down the side of the centrifuge tube so as to keep the pellet intact. For large pellets, the cells should be resuspended in the fixative by gentle shaking.

11. Scanning electron microscopy

The same standard fixatives as used for TEM can be used for preparing tissue for SEM. That is, primary fixation in 3% phosphate buffered glutaraldehyde following by secondary fixation in 1% osmium tetroxide.

In many instances, single fixation by glutaraldehyde can yield satisfactory results, but one can obtain comparatively better preserved surface membranes with the double-fixation method. The better results obtained by the double-fixation are chiefly attributed to osmium tetroxide which is capable of fixing neutral lipids and, phospholipids of cell membranes, thereby preventing their extraction during the process of dehydration.

After osmium fixation, one or more rinses to remove unreacted fixative is advisable. Osmium can be reduced by many of the agents used in dehydration causing a precipitate to form on the sample surface. Careful rinsing in low concentrations of alcohol can, however, obviate the need for rinsing with water prior to dehydration.

12. Artefacts

A number of artefacts may occur with tissue fixation. It is important to be able to recognize those in order to counter them. They may be classified as follows:

12.1 Agonal phase

That is, the time between the tissues blood supply being cut off and fixation taking place. This may be a few minutes at the shortest, where there is complete control over the system. At the other extreme, the time elapsed may be a day or more, for example, in post-mortem studies. This phase may be partly controllable. For example, with human tissue, if you explain to the surgeon the fixation problems, it may be possible to arrange delivery of the specimen more rapidly from the operating theatre.

12.2 Loss of substances

Different groups of substances may be lost from the tissues. Some, such as sugars, may be metabolized before fixation occurs. Small and reactive molecules such as biogenic amines and ATP may diffuse out and chemically break down. Lipids are poorly preserved and largely removed during the later stages of processing. Mucosubstances in general are often poorly preserved. A number of additives have been devised to try to overcome this problem (see Section 4 on additives). Some larger molecules such as enzymes may also be lost.

12.3 False localization

Related to loss of substances is false localization with cells or tissues. Probably, the best example of this is glycogen fugue in liver. Other substances such as iron may behave similarly

12.4 Loss of biological activity

A loss of biological activity in terms of enzyme or immunological activity or reactivity towards lectin may follow tissue fixation. This is probably due to the active centre or epitope no longer being accessible either by being covered or due to rigidity of the proteins brought about by cross-linking. Various methods have been proposed to overcome this problem:

- fixation in the presence of the substrate—for enzyme histochemistry (see Section 6)
- trypsinization (see Section 8 on immunoelectron microscopy)
- cryoultramicrotomy (see Section 8)
- microwave stabilization (see Section 3)

12.5 Non-specific reactions

Tissue fixation may produce various non-specific reactions. Fixation with the dialdehyde glutaraldehyde will induce an excess of aldehyde groups in tissues due to the reaction of some molecules at one end only. Formaldehyde fixation may cause non-specific binding of tritiated amino acids to tissues. Glutaraldehyde may bind horseradish peroxidase non-specifically to cell surfaces as another example of the same phenomena.

12.6 Changes in volume of tissues

Volume changes occurring during tissue fixation and processing have been recognized for a long time. This may occur from the gross to the ultramicroscopic level; in the latter, often noticeable with small vesicles. This artefact may be countered by using fixatives of differing molarity, if necessary, with the addition of substances such as dextrans. The major contribution to the effective osmotic pressure of the fixative solution probably comes from the buffers. It has been found in practice that slightly hypertonic fixatives give the best morphological results. This problem has been investigated histometrically by Mathieu and his colleagues (25).

Volume change due to processing is important in relation to any subsequent histometric analysis. It has also been noted that injured or diseased tissues behave differently volumetrically to normal control tissues towards fixatives (26).

13. Hazards

There are a number of hazards of tissue fixation that microscopists should be aware of, for their own well-being, as well as that of their colleagues. These fall under three main headings.

13.1 Biological

The specimen (in particular human tissue and the possibility of infection):

- tuberculosis
- hepatitis B
- AIDS

13.2 Chemical

- sensitivity to aldehydes
- toxic effects of osmium tetroxide
- toxic effects of arsenic in cacodylate buffer

13.3 Physical: microwave ovens

- burns from irradiated vessels
- formation of explosive gaseous mixtures by evaporation
- sparking from writing in lead pencil on (frosted) glassware

These can be countered by the proper handling of unfixed tissues, the thoughtful use of chemicals and microwave ovens.

References

1. Hopwood, D. (1990). In *Histochemistry in pathology*, 2nd edn (ed. M. I. Filipe and B. D. Lake), pp. 1–11. Churchill Livingstone, Edinburgh.
2. Bullock, G. R. (1984). *Journal of Microscopy*, **133,** 1.
3. Sleytr, U. B. and Robards, A. W. (1982). *Journal of Microscopy*, **126,** 101.
4. Hopwood, D., Slidders, W., and Yeaman, G. R. (1989). *Histochemical Journal,* **21,** 228.
5. McDowell, E. M. and Trump, B. F. (1976). Histologic fixatives suitable for diagnostic light and electron microscopy. *Archives of Pathology and Laboratory Medicine*, **100,** 405.

6. Boon, M. E. and Kok, L. P. (1989). *Microwave cookbook of pathology*, 2nd edn. Leyden Press, Leiden.
7. Login, G. R. and Dvorak, A. M. (1990). *Proceedings of the Microscopical Society of Canada, 17th Annual Meeting,* pp. 12–13. Microscopical Society of Canada, Toronto.
8. Eisenman, E. A. and Alfert, M. A. (1982). *Journal of Microscopy*, **125,** 117.
9. Maupin, P. and Pollard, T. D. (1983). *Journal of Cell Biology*, **96,** 51.
10. Mizuhira, A. V., Shiihasgi, M., and Futaesaku, Y. (1981). *Journal of Histochemistry and Cytochemistry*, **29,** 143.
11. Nuñez-Duran, H. (1980). *Stain Technology*, **55,** 361.
12. Stefanini, M., de Martino, C., and Zamboni, L. (1967). *Nature,* **216,** 173.
13. McLean, I. W. and Nakane, P. K. (1974). *Journal of Histochemistry and Cytochemistry,* **22,** 1077.
14. Hixson, D. C., Yep, J. M., Glenney, J. R. *et al.* (1981). *Journal of Histochemistry and Cytochemistry*, **29,** 615.
15. Karnovsky, M. J. (1971). *11th Annual Meeting of the American Society for Cell Biology*, p. 146. American Society for Cell Biology, New York.
16. De Bruijn, W. C. (1973). *J. Ultrastruct. Res.*, **42,** 29.
17. van Kuppevelt, T. H. M. S. M., Domen, J. G. W., Gremer, F. P. M. *et al.* (1984). *Histochemical Journal*, **16,** 671.
18. Luft, J. H. (1971). *Anatomical Record,* **171,** 369.
19. Cope, G. H. and Williams, M. A. (1968). *Journal of the Royal Microscopical Society,* **85,** 259.
20. Angermüller, S. and Fahimi, H. D. (1982). *Histochemical Journal*, **145,** 825.
21. Elbers, P. F., Ververgaert, P. H. J. T., and Demel, R. (1965). *Journal of Biology,* **24,** 23.
22. Fruhling, J., Penasse, W., Sand, G. *et al.* (1971). *Journal de Microscopie (Paris),* **12,** 83.
23. Hopwood, D. (1975). The reaction of glutaraldehyde with nucleic acids. *Histochemical Journal*, **7,** 267.
24. Bendayan, M. and Zollinger, M. (1983). Ultrastructural localisation of antigenic sites on osmium fixed tissues applying the protein-A–gold technique. *Journal of Histochemistry and Cytochemistry*, **31,** 101.
25. Mathieu, O., Claasen, H., and Weibel, E. R. (1978). *Journal of Ultrastructure Research*, **63,** 20.
26. Pentilla, A., McDowell, E. M., and Trump, B. F. (1975). *Journal of Histochemistry and Cytochemistry,* **23,** 251.

2

Embedding and thin section preparation

MICHAEL SMITH and SIMON CROFT

1. Introduction

The earliest applications of the transmission electron microscope (TEM) to the biological sciences involved the examination of whole samples and various replication techniques. However, a technique was required whereby thin sections of the sample could be examined in the microscope. The use of light microscopical techniques for section preparation produced disastrous results both for the sample and for the microscope, and the development of alternative techniques has been taking place for the past 50 years.

The ideal embedding medium to yield the maximum 'real-life' information for biological TEM should possess the following characteristics:

- high solubility in water
- low viscosity
- be capable of polymerizing or hardening uniformly with no shrinkage
- be capable of preserving antigenicity
- form a stable matrix for sectioning and microscopy
- allow easy and repeatable staining
- be readily available and consistent

There has been extensive experimentation with various media, but to date none of them fulfils all of the above criteria. The microscopist is therefore faced with a daunting range of embedding media, all of which are advertised as being ideal, and all of which have one or more failings.

2. Dehydration

None of the commercially available embedding media is soluble in water, although several types are water-miscible. The most common dehydrating agents are ethanol, methanol and acetone, with propylene oxide (1, 2-epoxy propane) being used as an intermediate solvent.

The major chemical problem associated with dehydration is the extraction of lipid, which may be as high as 95%, with the majority being lost in the higher solvent concentrations. Some proteins (up to 4%) are also lost at the lower solvent concentrations. The degree of extraction and loss varies markedly between different sample types, so it is not possible to lay down definite rules. The use of osmium tetroxide and uranyl acetate during processing reduces the rate of extraction, as does lowering the temperature to 4°C. In addition to the loss of chemical content this extraction leads to shrinkage of the sample, often up to 10%. However, these effects are slight in comparison to those which may be induced by poor fixation. A standard protocol is given below (*Protocol 1*).

Protocol 1. Dehydration

• Distilled water	2 × 30 min
• 30% Methanol	2 × 10 min
• 60% ,,	10 min
• 70% ,,	10 min
• 80% ,,	10 min
• 90% ,,	10 min
• 100% ,,	2 × 10 min
• Propylene oxide	2 × 10 min

Notes:
Ethanol can be used as an alternative to methanol. Propylene oxide can sometimes be omitted if the resin is easily miscible with alcohol.

Generally, dehydration should be fairly rapid, using several short changes (5–10 min each) rather than one long soak, carried out at room temperature and the samples should be agitated using a rotator during the dehydration process.

Other solvents, [xylene, toluene, ethylene glycol, and polyethylene glycol (carbowax)], have been used for dehydration, but none is particularly successful. Nonetheless, the use of alternative chemicals should not be dismissed if the 'standard' technique does not work.

Water-miscible embedding media can eliminate the use of alcohol, acetone, and propylene oxide but these media can themselves cause extraction of chemical components and modification of the sample. For example, water-miscible epoxy resins react with proteins and nucleic acids. Methacrylates on the other hand are powerful solvents for neutral lipids and some phospholipids, even at low temperatures.

Chemical dehydration using 2,2 dimethoxy-propane (DMP) was proposed in 1975 as an alternative (1). Acidified 2,2 DMP (0.05 ml concentrated hydrochloric acid per 100 ml 2,2 DMP) reacts with water to form anhydrous methanol and acetone. The sample should be transferred to 1–2 ml of 2,2 DMP (the reaction is extremely exothermic) and left for between 15 min and 16 h at room temperature (RT). Care must be taken to ensure that the sample is thoroughly washed in water prior to its dehydration since phosphate buffer precipitates in 2,2 DMP.

3. Safety

In general, care must be taken when handling all embedding medium components. Epoxy resins are mutagens and weak carcinogens. Many components can cause dermatitis, and their vapours may be toxic. In addition, some components are flammable and may be explosive, for example benzoyl peroxide (2).

Handling procedures:

(a) Use a fume cupboard at all times, including during the polymerization process.

(b) Wear gloves when handling resins and components.
Check that the gloves used are actually resistant to the chemicals; for example, nitrile gloves for epoxy resins, vinyl for acrylics.

(c) Consider barrier creams as an alternative to gloves when using a higher viscosity resin.

(d) Remove skin contamination by soap and water alone. Do not use solvents.

(e) Dispose of the waste resin mix by polymerization, and dispose of the waste components as solvent waste.

(f) Ideally, use disposable instruments, beakers, and pipettes when handling resins, but if necessary clean instruments thoroughly by sonication in acetone.

Care must also be taken to ensure that potentially hazardous mixes do not occur: propylene oxide and phosphotungstic acid can form an explosive complex, while epoxy monomers may explode if stored in glass bottles and exposed to heat. Amine accelerators (BDMA and DMP-30) give off toxic fumes when heated.

4. Acrylic resins

Acrylic resins were first introduced to electron microscopy in 1949. The first resins were based on a mixture of *n*-butyl and methyl methacrylate using benzoyl peroxide as a catalyst for heat polymerization, or benzoin as a catalyst for polymerization using ultraviolet light. This resin forms a two-dimensional polymer which is prone to shrink up to 20% on polymerization. In addition, up to 50% of the resin mass can be lost by electron bombardment and the resulting flow of the remainder can lead to distortion of the final image. Uneven polymerization can occur resulting from impurities in the resin or the sample, but this can be reduced by the addition of 0.01% uranyl nitrate or by extending the post-fixation time to 16 h with osmium tetroxide.

4.1 Glycol methacrylate

Glycol methacrylate (GMA) was introduced in 1960. This is a water-miscible resin, to which *n*-butyl methacrylate or styrene should be added to improve sectioning characteristics. Final infiltration can be improved by using a partially polymerized mix or pre-polymer (*Table 1*). Glycol methacrylate can also be used as a dehydrating agent leading to infiltration and embedding in an epoxy resin. It can be co-polymerized with epoxy resins.

Table 1. Glycol methacrylate

Mix A:	Glycol methacrylate	68 g
	Distilled water	2 g
Mix B:	*n*- Butyl methacrylate	29 g
	Benzoyl peroxide	1 g

Mix A and B prior to use.

Dehydration:	GMA: Water 8:2	20 min
	GMA: Water 97:3	20 min
	Mix A and B 1:1	20 min
	Pre-polymer [a]	30 min
	Embed in gelatin capsules (not polythene)	
	Polymerize with ultraviolet light (315 nm at a distance of 10 cm) for 24–48 h	

[a] Pre-polymer: Heat a small volume of A and B in a water bath until the viscosity is similar to thin syrup (i.e. a few minutes at 80°C) and cool rapidly to 3°C.

With the establishment of epoxy resins, methacrylate resins were out of favour until 1981, when a new generation of acrylic resins came on to the market: the 'London resins' and the Lowicryls.

4.2 London resins

Two resins are manufactured by the London Resin Company: LR Gold for use at low temperature, initially on unfixed material, and LR White for use at room temperature. These resins have been developed for use in both light and electron microscopy.

4.2.1 General considerations

The resins are single components, with the possible addition of accelerators for UV or chemical polymerization. Acetone interferes with the polymerization of the LR resins and should not be used. This means that the use of 2,2 DMP, which forms acetone, should also be avoided. The resins polymerize anaerobically and are capable of polymerizing in tissues containing up to 10% water. They have a low viscosity as well as low toxicity. They are less beam stable than epoxy resins and there is a significant loss of section mass in the electron beam. However, this loss is probably due to removal of water and sections show less distortion than the methacrylates. These resins are ideally suited for use with immunocytochemical techniques (see Chapter 4).

4.2.2 LR White (*Protocol 2*)

This resin is available as three grades: Hard, Medium, or Soft. The hard grade is probably most suitable for biological tissue samples. It can be polymerized by heat, UV light, or chemicals. Using heat, the resin is polymerized within 24 h at 50°C. This prevents complete cross-linking of the resin, which improves the tissue response to antibody labelling. For UV light polymerization, 0.5% benzoyl methyl ether is added to the resin and polymerization using reflected UV light occurs in 24 h. The resin cross-linking is more difficult to control, but in general the antibody response is similar to heat polymerization.

Protocol 2. LR White

• 40% Ethanol	2 × 10 min
• 60% Ethanol	10 min
• 70% Ethanol	10 min
• LR White: 70%	30 min
• Ethanol 2:1	60 min
• LR White	Overnight 4°C
• LR White	24 h 55°C
• Embed in gelatin capsules	

Chemical polymerization (*Protocol 3*) can be 20–40 times more sensitive for immunochemical detection than heat polymerization, although excess accelerator will damage antigenicity and cause over-cross-linking. Osmium tetroxide should not be used with chemical polymerization since the dark colour causes focal heat accumulation and the temperature inside the block can rise to over 60°C.

Unopened and unsectioned blocks in gelatin capsules can be stored indefinitely at room temperature.

Protocol 3. Chemical polymerization of LR White

1. Dehydrate in 70% ethanol 3 × 10 min.
2. Infiltrate in LR White 3 × 20 min at 50°C.

Protocol 3. *Continued*

3. Embed in BEEM capsules at RT in LR White + fresh accelerator (10 ml + 15 μl vigorously stirred immediately prior to use). Polymerization is complete within 30 min.

   The reaction is exothermic, although the temperature will only exceed 50 °C for a few seconds.

4.2.3 LR Gold

LR Gold was developed for use with unfixed material, being processed and embedded at low temperature using benzil as the catalyst for polymerization by visible light. The addition of polyvinyl pyrrolidone (PVP), molecular weight 44 000, will help to reduce the osmotic effects on unfixed tissue (*Protocol 4*).

Protocol 4. LR Gold

Step	Time	Temperature
• 50% Methanol (+ 20% PVP)	15 min	0 °C
• 70% Methanol (+ 20% PVP)	45 min	−25 °C
• 90% Methanol (+20% PVP)	45 min	−25 °C
• LR Gold: Methanol 1:1 (+ 10% PVP)	30 min	−25 °C
• LR Gold: Methanol 7:3 (+ 10% PVP)	60 min	−25 °C
• 100% LR Gold	60 min	−25 °C
• LR Gold + 0.1% Benzil	60 min and overnight	−25 °C
• LR Gold + 0.1% Benzil	24 h	−25 °C

- polymerize using a 12 V 100 W quartz–halogen projector bulb (operating at 7–9 V) at a distance of 10–15 cm

As with LR White, this resin can be cured at room temperature by the addition of 1% dry benzoyl peroxide (or 1.5% benzoyl peroxide paste in dibutyl phthalate). This curing is accelerated by the addition of 50 μl of LR White accelerator to 20 ml LR Gold/benzoyl peroxide. UV light may be used for polymerization by the addition of approximately 0.5% benzoin methyl ether in place of the benzoyl peroxide or benzil.

4.3 Lowicryl

The Lowicryl resins are highly cross-linked acrylate- and methacrylate-based media with low viscosity useable at low temperatures. There are four types of Lowicryl, summarized in *Table 2*.

4.3.1 K4M (3)

Being a polar, hydrophilic medium, this is capable of polymerization with up

Table 2. Lowicryl resins

Name	Type	Characteristics	Lowest useable temp.
K4M	Polar	Hydrophilic	−35°C
HM20	Non-polar	Hydrophobic	−40°C
K11M	Polar	Hydrophilic	−60°C
HM23	Non-polar	Hydrophobic	−80°C

to 5% water. Normally, small unpigmented blocks are polymerized anaerobically using indirect UV light (360 nm) at low temperature, although it can be polymerized at 60°C. The components should be weighed out and mixed carefully by bubbling nitrogen gas through the mix, or *stirred gently* with a glass rod. Excessive mixing will incorporate oxygen into the resin which will inhibit polymerization, especially at 60°C. Polymerization is carried out in gelatin capsules, *not* BEEM polythene capsules as the plasticizer tends to be leached out. UV polymerization, by, for example, two 15 W fluorescent tubes (Philips TLD 15W/05), is carried out at a distance of 35–40 cm using indirect irradiation for 24 h, often followed by a further 2–3 days at RT. Infiltration and sectioning may be improved by staining with 0.5% aqueous uranyl acetate for 30 min prior to dehydration.

Protocol 5. Lowicryl K4M

• 30% Ethanol	30 min	0°C
• 50% ,,	60 min	−20°C
• 70% ,,	60 min	−20 to −35°C
• 95% ,,	60 min	,,
• 100% ,,	2 × 60 min	,,
• K4M: Ethanol 1:1	60 min	,,
• K4M: Ethanol 2:1	60 min	,,
• K4M	Overnight	,,
• Polymerize	24 h	,,

Notes:

Resin mix:
- Cross-linker A 2.7 g
- Monomer B 17.3 g
- Initiator C 0.1 g

Temperatures: −20°C achieved using 3:1 ice:sodium chloride or a modified crystat
−35°C achieved using a household chest freezer

Sample size: large samples may suffer from excessive temperature rise during polymerization

See also Chapter 3, Section 5.

5. Epoxy resins

Epoxy resins were first introduced in 1956 as an alternative to methacrylates. They were initially produced by all the major chemical companies (for example Shell, Ciba, Dow, and Union Carbide) and much work has been carried out experimenting with different combinations of components. The final polymerized block is formed by the reaction between the epoxy resin and an anhydride hardener. The polymerization is speeded up by the addition of an amine accelerator. DMP-30 was initially used as the principle accelerator but is prone to inactivation by water and has a short shelf-life. In recent years it has generally been replaced by BDMA. Proportions of accelerator used do not normally need to be changed in recipes when switching from one accelerator to the other.

There are fundamentally three forms of epoxy embedding media based on different resin types, each having slightly different characteristics.

- Diglycidyl ether of bis phenol A — Araldite
- Triglycidyl ether of glycerol — Epon
- Vinyl cyclohexane dioxide — Spurr

All these resins polymerize uniformly, forming a three-dimensional cross-linked structure, with little volume change (typically 2% shrinkage). They are also stable in the electron beam, only losing up to 20% of their mass without any flowing of the remainder of the resin matrix.

The final block hardness can be reduced by the addition of a plasticizer, which also serves to improve the sectioning characteristics and prevents the block from becoming too brittle. The hardness of the block is also affected by the choice of anhydride hardener. DDSA will produce softer blocks and replacing some or all of the DDSA with MNA will increase the block hardness.

A water-miscible epoxy resin, Durcupan, has been developed, but in general this is less satisfactory than many of the acrylics.

The different epoxy types have widely varying viscosities, ranging from 7.8 cps for Spurr to 1300 cps for Araldite. Although the high viscosity of

Table 3. Araldite CY212

• Araldite CY212	10 ml
• DDSA	10 ml
• DBPth	0.25 ml
Mix for 20 min using a magnetic stirrer.	
Add BDMA	0.15 ml
Mix for a further 20 min.	

Note: For harder blocks, add 1 ml MNA to 9 ml DDSA.

Araldite may be considered to be a problem initially, longer infiltration and the use of smaller sample sizes should reduce the difficulties. Although the epoxy resins cross-link slowly at RT, giving them a long period of usefulness or 'pot-life', the viscosity increases slowly with time and care must be taken that the final blocks are embedded before the resin has become too viscous to handle easily. Media can be stored for several weeks at +4°C or for months at −20°C. Media stored below RT must be allowed to warm up before the containers are opened because of the hygroscopic nature of the epoxy resins.

Polymerization of epoxy blocks is usually carried out at 60°C for 16–24 h, although the standard resin mixes will polymerize at 50°C (48 h), 70°C (8 h), or 100°C (1–2 h).

5.1 Epon 812

Epon 812 (manufactured by Shell) was one of the first epoxy resins used for electron microscopy and was extremely popular. Unfortunately, when Shell ceased to manufacture it, several companies produced 'equivalents' which in general were not as reliable or predictable as the original. These problems are now being overcome and several good alternatives are now on the market (*Table 4*).

Table 4. Epon 812 (or equivalent)

• Epon 812 resin	15 ml
• DDSA	8 ml
• MNA	8 ml
Mix well and then add BDMA	0.6 ml

Notes:
The DDSA:MNA ratio is variable from 3:1 to 1:2 depending on the resin hardness deemed to be most suitable for the sample under examination.
To increase the rate of polymerization, increase the ratio in favour of MNA.
Epon is miscible with ethanol, so the intermediate solvent (propylene oxide, etc.) can be omitted.

Alternatively:

Mix A:

Epon 812 resin	6.2 ml
DDSA	10.0 ml

Mix B:

Epon 812 resin	10.0 ml
MNA	8.9 ml

Vary the ratio of the two mixes to adjust the hardness,

	Soft	–	–	Hard
Mix A	10	5	2	0 ml
Mix B	0	5	8	10 ml

and add 0.15 ml BDMA.

Many users have attempted to combine the best characteristics of Araldite and Epon by mixing them together, and on the whole this has been successful (*Table 5*).

Table 5. Epon/Araldite mixtures (4)

• Araldite CY212	22.3 g
• Epon 812	31.0 g
• DDSA	60.0 g
• BDMA	2.0 g

5.2 Spurr's resin

Spurr (5) developed a low viscosity epoxy resin (*Table 6*) to overcome many of the problems associated with high viscosity. This resin found great favour with many researchers, especially those working in the field of mycology, until it became evident that the epoxy component and the accelerator he had chosen were particularly volatile and carcinogenic. The resin has fallen out of favour, and extreme care must be taken when handling it; all stages of its use, including mixing and polymerization, must be carried out in an efficient fume cupboard.

Table 6. Spurr's resin

• ERL/VCD 4206	10.0 g
• DER 736	6.0 g
• NSA	26.0 g
• S1	0.4 g

Notes:
The resin should be mixed well and used the same day or stored at −20°C.
The resin is miscible with acetone and alcohol, so propylene oxide need not be used.
The DER component can be varied from 4–8 g. Increasing the amount decreases the hardness of the final block.
The above recipe is polymerized in 9 h at 70°C, with a pot-life of 3–4 days. For rapid curing (3 h) the S1 component should be increased to 1.0 g, reducing the pot-life to 2 days.

6. Other resin types

6.1 Melamine

Melamine resins were first suggested for EM in 1972 and became more readily available in 1983 (6). There are three versions available with different charac-

Table 7. Melamine resins

Type	Infiltration (temp. °C)	Hardening (temp. °C)
(A) FB101	40	60
(B) AME01	−82	60
(C) MUV116	−82	−82 (UV light)

teristics which are given in *Table 7*. All are water-miscible and non-toxic, although formaldehyde vapour may be released during polymerization.

Types B and C above are used in freeze-substitution protocols and so are beyond the scope of this chapter. FB101 (Nanoplast) has a viscosity of 48 cps and shrinks around 2% upon curing. Once polymerized, the resin is totally inert and cannot be removed, so this medium is useless for antibody studies.

Protocol 6. Melamine FB101 resin

1. Fix tissue in glutaraldehyde as usual.
   Omit osmium tetroxide.
2. Stain in 1.0% uranyl acetate for several hours in the dark.
3. Mix 10 g MME 7002 with 0.15–0.2 g B52, stir until dissolved, and allow to stand for 30 min.
4. Orientate small samples in pre-dried BEEM capsules or flat moulds, and cover with resin to a depth of no more than 5 mm.
5. Place in a dessicator in an oven at 40°C for 48 h.
6. Remove from dessicator and harden at 60°C for a further 48 h.
7. Allow block to cool before attempting to remove from mould.

Notes:
Section on to water.
Examine unstained. 0.5% uranyl acetate can be added to the resin for low contrast samples.
If the blocks are too small to mount, epoxy resin can be added to the moulds after polymerization of the melamine.

6.2 Polyester

These were introduced in 1958, the principal type being Vestopal. They are less generally available than the epoxys and the accelerator and initiator are less stable than other components, having only a two-month shelf-life. These two components are also explosive when mixed without the resin, so care must be taken when making up the medium. The final mix has a pot-life of a few hours and can be polymerized using UV light at RT in the presence of benzoin or by heating to 60°C. The resin is not soluble in ethanol and dehydration should be carried out using dry acetone or methyl methacrylate. Around 20% of the section mass will be lost upon exposure to the electron beam.

6.3 Others

6.3.1 Urea–aldehyde

50% Glutaraldehyde can be mixed with 4% urea using oxalic acid as a catalyst and as an embedding medium, although uniform infiltration throughout the block is often not easy to achieve and blocks produced with this medium are extremely hydrophilic and difficult to section.

6.3.2 Gelatin

Bacteriological grade gelatin mixed with 2% glycerine and dried in a dessicator at 37°C can be used to produce blocks, but they are difficult to section and suffer from extreme shrinkage.

6.3.3 Glutaraldehyde/carbohydrazide (GACH) (*Protocol 7*)

A water-miscible, lipid-retaining embedding mix was developed in 1973 using glutaraldehyde and carbohydrazide (7). Although this mix avoids the use of organic solvents, it has failed to achieve widespread acceptance since its sectioning characteristics and beam stability are far from ideal.

Protocol 7. Glutaraldehyde–carbohydrazide (GACH)

1. Fix and wash as normal.

2. Dehydrate

20%	GACH[a]	3–16 h	1°C
50%		2–4 h	
80%		2–4 h	
100%		1–2 h	

3. Embed sample in a small droplet of fresh 100% GACH on a wax plate at RT and transfer to an incubator at 37°C for 8–14 h.

4. Mount droplets on dowels using a rapid-setting epoxy resin.

5. Trim, section, and stain as normal.

Note: Small samples MUST be used.

[a] 100% GACH: Add carbohydrazide in three portions to 50% glutaraldehyde to give a final concentration of 150 mg/ml.
Mix in a beaker in an ice bath with a magnetic stirrer.
Stir rapidly during each addition and for 15 min between additions.
Adjust pH to 5.5 with 1 N NaOH to maximize lipid retention.
Use immediately or store in tight-capped vials at −20°C.

---

7. Embedding

7.1 General considerations

If small samples are being examined, it is essential that the sample is correctly orientated in the mould prior to polymerization. Failure to do this will lead to a prolonged period of trimming or sectioning or, in the worst case, to a wasted experiment.

All blocks should be labelled with the sample code or reference number. The best way to do this is to place a small piece of paper bearing the code in the end of the mould away from the sample. This label can be written using a typewriter, pencil, or indian ink, but biros must be avoided.

Polyethylene moulds must not be used with polyester resins, since the resin accelerator can be leached out into the mould plastic. Moulds should be pre-dried in an oven to prevent traces of water on the mould surface interfering with the polymerization of the resin.

7.2 Tissue blocks and cell pellets

7.2.1 Capsules

These are made of gelatin or polyethylene (BEEM capsules). They are available in a variety of sizes. They must be supported either using special holders supplied by the capsule manufacturers or by simply punching holes in small cardboard boxes.

7.2.2 Flat embedding

Moulds for flat embedding can be made from polyethylene, aluminium foil, or silicon rubber. They are available from most EM suppliers or can be manufactured in the laboratory using modellers rubber solution; smaller samples can be embedded in the lids from BEEM capsules. Transparent polyethylene moulds can be used with LR and Lowicryl resins, if they are covered with cling film to exclude air. Care must be taken to ensure that the materials used are UV-transparent.

7.3 Monolayers

7.3.1

Where the orientation of cells is not important, cell monolayers can be scraped off their substrate after fixation and prior to embedding and handled as a pellet.

7.3.2

Cells grown on glass slides can be embedded *in situ* by covering them with resin and positioning inverted capsules full of resin over selected areas of the cells. After polymerization, excess resin can be trimmed away and the capsules removed from the slide. Dry ice or a freezing spray can be applied to the back of the slide to encourage the release of the resin from the slide. Removal of the sample is also eased if the slides are coated with collagen. The monolayer samples can be re-embedded in moulds for sectioning perpendicularly to the plane of the culture (8).

7.3.3

Cells grown on Falcon plastic will be released from their subtrate by propylene oxide. Care must be taken to ensure that the propylene oxide is not left in contact with the plastic for more than a minute as this leads to the transfer of some of the plasticizer from the plate into the cells which will in turn lead to poor polymerization of the resin.

7.3.4

Cells can be grown on substrates which are sectionable, for example Millipore or Nucleopore filters, cling film, Thermonox plastic coverslips, silicon rubber membranes, or methacrylate or epoxy resin sheets (NB: beware of the cytotoxic effects of resin sheets as a substrate). For all of these substrates, the cells can be embedded *in situ*, cut up, and either mounted on the end of wooden or Perspex dowels or re-embedded in moulds or capsules.

8. Sectioning

Early attempts to section resin embedded material with steel knives, even when specially sharpened, proved fruitless. Sections are now cut using glass, diamond, and most recently, sapphire knives. They are generally cut on to a water bath, formed on the back face of the glass knife, either using a preformed boat sealed into position with glue or melted dental wax, or made from waterproof surgical strapping tape. The bath should be filled with distilled water (NOT from ion-exchange systems) containing up to 5% acetone.

Glass for knives should be cleaned with soap and distilled water, preferably without the use of organic solvents, and dried using a lint-free or nylon cloth. The glass knives have an almost indefinite shelf-life if stored in clean conditions. Under good sectioning conditions, a glass knife will cut a ribbon of sections 15 mm long. In use, a knife will deteriorate at a constant rate regardless of the section thickness being cut, below 1 μm.

The requirements for good sectioning are listed in *Table 8*.

The block should be trimmed to give a trapezoid face using small cutting strokes with a single-edged, backed razor blade, with the block sides sloping away at an angle of 30–45 degrees. The razor blade should be changed frequently to ensure that a blunt edge is not used, and the trimming action should always be away from the block face to reduce the possibility of the face being scratched or marked. The block should be positioned so that the shorter of the parallel edges of the block face trapezium should contact the knife first.

Table 8. Requirements for good sectioning

- uniform embedding
- adequate trimming
- correct knife angle
- even section flattening
- good mounting/recovery of sections from water bath
- high level of cleanliness

This will reduce the shock and hence the compression. Mesa trimming (that is, trimming the block so that the initial area of interest forms a raised plateau on the block face) will increase stability of the block while being sectioned, as long as the block itself is securely clamped.

Although the optimum setting for an ultramicrotome will vary from machine to machine, as well as from sample to sample, the settings given below will be useful as a starting point.

- knife angle 45 degrees
- cutting angle 7 degrees
- cutting speed 2–5 mm/sec for a glass knife
  0.5–2 mm/sec for a diamond knife

Even under the most favourable conditions, sections will have suffered 5–20% compression. In order that the sample detail is not obscured, they must be flattened carefully and evenly before being picked up from the water-bath surface. A thin artist's brush or strip of filter paper dipped in chloroform can be waved over the sections, or a heat pen can be used. Acrylic sections may be rather slow to flatten. Care must be taken not to actually touch the sections or the water surface.

To minimize wrinkling and distortion, sections should be picked up from below using a suitable grid held carefully by a pair of watchmakers' forceps and dried immediately with a piece of filter paper.

9. Staining

While many stains can be applied to samples during fixation and dehydration, the majority of electron microscopical staining techniques will be applied to the sections. However, there may be occasions when the less selective more intense *en bloc* staining may have distinct advantages. Different chemical stains and fixatives have varying affinities for constituents within the cell and these are summarized in *Table 9*.

There is often little difference between the fixative qualities and staining characteristics of any one chemical. It must always be borne in mind that the

Table 9. Chemical affinities

	Nucleic acids	**Proteins**	**Phospho-lipids**	**Polysac-charides**	**Fats**
Glutaraldehyde	weak/none	strong	weak/none	weak/none	weak/none
Osmium tetroxide	moderate	moderate	strong	moderate	variable
Potassium permanganate	none	strong	strong	moderate	moderate
Lead citrate	strong	strong	strong	moderate	moderate
Uranyl acetate	strong	strong	moderate	none	none
Phosphotungstic acid	none	strong	none	none	none

different chemicals often work together and omitting any one step will affect the reactions of the remainder.

Stains can be divided into several classes:

(a) General stains which will be most widely applied, e.g. uranyl acetate and lead citrate;

(b) Less frequently used general stains, e.g. potassium permanganate and phosphotungstic acid;

(c) Stains for specific cellular components, e.g. for polysaccharides;

(d) Stains for enzymes and immunological probes. (These are covered by Chapters 4 and 6.)

Table 10. Uranyl acetate (9)

Saturated aqueous solution of uranyl acetate, centrifuged or filtered before use.

Staining times: 30–120 min at RT.

Notes:
As an alternative, 2% uranyl acetate in 30% methanol can be used for 25 min during tissue dehydration.
The staining can be carried out at 60°C but care must be taken to stop the stain drying out and damaging the sample.
Uranyl acetate is toxic and mildly radioactive and should be handled with care, using gloves.

Section staining is generally carried out on a piece of clean dental wax or parafilm, using small drops of the solutions covered by a Petri dish lid during staining. Grids must be handled carefully during staining procedures using fine watchmaker's forceps, holding the edge of the grids only. After each stain, the grids must be washed thoroughly in distilled water either by immersion in several beakers of water or by carefully running water down the forceps and across the grid faces. The grids should be dried before being replaced in their storage Petri dishes or grid boxes.

Table 11. Lead citrate (10)

- lead nitrate 1.33 g
- distilled water 30.0 ml

Mix well and add;

- sodium citrate 1.76 g

Shake the resulting suspension for 1 min and allow to stand for a further 30 min, shaking briefly every 5 min.

- 1 N Sodium hydroxide 8.0 ml
- distilled water to make up total volume to 50 ml

Staining times:	Araldite, Spurr, and Acrylics	5 min at RT
	Epon	15–30 min at RT

Notes:
A small number of sodium hydroxide pellets should be placed in the Petri dish during staining to absorb atmospheric carbon dioxide.
The distilled water used for making this stain should be boiled to remove the carbon dioxide and cooled, covered with aluminium foil or cling film to prevent the reabsorption of atmospheric carbon dioxide.
After staining, the grids can be washed in 0.2 N sodium hydroxide, before the distilled water.

Generally, more than one stain will be applied to sections to give optimal staining. If this is done, then the sequence given in *Protocol 8* should normally be followed.

Protocol 8 Overall staining sequence

1. Block stain during dehydration (uranyl acetate in 30% methanol/ phosphotungstic acid in 100% methanol)
   Embedding and sectioning
2. Potassium permanganate
   Citric acid
   Distilled water
3. Specific stains: cytochemical/enzyme/immunocytochemical
   Distilled water
4. Uranyl acetate: unless used as a block stain [(a) above]
   Distilled water
5. Lead citrate
   Sodium hydroxide
   Distilled water

Table 12. Potassium permanganate (11)

1% Aqueous potassium permanganate

Staining times: Spurr 3–15 min
Araldite 10–120 min

Notes:
Stain, then briefly wash in distilled water, place grid on drop of diluted citric acid (5% stock solution diluted 1 drop/ml water) for 30 sec, then wash again in water.

Table 13. Phosphotungstic acid (12)

1% Phosphotungstic acid in 70–100% acetone or alcohol

Staining time: 30 min at RT

Notes:
This stain is only used during dehydration, and will not be used in combination with any other stains.
Note that phosphotungstic acid and propylene oxide can form an explosive mixture.

In the process of section staining only the section surface is actually being stained. This is primarily due to the hydrophobic nature of the resins interacting with the hydrophilic stains. In practice, there is found to be little difference in stain penetration between the hydrophilic and hydrophobic resins, and it may be reasonable to think of the resin differences to be due to degrees of hydrophobicity rather than the fact that they are hydrophobic or hydrophilic.

Etching techniques which soften the resin to enable greater stain penetration are generally to be avoided, since they may well alter the tissue morphology; they certainly weaken the sections, and there is little evidence that they do, in fact, improve general stain penetration to any marked degree.

Protocol 9. Polysaccharide stain (13)

1. 20 min — 1% aqueous periodic acid at RT.
2. 2–3 changes of distilled water
3. 12–72 h — 0.2% thiocarbohydrazide (TCH) in 20% acetic acid
4. 2 × 1 min — 10% acetic acid
5. 20 min — 5% acetic acid
6. 20 min — 2.5% acetic acid
7. 20 min — 1% acetic acid

8. 2 × 15 min distilled water

9. 30 min 1% aqueous silver proteinate in the dark at RT.

10. 2 × 5 min distilled water in normal light

Notes:

Section on to gold grids (copper and nickel will be dissolved).

1.0% Thiosemicarbohydrazide in 10% acetic acid can be used in place of TCH.

The silver proteinate should be prepared 30 min before use with a magnetic stirrer under a dark red safelight (Kodak Wratten No. 1), with acid-cleaned glassware and filtered immediately before use.

Steps 1 or 3–7 should be omitted for control sections.

Exposure to osmium tetroxide vapour for 1–2 h can be used as an alternative to step 9 if necessary.

---

10. Trouble-shooting

Various problems may occur during embedding and sectioning, and the following notes may give some indication as to what might be done to rescue the situation or at least try to prevent it recurring.

10.1 Hardness variation

Variation in the resin hardness is often caused by incomplete mixing of the resin components. Poor infiltration of the specimen causes it to appear softer than the surrounding resin. This may be caused by incomplete mixing of the resin or as a result of the different components penetrating the sample at different speeds. Infiltration can be improved by:

(a) Increasing infiltration time, often to as much as several days, taking care to ensure that the samples are finally embedded before the resin thickens too much to allow easy orientation.

(b) Using a lower viscosity resin.

(c) Using a 1:1 resin: solvent mix prior to full infiltration.

(d) Increasing the infiltration temperature to up to 50°C, taking care that prepolymerization of the resin does not start.

10.2 Instability in the electron beam

This is caused by three factors:

(a) Incomplete polymerization, which can be completed by returning the block to the oven (or UV light).

(b) Loss of volatile resin components during examination.

(c) Radiation damage caused by the electron beam. (Although this may be reduced by reducing the microscope's accelerating voltage, the use of an alternative medium is probably preferable.)

11. Future trends

As biological research progresses there is an increasing interest in the biochemical, immunochemical, and molecular aspects of cells and tissues and their interactions, as well as an awareness of the need for greater integration of light and electron microscopical techniques. It would seem reasonable therefore to envisage that these are the two paths along which biological TEM sample preparation is likely to advance. New dehydration and staining techniques, and particularly embedding media, will be developed which will allow more complete integration of ultrastructural, biochemical, immunocytochemical, and molecular examination in both light and electron microscopes.

12. Resin names and synonyms

Many companies manufacture resin components and a wide variety of abbreviations are used. The confusion this may cause will be reduced by the information given below.

12.1 Hardeners or cross-linkers

DDSA	Dodecenyl succinic anhydride	HY964/964B
MNA	Methyl nadic anhydride	NMA
NSA	Nonenyl succinic anhydride	

12.2 Catalysts or accelerators

BDMA	Benzyl dimethylamine	DY062
DMP-30	Tridimethylaminomethylphenol	DY064/964C
S1	Dimethylaminoethanol	DMAE
B52	*p*-Toluene sulphonic acid	

12.3 Modifiers or plasticizers

DBPth	Dibutyl phthalate
DER732	Polyglycol diepoxide

12.4 Resins

DER736	Diglycidyl ether of polypropylene glycol	
ERL4206	Vinyl cyclohexene dioxide	VCD

MME7002	Hexamethylol–melamine–methyl ether
GMA	Glycol methacrylate/hydroxy ethyl methacrylate
CY212	Diglycidyl ether of bis-phenol A
Epon 812	Triglycidyl ether of glycerol

References

1. Muller, L. L. and Jacks, T. J. (1975). *Journal of Histochemistry and Cytochemistry*, **23,** 107.
2. Causton, B. E. (1981). *Proceedings of the Royal Microscopical Society*, **16,** 265.
3. Carlemalm, E., Garavito, R. M., and Villiger, W. (1980). *Electron Microscopy Proceedings of the 7th European Congress*, Vol. 2, p. 656.
4. Cross, R. H. M. (1989). *Micron and Microscopica Acta*, **20,** 1.
5. Spurr, A. R. (1969). *Journal of Ultrastructure Research*, **26,** 31.
6. Buchhuber, K. and Forsch, D. (1983). *Journal of Microscopy*, **130,** 1.
7. Heckman, C. and Barrnett, R. (1973). *Journal of Ultrastructure Research*, **42,** 156.
8. Jones, H. B. and Gray, T. J. B. (1985). *Journal of Microscopy*, **139,** 321.
9. Watson, M. L. (1958). *Journal of Biophysical and Biochemical Cytology*, **4,** 475.
10. Reynolds, E. S. (1963). *Journal of Cell Biology*, **17,** 208.
11. Sutton, J. S. (1968). *Journal of Ultrastructure Research,* **21,** 424.
12. Mazinozzi, V. (1962). *Journal of The Royal Microscopical Society*, **81,** 141.
13. Ryter, A. and DeChastellier, C. (1977). *Journal of Cell Biology*, **75,** 200.

3

Freeze-substitution and other low temperature embedding methods

NORBERT ROOS

1. Introduction

The typical living cell is an extremely complex and usually dynamic entity. It is surrounded by aqueous domains and contains a variable amount (usually more than 70%) of non-uniformly distributed water, which is an important structural component of most biological systems. Unfortunately, the environment in an operational electron microscope does not allow living specimens to be examined. The high vacuum (less than 1.33×10^{-3} Pa) would lead to explosive evaporation of a fraction of the tissue water. Subsequently, extraction of the latent heat of evaporation would cause the remaining water to freeze, and finally the frozen water would evaporate. The very high surface tension of water and the forming ice-crystals would destroy the ultrastructural details in the cells. Therefore, the extra- and intracellular water has to be either removed (chemically or physically) or immobilized (by freezing).

Conventional preparation methods employ a series of aqueous and organic reagents at ambient temperatures. The specimen is chemically fixed (using cross-linkers such as, for example, glutaraldehyde or formaldehyde) in order to arrest all cellular processes and precipitate and cross-link proteins, lipoproteins, etc. The chemical fixation process is slow and happens within minutes rather than seconds. The water is subsequently removed by organic solvents (e.g. ethanol or acetone), which in turn are replaced by resins for thin-sectioning (1, 2) (see Chapter 2).

It is clear that exposure to aqueous organic media will alter both the structure and composition of the specimen rendering a sample that bears little resemblance to the original *in vivo* state. Artefacts imposed by conventional preparation methods include leaching and translocation of soluble molecules, extraction of phospholipids (3), enzyme inhibition, and denaturation of antigenic groups.

2. Cryofixation

Cellular processes can be arrested using chemical fixatives as described above or by immobilizing the tissue water by freezing; that is, cryofixation. This method offers a series of advantages, because it:

- maintains the original elemental composition of the specimen (4, 5)
- arrests transient morphological changes since tissues are immobilized within 0.1–1 ms (6, 7)
- avoids denaturation of proteins
- allows cells to be processed, transferred, studied, and imaged in their native environment (8)
- provides a hard specimen that can be sectioned without the introduction of extraneous materials (9)

In order to utilize these advantages optimally, considerable care has to be taken before, during and after freezing to ensure that the artefacts introduced by the freezing method employed are less damaging than those the technique seeks to avoid.

It is important to keep the sample as close to the *in vivo* state as possible and one should be aware of the fact that sample handling before the actual freezing step can lead to artefacts. Anaesthetics can interfere with the physiological state of cells, and subsequent dissection of the sample may lead to mechanical deformation, reduced oxygen supply and falling blood pressure. To avoid gross post-mortem damage, tissue pieces should be frozen *in situ* (10–12). Where this is not possible, the organ(s) should be dissected carefully, placed in a bath containing physiological medium, and prepared for freezing.

The freezing process itself (that is, the transition from liquid into solid water) is complex and can obviously have significant effects on other cellular constituents. It is beyond the scope of this chapter to elaborate on these complexities, and we would like to restrict ourselves to a simple description of the main processes involved in freezing biological specimens. For more comprehensive reviews the reader is referred to articles by Dubochet (8) and Mazur (13); or books by Robards and Sleytr (14) and Roos and Morgan (15).

Cells consist of compartments that are separated by semi-permeable membranes and differ in protein, phosopholipid, and probably ion and water composition. A proportion of the intracellular water (up to 15% of the total cell water; ref. 16) is associated with the surface of molecules ('hydration' water) and will not be available during freezing of the specimen for the formation of ice crystals.

Slow cooling of the specimen in the absence of cryoprotectants will result in ice-crystal growth (hexagonal ice) in the extracellular space. The growing ice-crystal will exclude solutes thus concentrating salts in the remaining liquid. In order to maintain the osmotic equilibrium, intracellular water will be forced out of the cell. Subsequently the cell will shrink, and the intracellular solute concentration will increase, possibly acting as a cryoprotectant. Intracellular ice-crystal damage can be avoided in such specimens.

Higher cooling rates will also result in extracellular ice-crystal growth, but the intracellular water would not be able to diffuse out of the cell quickly enough. Intracellular ice-crystal growth will commence and the crystal will grow 'dendrite like' (8) in the cell, destroying cellular structure.

Extremely high cooling rates ($>10^4$ K/s) can vitrify a biological specimen. Water molecules are arrested instantly and are not able to arrange themselves in a crystal structure and it is therefore unlikely that any redistribution or destruction of cellular components will occur. The cells are preserved fully hydrated and as close to the *in vivo* state as possible (see Chapter 10).

The goal of cryofixation is to achieve this state. Unfortunately, the high cooling rates required make it extremely difficult to obtain vitrified bulk specimens. Most of the freezing methods available produce frozen bulk specimens with only a thin (up to 20 μm) vitrified surface layer with the possible exception of high-pressure freezing (14, 17). This implies that the specimen to be vitrified be kept small unless the use of a cryoprotectant is permissible. Cryoprotectants are compounds that interact with water in such a way that its ability to provide nucleation sites is greatly reduced. One can distinguish between non-penetrating (e.g. polyvinyl pyrrolidone) and penetrating (e.g. glycerol and dimethylsulphoxide) cryoprotectants.

A detailed description of different freezing techniques such as plunge-freezing (18–20), spray-freezing (21, 22), metal mirror freezing (11, 23–26), propane jet freezing (27–30), high-pressure freezing (17), etc., can be found in various books (14, 15,31).

A frozen sample can be processed in many different ways but one has to be aware of the fact that a vitrified non-cryoprotected specimen should never be allowed to warm up above 135 K/−138°C, the recrystallization point of water. Warming of the specimen above its recrystallization point will lead to ice-crystal growth even at low temperatures, thus ruining a previously well-vitrified specimen.

2.1 Fully hydrated specimens

A piece of frozen tissue is very hard, which allows ultra-thin sectioning at low temperatures (cryosectioning) using a cryoultramicrotome (32). Sections can than be transferred to and imaged in the electron microscope using a cold transfer stage. Imaging of frozen-hydrated sections has to be done under minimal dose conditions at very low temperatures (>110 K/−163°C) in order to avoid beam damage. This method avoids the usual artefacts due to chemical fixation, dehydration, staining, and embedding. Unfortunately, sectioning below the recrystallization point of the specimen is difficult and introduces a series of artefacts that are more pronounced than with conventional plastic sections (33). Nevertheless, impressive results have been obtained using this sophisticated approach (8, 32).

2.2 Frozen-dried cryosections

Analytical electron microscopists make use of fully-hydrated specimens in a different way. Ultra-thin cryosections are transferred to the microscope using a cold transfer stage and frozen-dried in the column. This greatly reduces the risk of elemental redistribution, the sections are stable under the beam and can subsequently be used for X-ray microanalysis (4, 34, 35) to localize elements in subcellular compartments.

2.3 Cryosections for immunocytochemistry

Tissue samples can be vitrified after chemical fixation followed by infusion of, for example, high (2.1–2.3 M) concentrations of sucrose. Sections can easily be cut from these frozen blocks of tissue. The sections are thawed, exposed to a series of immunolabelling reagents in aqueous media, and the water finally replaced by a matrix of (for example) methyl cellulose, which maintains the ultrastructural identity of the sample by avoiding air-drying artefacts and is relatively stable in the electron beam (36–38). Whereas the sample is sectioned at low temperatures, all the subsequent steps in the specimen preparation chain are performed either on ice or at ambient temperatures. The technique could therefore be called a hybrid technique.

Different techniques have been developed to bridge the gap between quench-freezing and conventional sectioning at ambient temperatures. They combine some of the advantages of the pure cryopreparative methods with the ease of sectioning at room temperature (14, 15, 39). These techniques are the freeze-substitution method and the freeze-drying method (*Figure 1*). Since they rely heavily on quench-frozen fresh tissue, these methods did not have a breakthrough until suitable freezing methods were introduced. Common for both methods is the removal of frozen water from the specimen. As mentioned above, water can either be 'free' or 'bulk' water or tightly associated with macromolecules as 'bound' water. This implies that it may be easier to remove the free water and that removal of the bound water might damage or change ultrastructural details. Only a few studies addressed this problem and it has been shown that freeze-drying below 210 K/−63°C does not remove water completely from a specimen (40, 41). First, upon further warming under vacuum the residual water will be removed. However, the critical temperature above which the residual water can be removed seems to depend on the composition of the specimen (40). It is therefore likely that the removal of water from a frozen specimen using either freeze-substitution or freeze-drying happens in two distinct steps.

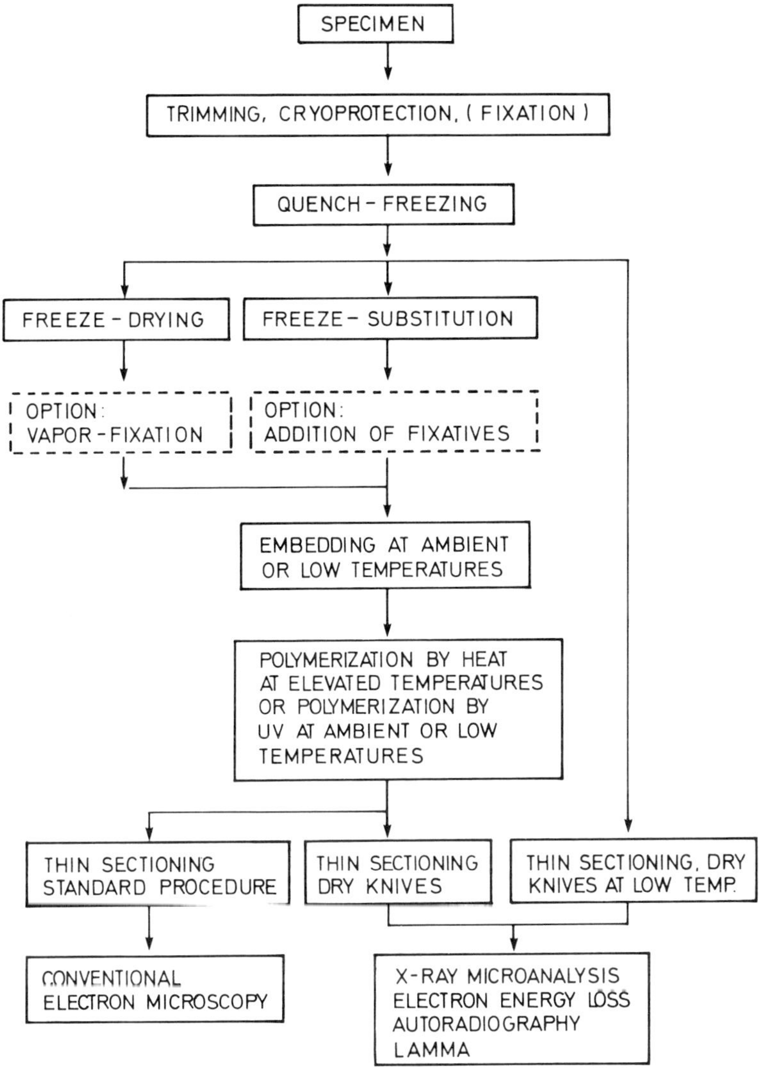

Figure 1. Flow chart comparing the main steps in different low-temperature specimen preparation procedures described in this chapter.

3. Freeze-substitution

3.1 General principles of freeze-substitution

Freeze-substitution is a procedure whereby the ice in a quench-frozen specimen is slowly replaced at low temperatures by anhydrous organic solvents usually containing a fixative, followed by infiltration with a resin. The method has been used in electron microscopy since 1957 (42) and has been modified and improved ever since. In common with all cryopreparative procedures, the final result is very much dependent on the quality of the freezing method employed. Assuming that the quality of freezing is good and vitrification of at least a thin surface layer of the sample has been achieved, substitution of the ice has to be performed at temperatures below the recrystallization temperature of the specimen. The major problem of the method is that it is very difficult to establish the recrystallization temperature of a given specimen. However, it is generally assumed that the recrystallization temperature of a biological specimen is higher than the recrystallization temperature of pure water (about 138 K). In addition, Steinbrecht (43) performed an elegant study on ice-crystal damage in freeze-substituted biological material which showed that recrystallization is a fairly slow process compared to the time it takes to substitute ice with an organic solvent. Again, the time it takes for recrystallization to occur is obviously dependent on the nature of the specimen studied (39). However, the substitution medium has to be liquid at low temperatures and should have the capacity to dissolve additives such as OsO_4 for fixation and staining of specimen structures. Some frequently used freeze-substitution liquids and their properties are listed in *Table 1*. Ethanol containing glutaraldehyde has been used by Nagele *et al.*, (44), diethylether containing 20% acrolein was recommended by Harvey (45) for botanical specimens, methanol with additives such as acrolein, uranyl acetate, and osmium tetroxide has been used by Zalokar (46) and Müller *et al.*, (47), but the most widely used solvent is acetone, with or without additives (48–50).

The quality of the freeze-substitution, however, is not only dependent on the choice of the substitution medium, but also on the conditions during the substitution process. For some applications (for example, analytical microscopy) water has to be removed from the substitution liquid before specimen immersion and constantly during the substitution phase (51). This is achieved by adding molecular sieves to the systems. Another parameter affecting the quality of the substitution process is the time-course of the warming up after the substitution has taken place. Using model systems labelled with tritiated water, and methanol, diethylether, and acetone respectively as substitution liquids, Humbel *et al.*, (52) showed that methanol is the fastest, and diethyl-

Table 1. A selection of freeze-substitution liquids (from ref. 14)

Solvent	Melting point (K)	Melting point (°C)
Ethanol	156	−117
Diethyl ether	157	−116
Propylene oxide	161	−112
Acetone	178	−95
Methanol	179	−94
Acrolein	186	−87

ether the slowest, substitution liquid. The capacity of the substitution liquid to take up water depends on the substitution temperature. Of the substitution liquids tested methanol had the highest capacity to take up water and diethylether the lowest (39). Whatever substitution medium is chosen, for all critical purposes where the localization of soluble constituents is to be performed, it is recommended that molecular sieves are used to capture the water released during substitution and embedding.

Usually chemical fixatives are added to the substitution liquid in order to stabilize ultrastructure and retain macromolecules. Whereas cross-linkers such as glutaraldehyde work only reasonably fast (1 h) at 263 K/−10°C (52), OsO_4 may react with double bonds of unsaturated acids already at 200 K/−73°C (53). Preservation of fine structure can possibly be improved by infiltration of resin monomers at low temperatures.

3.2 Instrumentation for freeze-substitution

The required hardware is easily made using basic workshop facilities and/or some imagination. A simple freeze-substitution set-up consists of a metal container filled with substitution liquid cooled by liquid nitrogen. The frozen specimen is placed on the frozen surface of the substitution liquid. The metal container is insulated, with the insulator determining the rate of warming. Alternatively, plastic vials containing the frozen-substitution liquid and the specimen can be placed in a pre-cooled metal block. The rate of warming is determined by the heat capacity of the block. It is not possible to keep a defined temperature for a period of time with this system. In order to obtain reproducible results it is fairly important that the device permits the temperature of the specimen and the substitution medium to be recorded and controlled in the range 173 K/−100°C to 294 K/21°C. In addition, defined temperatures should be stable over a set period of time (usually several hours, or a few days) and the whole arrangement should be sealed off from the ambient atmosphere, to ensure that the water content of the substitution medium is maintained as low as possible.

More sophisticated methods including temperature control (from 77 K/−96°C to 273 K/0°C) and pre-setting of times (up to 100 days) are described by Müller (54) and Sitte (55). Additionally, UV light sources can be fitted to allow polymerization of resins at low temperatures.

Freeze-substitution devices that are almost operator independent are also commercially available: Reichert-Jung (Cambridge Instruments) sell a device, the 'CS auto', manufactured according to principles defined by Sitte and Edelmann (55). RNC market freeze-substitution systems called MS 6000 and MS 6200, and Balzer's market a system designated FSU 010.

3.3 Freeze-substitution procedures for thin sectioning

For the preservation of morphology by freeze-substitution the major problem is to fix cellular structures after the ice has been replaced, to prevent excessive extraction of cellular material. However, fixatives dissolved in the substitution medium (for example, aldehydes and OsO_4) are usually not active at very low temperatures. Substitution schedules therefore include a slow stepwise temperature increase after the actual substitution has taken place. A general freeze-substitution regime is presented in *Table 2*.

Table 2. A general freeze-substitution regime for morphological studies

(a) Rapid freezing	of preferably very small tissue specimens in order to achieve vitrification
(b) Transfer	of the quench-frozen specimen onto a frozen substitution liquid at 77 K/−196°C (liquid nitrogen temperature)
(c) Substitution	in acetone, while shaking at 183 K/−90°C for 8–12 h depending on specimen size and volume (possible additives: osmium tetroxide, acrolein); for dehydration add molecular sieve
(d) Warming	to 213 K/−60°C for 8 h to 243 K/−30°C for 8 h
(e) Washing	twice in substitution liquid kept at 243 K/−30°C
(f) Embedding	using conventional epoxy resins (such as Spurr's or Epon)

Morphological preservation with this technique can be impressive, particularly if a fixative is incorporated into the substitution medium (*Figure 2*) (50). A freeze-substitution regime for plant tissue containing dimethyloxypropane has been introduced recently (56).

If thin sections of freeze-substituted material are to be used for the localization of electrolytes, a more complex schedule has to be employed to minimize the risk of redistribution and loss of elements (45, 51).

However, the utility and validity of the technique for X-ray microanalytical purposes are not been unequivocally proven (4), so it would not be reasonable to recommend it in preference to cryomicrotomy for quantitative purposes. Freeze-substitution is frequently used as an adjunct to other more definitive cryoprocedures for morphologically assessing the quality of freezing and the extent of freezing damage. (Note, however, that the only approach for unambiguously identifying the quality of the freezing is to record an electron diffraction pattern of a section of the fully hydrated biological specimen!). It is also frequently used for botanical studies (45), where the plant material often does not permit cryosectioning because of large water and air spaces. Recent studies show, however, that certain plant tissues can be cryosectioned (57). Improved cryoultramicrotomes, better glass knives (58), or the use of 'cryo'-diamond knives will make cryosectioning of difficult materials feasible.

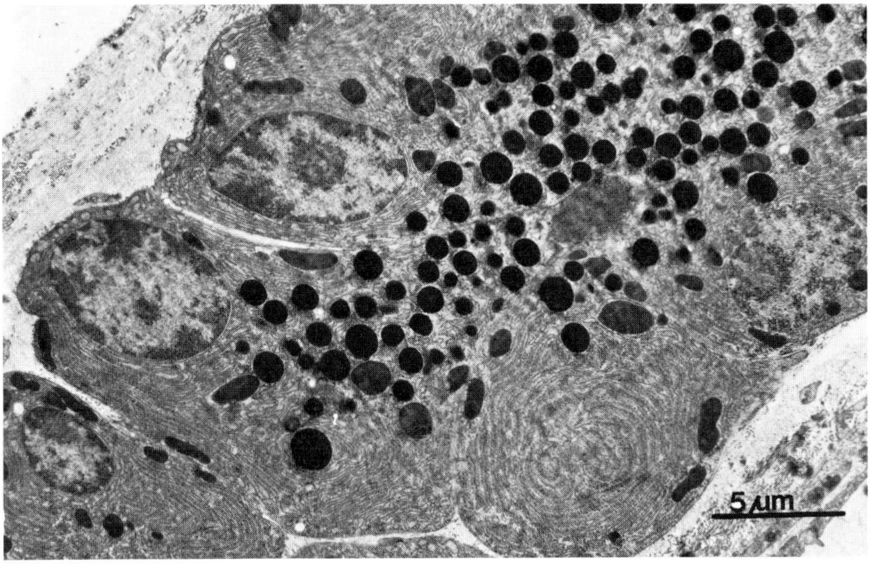

Figure 2. Transmission electron micrograph of an ultra-thin section of rat exocrine pancreas, quench-frozen and freeze-substituted in acetone/osmium. Section post-stained with uranyl acetate/lead citrate.

4. Freeze-drying

4.1 General principles of freeze-drying

For the examination of biological material in the electron microscope at ambient temperatures, water has to be removed from the specimen. This can be achieved in several ways; for example, by substituting the water in the specimen with a highly volatile solvent which then in turn can be evaporated, by air-drying, or freeze-drying of the hydrated specimen (lyophilization) (for details see ref. 14). Freeze-drying, a well-established and versatile method, is defined as the removal of water vapour from ice. The sublimation of water molecules from the solid phase (ice) depends on the temperature of the ice and can only occur when the partial pressure of the water vapour in the frozen surface exceeds that of the atmosphere immediately adjacent to it; the reverse process, the condensation of water molecules on to the ice, depends on the pressure and temperature of the vapour phase. It is very difficult to calculate theoretical sublimation rates, although Newbury *et al.* (58) provide a useful expression:

$$Q_n = 0.25(P_T\text{-}P_A)T^{0.5} \qquad [1]$$

where Q_n is the amount of water lost from the specimen (g/cm^2/s), P_A is the ambient vapour pressure (Pascals) of water, P_T is the saturated vapour pressure (Pascals) of water at the specimen temperature, T (K). [NB: when the vacuum pressure is sufficiently low, or when the specimen is surrounded by a sufficiently dry atmosphere Q_n will be determined by T. For example, at 273 K/0°C the maximum evaporation rate from pure ice is 8.3×10^{-2} g/cm^2/s, whilst at 173 K/−100°C the rate is 5×10^{-7} g/cm^2/s; that is, 100 000 times slower.] Umrath (60) used this expression to estimate the decrease of a pure ice layer in relation to the sublimation time, and *Table 3* provides an edited version of his results. It is clear that freeze-drying at very low temperatures leads to very long drying times.

Table 3. Comparison of freeze-drying times (from ref. 60)

Temperature (K/°C)	Vapour pressure (Pa) $\times 10^{-3}$	Drying time (s/μm)	Drying time (per μm) days	h	min	s
213/−60	1080	0.676				0.68
203/−70	259	2.75				2.75
193/−80	53.6	13				13
183/−90	9.32	78.2			1	18
173/−100	1.32	499			8	19
163/−110	0.148	4340		1	12	20
153/−120	0.0124	50200		13	56	40
143/−130	0.000739	817000	9	10	56	

These data represent maximum sublimation rates for pure frozen water and are probably only of value for biological specimens with very high water contents (such as dilute suspensions of viruses, etc.).

4.2 Freeze-drying of biological specimen and resin embedment

In biological tissue specimens the situation is more complex. Sublimation of water leads to a layer of dry biological material which inhibits further sublimation. Experiments using tissue models (61) clearly show that the sublimation rate decreases exponentially with the thickness of the already dry surface layer. In addition, bound water may be extremely difficult to remove from biological specimens at low temperatures. For these reasons the drying times for a tissue specimen of a given thickness are from 20 to 100 times longer than for solid pure water of similar thickness (60). Furthermore, biological specimens vary in size, structure, and composition, and it is therefore difficult to establish a general freeze-drying regime. However, since the specimen is preferably vitrified, recrystallization has to be avoided. As discussed above, recrystallization of water is temperature- and time-dependent and short drying times at 'safe' temperatures (preferably below 193 K/−80°C) are therefore mandatory. Short drying times can be achieved by using thin specimens, with a high surface-to-volume ratio; for example, ultra-thin cryosections, which after drying can be used for X-ray microanalytical studies (18). As a matter of fact, freeze-drying of biological material may not yield dry specimens at low temperatures! A fraction of the tissue water seems to be removed from the specimen at elevated temperatures only (39, 41).

Biological specimens can be frozen-dried and subsequently embedded in resin either at ambient or low temperatures, depending on the nature of the resin. Even though this method undoubtedly introduces its own suite of artefacts, it has been developed as an alternative to the conventional wet-chemical preparation methods, since it offers a more valid (4) but possibly not the best preservation of the original specimen composition (50).

4.3 Freeze-drying apparatus

Most commercially available freeze-driers (lyophilizers) operate at temperatures far above the presumptive recrystallization temperature (193 K/−80°C) of biological specimens and are therefore not adequate for ultrastructural studies.

Best freeze-drying results are obtained when the partial pressures of the gases around the specimen are kept much lower (preferably by a factor of about a 1000 times) than the saturation vapour pressure for ice at the specimen temperature. Three different approaches to remove the released water vapour and other gases from the system are available:

- *condensation* of the condensable gases on to a surface (condenser, cold-finger) kept at much lower temperatures than that of the specimen;
- *pumping* the gases out of the freeze-drier, using rotary, oil diffusion, or turbo molecular pumps;
- *adsorption* of the water vapour to chemicals (phosphorus pentoxide, calcium chloride, copper sulfate, or molecular sieves)

Most 'home-made' and commercially-available freeze-drying units combine at least two of the strategies. Whatever system is chosen, the condensers or chemical dessicants should be closer to the specimen than the mean free path of the residual gases (α) at a given pressure (equation 1):

$$\alpha = 1/\sqrt{(2\pi n d^2)} \qquad [2]$$

where α is the mean free path (the average distance a molecule would have to travel before striking another molecule), d is the diameter of the molecule and n the number of molecules per cubic metre.

This will ensure maximum removal of molecules from the gas phase. If molecular sieves are used at low temperatures they perform the functions of a vapour trap (especially water vapour) and vacuum pump, and they can be re-activated simply by heating. Edelmann (62) describes a simple system based on the use of molecular sieve that operates without additional vacuum pumps.

All vacuum-pumping freeze-drying systems should have the following features

- a vacuum gauge
- temperature control of the specimen holder
- a thermocouple close to the specimen to monitor the actual specimen temperature
- the possibility to introduce dry nitrogen gas in order to prevent rehydration
- the possibility to introduce vapour fixatives
- access for vacuum infiltration of resins

A general freeze-drying and resin-embedding regime is summarized in *Table 4*.

Table 4. A typical freeze-drying schedule for quench-frozen soft biological tissue (62)

	Accumulated time (h)	Drying time (h)	Temperature (K/°C)
(a)	1	1	138/−135
(b)	3	2	138–203/−135 to −70
(c)	33	30	203/−70
(d)	57	24	203–233/−70 to −40
(e)	63	6	233–297/−40 to −24
(f)	80	17	297/24
(g)	Formaldehyde–vapour fixation	(3)	333/24
(h)	Osmium–vapour fixation	(3–18)	297/24
(i)	Plastic-infiltration	(2–4)	297/24

Most conventional resins are suitable provided their viscosity is low enough to ensure complete infiltration of the specimen. Some resins allow infiltration and polymerization at low temperatures (see below). The morphological preservation of tissues by the freeze-drying/resin embedding method can be very good (*Figure 3*), but the technique is subject to a number of artefacts:

(a) The formation of ice-crystals during freezing will cause severe morphological changes, and secondary ice-crystal growth due to freeze-drying temperatures necessarily higher than the critical recrystallization temperature may damage the fine structure.

(b) Structures subjected to freeze-drying may shrink and collapse.

(c) Frozen-dried specimens are usually extremely hygroscopic. Rehydration

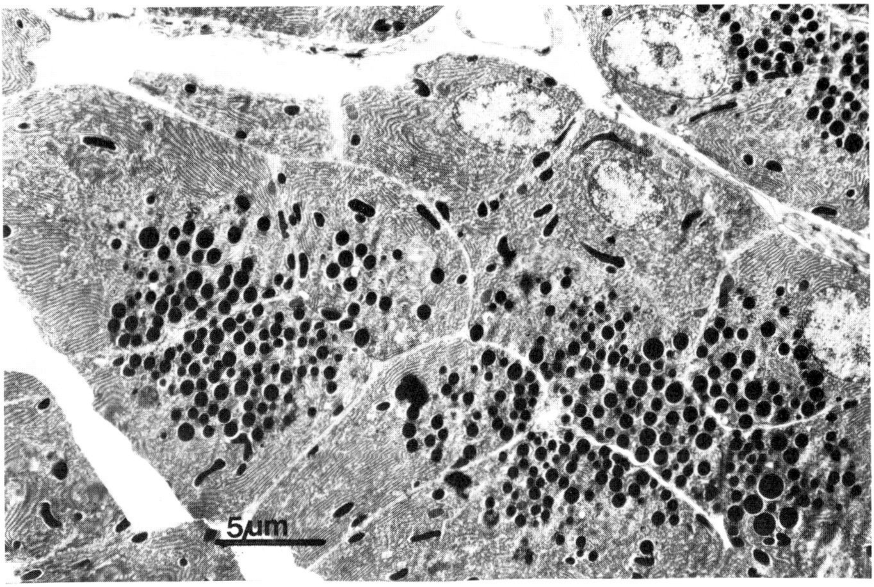

Figure 3. Transmission electron micrograph of an ultra-thin section of quench-frozen, frozen-dried, osmium/formaldehyde-vapour-fixed, resin-embedded rat exocrine pancreas. Section post-stained with uranyl acetate/lead citrate.

will occur instantly and can only be avoided by keeping the specimen at temperatures slightly elevated compared to the environment.

(d) Infiltration of the dry specimen with a resin at low temperatures will avoid the problem of rehydration but may cause redistribution of elements.

5. Low temperature embedding

Methods combining freeze-drying and resin-embedding have been developed in order to obtain specimens that reflect the *in vivo* structure and elemental composition better than the standard chemical methods.

The development of resins (methacrylate and acrylate formulations) with low enough viscosity to allow for infiltration and polymerization at temperatures below 273 K/0°C (64, 65) opened a new field of applications for the freeze-substitution and freeze-drying methods. The resins used mostly are the relatively electron beam-stable methacrylates Lowicryl K4M and K11M (polar resins usable at 237 K/0°C and down to 190 K/−83°C respectively) and Lowicryl HM20 and HM23 (non-polar resins applicable at temperatures as low as 190 K/−83°C (66), due to reduced viscosity). Both resins can either be polymerized by heat at 333 K/60°C or by UV light (360 nm) at low temperatures (e.g. in a conventional refrigerator). If UV light is used, vapour fixation with OsO_4 has to be avoided, since osmium blocks the penetration of the UV rays. Providing that the penetration of the resin into the specimen and the polymerization of the resin is complete, the plastic can be sectioned with either glass or diamond knives. In the case of the polar K4M and K11M resins contact between the trough liquid and the block face has to be avoided.

The resins have mainly been used in conjunction with the progressive lowering of temperatures, thus replacing conventional embedding techniques (for a schedule see *Table 5*) and to perform immunocytochemical labelling on sectioned material (67–69).

Table 5. Embedding[a] with Lowicryl K4M resin

Technique	Medium	Temperature (K/°C)	Time
Dehydration	(a) 30% ethanol	273/0	30 min
	(b) 50% ethanol	258/−15	60 min
	(c) 70% ethanol	248/−25	60 min
	(d) 100% ethanol	238/−35	120 min
Infiltration	(a) 1:1 100% ethanol and K4M	238/−35	60 min
	(b) 1:2 100% ethanol and K4M	238/−35	60 min
	(c) 100% K4M	238/−35	60 min
	(d) 100% K4M	238/−35	overnight
Polymerization (UV-lamp)	(a) 100% K4M	238/−35	24 h
	(b) 100% K4M	295/22	72 h

[a] Follow the manufacturer's guide-lines, or those provided by Glauert and Young (70)

It is claimed that these embedding procedures reduce the extraction of biological material and the denaturation of antigenic sites. However, for some antigens the yield of labelling sites was not as good as expected after embedding in K4M (71), and the results were often not reproducible. This could be the result of a rise in temperature during polymerization, which is an exothermic reaction (72), although the temperature rise can be controlled or eliminated completely if certain precautions are taken (70). The more important precautions are (a) the use of diffuse UV radiation, and (b) the provision of heat 'sinks' (that is, containers for flat embedding are placed in recessed aluminium blocks, and gelatin capsules are held in an ethanol bath). In a few studies freeze-substituted material (73, 74) has been embedded at low temperatures. A modified procedure (75) is summarized in *Table 6*. It avoids chemical fixatives in the substitution liquid. A freeze-substitution, low temperature embedding method, using methanol as substitution liquid, has just recently been described by Monaghan and Robertson (76).

Table 6. Low temperature embedding of freeze-substituted biological material

(a) Rapid freezing	vitrify very small tissue specimens
(b) Transfer	the quench-frozen specimen into the freeze-substitution device
(c) Substitution	in pure acetone at 193 K/−80°C for 8 days
(d) Infiltration	with 70% K11M + 30% acetone at 193 K/−80°C for 5 h with 100% K11M at 213 K/−60°C for 5 h with 100% K11M at 213 K/−60°C for 12 h
(e) Embedding	in 100% K11M at 213 K/−60°C in flat embedding moulds
(f) Polymerization	with UV light at 213 K/−60°C for 24 h

Freeze-drying avoids contact of the specimen with solvents and low-temperature embedding of frozen-dried specimens could theoretically reduce artefacts introduced by embedding procedures performed at ambient temperatures. Unfortunately, only a few groups have made attempts to evaluate this preparation method (77–79). An alternative to ambient temperature embedding (*Table 4*) is summarized in *Table 7*.

The author states that such a low temperature embedding procedure will preserve the original electrolyte distribution of the frozen-dried biological specimen.

The new generation of acrylic resins, LR White and LR Gold (80, 81), have certain very attractive features (65) in the context of sub-ambient embedding temperatures and the immunolocalization of antigens:

- they are electron-beam stable
- they are hydrophilic, so that background staining with equally hydrophilic immunoreagents should be significantly reduced
- LR White is compatible with up to 12% tissue water, so partial dehydration is permissible, thus possibly improving antibody–antigen reaction

Table 7. A low temperature embedding procedure for frozen-dried specimens (78)

	Accumulated time (h)	**Time (h)**	**Temperature (K/°C)**
(a)	1	1	133/−140
(b)	73	72	193/−80
(c)	217	144	213/−60
(d)	Resin infiltration K11M	24	213/−60
(e)	UV-polymerization	24	213/−60
(f)	UV-polymerization	24	273/0

- LR White, being less lipophilic than epoxides, is less likely to disrupt membrane systems when osmication is omitted

Another new formulation, the LR Gold resin has been introduced recently. It allows infiltration of the unfixed specimen at 248 K/−25 °C. Polymerization can take place at the same temperature using visible light, and is complete after 24 h. (See also Chapter 2, Section 4.)

6. Applications

Methods which combine quench-freezing of biological specimens with subsequent removal of water either by freeze-substitution or freeze-drying have been developed to achieve a better preservation of the native specimens, since they reduce the extraction and to some extent the displacement of cellular components. Until recently these methods had their place mainly in special electron-microscopical applications, such as autoradiography of water-soluble substances (82, 83), laser microprobe mass analysis (83) and X-ray microanalysis (4, 26, 45, 51). However, with the growing interest of cell biologists, in particular, in immunocytochemical localization studies, frozen-substituted and frozen-dried specimens that are subsequently infiltrated with suitable resins at low temperatures are more frequently used (76, 84). At the same time it was realized that such specimens are extremely useful for ultrastructural studies as well (50, 85, 86).

7. Artefacts

It was soon realized that the freeze-drying and freeze-substitution procedures introduced their own set of artefacts. The initial freezing step is very crucial for the quality of the specimen obtained. Any ice-crystal formation during and after freezing will destroy at least ultrastructural details. Freeze-substitution exposes the specimen to solvents, albeit at low temperatures, leading to partial extraction of hydrophilic substances. Freeze-drying of specimens will lead to shrinkage (87, 88) and subsequent infiltration of resins may create surface tension forces which could cause considerable structural damage. Dry specimens are therefore often vapour-fixed in order to stabilize structures, unfortunately thereby introducing some of the fixation artefacts that the method tried to avoid in the first place.

References

1. Glauert, A. M. (1974). In *Fixation, dehydration and embedding of biological specimens: Practical methods in electron microscopy*, Vol. 4 (ed. A. M. Glauert). North-Holland Publishing Company, Amsterdam, New York, and Oxford.
2. Hayat, M. A. (1981). *Fixation for electron microscopy*. Academic Press, New York.

3. Coetzee, J. and Van der Merwe, C. F. (1984). *Journal of Microscopy*, **135,** 147.
4. Roos, N. and Barnard, T. (1986). *Scanning Electron Microscopy*, **2,** 703.
5. Zglinicki, T. von, Rimmler, M., and Purz, H.-J. (1986). *Journal of Microscopy*, **141,** 79.
6. Plattner, H. and Bachmann, L. (1982). *International Reviews in Cytology*, **79,** 237.
7. Plattner, H. and Knoll, G. (1987). *Scanning Microscopy*, **1,** 1199.
8. Dubochet, J., Adrian, M., Chang, J.-J., Homo, J.-C., Lepault, J., McDowall, A. W., and Schultz, P. (1988). *Quarterly Reviews in Biophysics*, **21,** 129.
9. McDowall, A. W., Chang, J.-J., Freeman, R., Lepault, J., and Walter, C. A. (1983). *Journal of Microscopy*, **131,** 1.
10. Hearse, D. J., Yellon, D. M., Chapell, D. A., Wyse, R. K. H., and Ball, G. R. (1981). *Journal of Molecular and Cellular Cardiology*, **13,** 197.
11. Hagler, H. K. and Buja, L. M. (1984). In *The science of biological specimens preparation* (ed. J.-P. Revel, T. Barnard, and G. H. Haggis), pp. 161–166. SEM AMF O'Hare, Il 60666, USA.
12. Somlyo, A. P., Bond, M., and Somlyo, A. V. (1985). *Nature*, **314,** 622.
13. Mazur, P. (1984). *American Journal of Physiology*, **247:16,** C125–C142.
14. Robards, A. W. and Sleytr, U. B. (1985). *Low temperature methods in biological electron microscopy: Practical methods in electron microscopy*, Vol. 10 (ed. A. M. Glauert). Elsevier, Amsterdam, New York and Oxford.
15. Roos, N. and Morgan, A. J. (1990). *Cryopreparation of thin biological specimens for electron microscopy: Methods and applications*. Royal Microscopical Society/ Oxford University Press, Oxford.
16. Clegg, J. S. (1982). In *Biophysics of water* (ed. F. Franks and S. Mathias) pp. 365–83. Wiley, Chichester, UK.
17. Moor, H. (1987). In *Cryotechniques in biological electron microscopy* (ed. R. A. Steinbrecht and K. Zierold), pp. 175–91. Springer-Verlag, Berlin and Heidelberg.
18. Barnard, T. (1982). *Journal of Microscopy*, **126,** 317.
19. Elder, H. Y., Gray, C. C., Jardine, A. G., Chapman, J. N., and Biddlecombe, W. H. (1982). *Journal of Microscopy*, **126,** 45.
20. Ryan, K. P., Purse, D. H., Robinson, S. G., and Wood, J. W. (1987). *Journal of Microscopy*, **145,** 89.
21. Bachmann, L. and Schmitt, W. W. (1971). *Proceedings of the National Academy of Sciences, USA*, **68,** 2149.
22. Dubochet, J., Groom, M., and Mueller-Neuteboom, S. (1982). In *Advances in optical and electron microscopy*, Vol. 8 (ed. R. Barer and V. E. Cosslett), pp. 107–35. Academic Press, London.
23. Heuser, J. E., Reese, T. S., Dennis, M. J., Jan, Y., Jan, L., and Evans, L. (1979). *Journal of Cell Biology*, **81,** 275.
24. Escaig, J. (1982). *Journal of Microscopy*, **126,** 221.
25. Philips, T. E. and Boyne, A. F. (1984). *Journal of Electron Microscopy Technique*, **1,** 9.
26. Ingram, M. J. and Ingram, F. D. (1984). In *The science of biological specimen preparation* (ed. J.-P. Revel, T. Barnard, and G. H. Haggis), pp. 167–74. SEM Inc., A.M.F. O'Hare, IL 60666, USA.
27. Moor, H., Kristler, I., and Müller, M. (1976). *Experientia*, **32,** 805.
28. Müller, M., Meister, N., and Moor, H. (1980). *Mikroskopie*, **36,** 129.

29. Pscheid, P., Schudt, C., and Plattner, H. (1981). *Journal of Microscopy*, **121,** 149.
30. Knoll, G., Oebel, G., and Plattner, H. (1982). *Protoplasma*, **111,** 161.
31. Sitte, H., Edelmann, L., and Neumann, K. (1987). In *Cryotechniques in biological electron microscopy* (ed. R. A. Steinbrecht and K. Zierold), pp. 87–113. Springer-Verlag, Berlin and Heidelberg.
32. McDowall, A. W., Hofmann, W., Lepault, J., Adrian, M., and Dubochet, J. (1984). *Journal of Molecular Biology*, **178,** 105.
33. Chang, J.-J., McDowall, A. W., Lepault, J., Freeman, R., and Walter, C. A. (1983). *Journal of Microscopy*, **132,** 109.
34. Hagler, H. K. and Buja, L. M. (1986). *Journal of Microscopy*, **141,** 311.
35. Roos, N. (1988). *Scanning Microscopy,* **2,** 323.
36. Tokuyasu, K. T. (1973). *Journal of Cell Biology,* **57,** 551.
37. Griffiths, G., Simons, K., Warren, G., and Tokuyasu, K. T. (1983). In *Methods in enzymology*, Vol. 96 (ed. S. Fleischer and B. Fleischer), pp. 435–50. Academic Press, New York and London.
38. Griffiths, G., McDowell, A. W., Back, R., and Dubochet, J. (1984). *Journal of Ultrastructure Research*, **89,** 65.
39. Steinbrecht, R. A. and Müller, M. (1987). In *Cryotechniques in biological electron microscopy* (ed. R. A. Steinbrecht and K. Zierold), pp. 149–72. Springer-Verlag, Berlin and Heidelberg.
40. MacKenzie, A. P. (1972). *Scanning Electron Microscopy*, 1972/II, 273.
41. Wildhaber, I., Gross, H., and Moor, H. (1982). *Journal of Ultrastructure Research*, **80,** 367.
42. Fernández-Morán, H. (1957). In *Metabolism of the nervous system* (ed. D. Richter), pp. 1–34. Pergamon Press, Oxford.
43. Steinbrecht, R. A. (1985). *Journal of Microscopy*, **140,** 41.
44. Nagele, R. G., Kosciuk, M. C., Wang, S. M., Spero, D. A., and Lee, H. (1985). *Journal of Microscopy*, **139,** 291–301.
45. Harvey, D. M. R. (1982). *Journal of Microscopy*, **127,** 209.
46. Zalokar, M. (1966). *Journal of Ultrastructure Research*, **15,** 469.
47. Müller, M., Marti, T., and Kriz, S. (1980). In *Electron microscopy 1980*, Vol. II. *Proceedings of the European Congress on Electron Microscopy, Leiden* (ed. P. Brederoo and W. de Priester), pp. 720–1.
48. Benshalom, G. and Reese, T. S. (1985). *Journal of Neurocytology*, **14,** 943.
49. Heath, B., Rethoret, K., Arsenault, A. L., and Ottensmeyer, F. P. (1985). *Protoplasma*, **128,** 81.
50. Roos, N., Kinde, U., and Morgan, J. (1990). *Journal of Electron Microscopy Technique*, **14,** 39.
51. Marshall, A. T. (1980). *Scanning Electron Microscopy*, 1980/II, 395.
52. Humbel, B. M., Marti, T., and Müller, M. (1983). *Beiträge zur Elektronenmikroskopischen Direktabbildung von Oberflächen*, **16,** 585.
53. White, D. L., Andrews, S. B., Faller, J. W., and Barrnett, R. J. (1976). *Biochimica et Biophysica Acta*, **436,** 577.
54. Müller, M. (1981). In *Membrane proteins* (ed. A. Azzi, U. Brodbeck, and Z. Zahler), pp. 252–6. Springer-Verlag, Heidelberg.
55. Sitte, H. (1984). *Zeiss Information Magazine for Electron Microscopists, MEM*, **3,** 25.
56. Kaeser, W. (1989). *Journal of Microscopy*, **154,** 273.

57. Parsons, D., Bellotto, D. J., Schulz, W. W., Buja, M., and Hagler, H. K. (1984). *EMSA Bulletin*, **14,** 49.
58. Stang, E. and Johansen, B. V. (1988). *Journal of Microscopy*, **149,** 77.
59. Newbury, D. E., Joy, D. C., Echlin, P., Fiori, C. E., and Goldstein, J. I. (1986). *Advanced scanning electron microscopy and X-ray microanalysis*. Plenum Press, New York.
60. Umrath, W. (1983). *Mikroskopie*, **40,** 9.
61. Stumpf, W. E. and Roth, L. J. (1967). *Journal of Histochemistry and Cytochemistry*, **15,** 243.
62. Edelmann, L. (1978). *Journal of Microscopy*, **112,** 243.
63. Roos, N. and Barnard, T. (1985). *Ultramicroscopy*, **17,** 335.
64. Carlemalm, E., Garavito, R. M., and Villiger, W. (1982). *Journal of Microscopy*, **126,** 123–143.
65. Newman, G. R. and Hobot, J. A. (1987). *Journal of Histochemistry and Cytochemistry*, **35,** 971.
66. Carlemalm, E., Villiger, W., Hobot, J. A., Acetarin, J. D., and Kellenberger, E. (1985). *Journal of Molecular Biology*, **140,** 55.
67. Roth, J. (1983). In *Techniques in immunocytochemistry,* II (ed. G. R. Bullock and P. Petrusz), pp. 217–84. Academic Press, London and New York.
68. Causton, B. E. (1984). In *Immunolabelling for electron microscopy* (ed. J. M. Polak and I. M. Varndell), pp. 29–36. Elsevier, Amsterdam, New York and Oxford.
69. Bénichou, J. C., Fréhel, C., and Ryter, A. (1990). *Journal of Electron Microscopy Technique,* **14,** 289.
70. Glauert, A. M. and Young, R. D. (1989). *Journal of Microscopy*, **154,** 101.
71. Roth, J., Bendayan, M., Carlemalm, E., Villiger, W., and Garavito, M. (1981). *Journal of Histochemistry and Cytochemistry*, **29,** 663.
72. Ashford, A. E., Allaway, W. G., Gubler, F., Lennon, A., and Sleegers, J. (1986). *Journal of Microscopy*, **144,** 107.
73. Ebersold, H. R., Lüthy, P., Cordier, J. L., and Müller, M. (1981). *Journal of Ultrastructure Research*, **76,** 71.
74. Humbel, B. M. and Muller, M. (1986). In *The science of biological specimen preparation* (ed. M. Müller, R. P. Becker, A. Boyde, and J. J. Wolosewick), pp. 175–83. SEM, AMF O'Hare, IL 60666, USA.
75. Edelmann, L. (1989). In *Electron probe microanalysis: Applications in biology and medicine* (ed. K. Zierold and H. K. Hagler), pp. 33–41. Springer Series in Biophysics, Vol. 4. Springer-Verlag, Berlin and Heidelberg.
76. Monaghan, P., and Robertson, D. (1990). *Journal of Microscopy*, **158,** 355.
77. Wroblewski, R., Wroblewski, J., Anniko, M., and Edström, L. (1985). *Scanning Electron Microscopy*, 1985/I, 447.
78. Edelmann, L. (1986). *Scanning Electron Microscopy*, 1986/IV, 1337.
79. Sitte, H., Neumann, K., and Edelmann, L. (1986). In *Science of biological specimen preparation* (ed. M. Müller, R. P. Becker, A. Boyde, and J. J. Wolosewik), pp. 103–18. SEM, AMF O'Hare, Il 60666, USA.
80. Newman, G. R., Jasani, B., and Williams, E. D. (1982). *Journal of Microscopy*, **127,** RP5–RP6.
81. Newman, G. R., Jasani, B., and Williams, E. D. (1983). *Histochemical Journal*, **15,** 543.

82. Stirling, C. E. and Kinter, W. B. (1967). *Journal of Cell Biology*, **35,** 585.
83. Edelmann, L. (1980). *Physiological and Chemical Physics*, **12,** 509.
84. Dürrenberger, M., Bjornsti, M.-A., Uetz, T., Hobot, J. A., and Kellenberger, E. (1988). *Journal of Bacteriology*, **170,** 4757.
85. Kaeser, W., Koyro, H.-W., and Moor, H. (1989). *Journal of Microscopy*, **154,** 279.
86. Horowitz, R. A., Giannasca, P. J., and Woodcock, C. L. (1990). *Journal of Microscopy*, **157,** 205.
87. Zierold, K. (1984). *Ultramicroscopy*, **14,** 201.
88. Zierold, K. (1988). *Journal of Electron Microscopy Technique*, **9,** 65.

4

Immunogold labelling techniques

ALEX D. HYATT

1. Introduction

The use of colloidal gold as a marker in immunoelectron microscopy (IEM) is now commonplace. The technique is generally referred to as immunogold staining or the IGS method. The success of the technique stems from the various characteristics of colloidal gold; for example (a) its electron-dense and particulate nature, (b) the range of sizes and the monodisperse size distributions to which it can be prepared, (c) the ease with which it can be complexed to a wide range of ligands, (d) the maintenance of the bioactivity of the complexed macromolecules, and (e) the ability of gold probes to produce strong secondary and backscattered signals when bombarded with electrons and thus its suitability to scanning electron microscopy (SEM) in addition to transmission electron microscopy (TEM). It is therefore not surprising that the IGS technique is now used routinely as opposed to other techniques which use either ferritin as a tag or peroxidase which leads to a diffuse electron dense substrate over the antigenic region of interest.

Over the past one to two decades the IGS method has been applied to the detection of micro-organisms, cell surface associated and intracellular antigens. The techniques utilized in these applications include negative contrast immunoelectron microscopy (NCIEM), pre- and post-embedding IGS, freeze-fracture immunoelectron microscopy, and scanning immunoelectron microscopy (SIEM). It is beyond the scope of this chapter to discuss each IGS technique which has been utilized over the past decade, thus only the more popular IGS procedures which can be performed in routine electron microscopy laboratories will be covered. It is therefore the aim of this chapter to present laboratory protocols for NCIEM, pre- and post-IGS of plastic sections, SIEM, and briefly the labelling of cryosections. These protocols will be based on methods which have been successfully implemented in this and other laboratories.

2. Background

IGS is a term which covers a broad range of techniques. The success of these techniques depends upon a standard approach; that is, the basic understanding of each reagent/step used in the protocol. The standard reagents/steps which influence the success of IGS are discussed below.

2.1 Antibodies and antigens

The major questions which must be asked prior to any IGS include the following.

(a) What is the antigen to which the antibody is directed? This is important when investigating particulate matter; for example, viruses, bacteria or isolated organelles. For example, if the antibody is directed against an organelle membrane then the thin-section freeze-fracture technique may be more applicable than post-embedding IGS. Alternatively, if the antibody is directed against an internal structure such as a capsid protein of an enveloped virus then either the envelope could be removed with mild detergent (NP40 or Triton X-100) and processed for NCIEM or post-embedding IGS or cryosections could be employed. The important point to note is that if the structure and general localization of the antigen is known then an appropriate protocol can be selected which will optimize its detection.

(b) What is the nature of the antibody? Polyclonal antiserum comprises a range of antibodies directed against a collection of antigenic determinants. Monoclonal antibodies (MAbs) on the other hand comprise a single class of antibody directed against a single epitope. The ramifications of this become apparent when the effects of fixation are considered. Aldehyde fixative such as glutaraldehyde produce strong cross-linking of proteins which in turn may result in direct modification of, or confirmational changes in the epitope(s). These effects may be pronounced in monoclonal antibodies where modification of one antigenic site, may effect the accessibility and binding of the antibody.

(c) From what species are the antibodies derived? The species in which an antibody is produced is another important factor in immunolabelling. If, for example, indirect or bridge methods (see below) are to be used, then attention must be given to ensure that the antibody from one species will interact with that from another. If protein A or protein G complexed gold is used then care should be exercised to ensure that these complexes are compatible with the specific antibodies. The form of the antigen to which the antibody has been raised must also be known. If, for example, the antigen was denatured with sodium dodecylsulphate then the resultant antibody may not react or may have limited reactivity with native cell-associated antigens. IGS should ideally be performed with antibodies raised to pure un-denatured antigens produced in a species compatible with the overall labelling system.

2.2 Fixation

Most specimens intended for IGS require some type of fixation. Fixation is required for ultrastructural preservation and to render the antigen insoluble (an exception to this rule can apply to IGS of viruses by NCIEM; see below). Whilst fixatives maintain specimen ultrastructure, which is imperative for morphological examination, they can have a deleterious effect on antigen reactivity. The general acceptance of the deleterious effect of fixatives is not always correct; for example, some antigens can survive severe fixation such as 2.5% glutaraldehyde (1 h) and 1% osmium tetroxide (1 h). Furthermore, these same antigens may also survive the extractive effects of dehydrating agents and infiltration and polymerization with epoxy resins (48 h, 60°C). Thus the choice of fixative, its concentration, vehicle, and time-course must in the end depend on the survival of sample antigenicity.

Overall the above information indicates the necessity for preliminary experiments. These experiments can be performed using fluorescence, immunoenzyme, or gold–silver methods at the light microscopy level thereby enabling a large number of variables to be tested in an efficient manner. A matrix can be designed whereby the possible adverse effects of fixatives (type, concentration, time-course) and buffers can be determined. Furthermore, if extracted or partially extracted specimens are to be used then the accessibility of the antigenic determinants to antibodies can be assessed. Other factors such as compatibility of antibodies and the effects of heat inactivation can also be quickly monitored. Once the fixation and labelling regimes have been selected, specimens can be processed for electron microscopy (EM) and the system 'fine-tuned'.

2.3 Labelling procedures

2.3.1 Basic procedures

Direct, indirect, and bridge techniques can be used in IEM. Today, many researchers use either the direct or indirect labelling techniques as opposed to the bridge technique which is usually restricted to applications where the signal requires amplification. Briefly, the direct method involves the use of labelled primary antibodies which bind to antigenic sites. The indirect method utilizes unlabelled primary antibodies and secondary labelled immunocomplexes; this is more sensitive than the direct method. The bridge technique, in its simplest form, involves unlabelled primary antibody (generated, for example, in mice), a secondary antibody (for example, rabbit anti-mouse IgG) and a third labelled antibody (for example, gold labelled mouse anti-rabbit IgG); thus the secondary antibody forms a bridge between the primary and final antibody. One of the more popular variations of this procedure is the streptavidin–biotin bridge technique.

2.3.2 Probe size

One of the advantages of colloidal gold is the range of monodisperse sizes to which it can be produced (3–>100 nm). The size of the gold probe influences the resolution and intensity of labelling. Resolution (distance from gold marker to the site of antigen–antibody interface) is influenced by the selection of immunolabelling procedure (direct indirect or bridge) and/or size of the gold probe. The best resolution is obtained with the direct labelling procedure which incorporates a small gold probe complexed antibody. When other procedures are used the resolution varies according to the size of the antibodies and gold complexes. Probe size also effects intensity of labelling. The reasons for this include the ability of smaller probes to penetrate specimens (for example, extracted or open matrixed specimens) and the reduced effect of steric hindrance.

2.3.3 Non-specific interactions

In any immunocytochemical reaction non-specific reactions and electrostatic interactions can occur. The sources of non-specific binding are many and are discussed in detail elsewhere (1, 2). Briefly, the non-specific interactions can be attributed to unreacted aldehyde groups, non-specific binding sites on the specimen, and possibly incomplete protein coating of colloidal gold (1, 3) and the resultant electrostatic interactions with the specimen. To reduce or abolish such reactions the following steps should be followed:

(a) If a fixation step is involved (such as 2.5% glutaraldehyde, 40–60 min) then free aldehyde groups must be quenched. For example, following fixation the sample should be exposed to NH_4Cl (50 mM, 1 h) glycine (50 mM, 1 h) or other amino acids.

(b) Condition the specimen to prevent or minimize non-specific binding of antibodies to 'reactive groups' within the specimen. This can be achieved by incubating the specimen with either high ionic strength buffers (for example, buffers containing up to 500 mM NaCl) or competing IEM inert proteins. These competing proteins or blockers are dissolved in buffers, the most common of which are phosphate-buffered saline (PBS) and 20 mM, Tris (pH 7.6–8.2). We have also had considerable success with 0.1 M cacodylate buffer (pH 7.2) and 0.1 M Sörensen's phosphate buffer (pH 7.2). Blockers which are routinely and successfully used are listed below:
  - ovalbumin (1%)
  - bovine serum albumin (BSA) (1%)
  - gelatin (including fish gelatin) (1–5%)
  - defatted (skim) milk (2–4%)
  - acetylated BSA (0.01%)

(c) If the colloidal gold was prepared by the tannic acid–lead citrate method (4) then some non-specific binding can occur due to tannic acid residues on the gold particles (1, 3). This problem can be rectified by diluting the gold probe in a buffer containing either 1–5% fish gelatin or a mixture of 1% BSA, 0.05% Triton X-100, and Tween-20 (3). If Triton X-100 or Tween-20 are used on unfixed specimens, then care should be exercised as these reagents are detergents and will have a disruptive effect on lipid containing structures.

(d) Select the optimum dilution of antibody or antibodies and/or gold complexes because as too concentrated dilution can lead to poor signal-to-noise ratio. A general rule is to select the largest dilution which maintains a discernible signal. Roth (3) recommends $OD_{525} = 0.44$ (15 nm protein A–gold) and $OD_{525} = 0.06$ (6–10 nm protein A–gold). Antibody dilutions are recommended at 2–20 μg/ml for low to medium titre antiserum.

2.3.4 Incubation times

Incubation times vary greatly. Some researchers use times as short as 10–15 min at room temperature (RT) whilst others use 2–4 h (RT), overnight (4°C) or 30–60 min (37°C). We have found that the cleanest preparations are obtained by overnight incubations at 4°C or 1–2 h (RT).

An indication of the effect of incubation times on intensity of labelling is illustrated in *Figure 1*. *Figure 1* shows that (for this example) incubation at 37°C for 1 h was excessive as 60–70% of optimal binding could be achieved within 20 min at RT. It should be noted, however, that if experiments are being performed which required saturation of epitopes or maximum gold-probe–antigen interactions then a longer incubation period would be required. Incubation times should be selected according to (a) the objective of the procedure, and (b) to minimize background if indeed this is a problem.

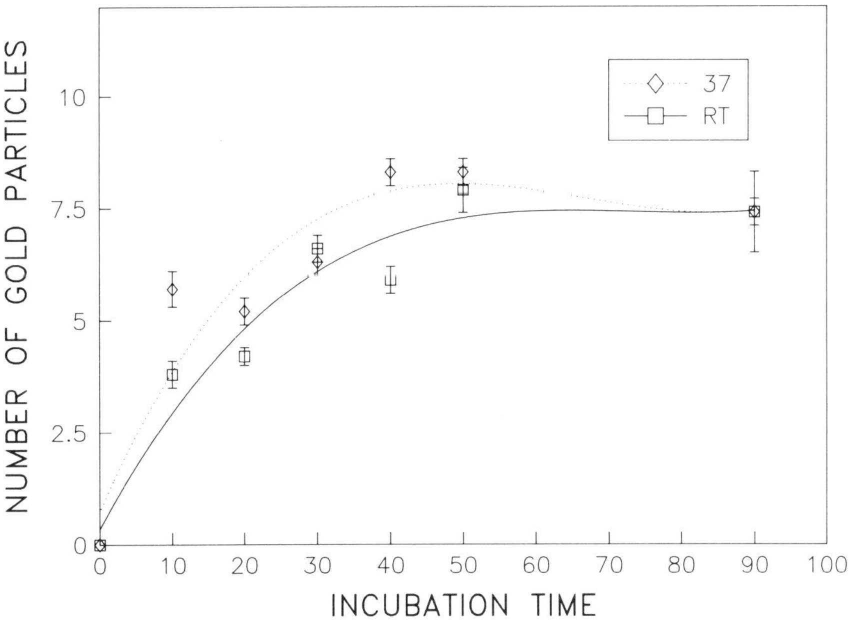

Figure 1. Time of incubation (RT and 37°C) vs. intensity of gold labelling. Standard error of means are indicated, $n = 200$ for each sample point.

3. Negative contrast immunoelectron microscopy (NCIEM)

NCIEM is mainly restricted to virological and bacterial applications. The method is, however, applicable to any specimen which is particulate and can be adsorbed to an EM grid.

3.1 Preparation of grids

Protocol 1. Preparation of grids

1. Select gold or nickel grids, the mesh size depends upon the application. Note that gold or nickel grids should be used for all IGS procedures.
2. If gold grids are selected then wash in concentrated HNO_3 followed with acetone (100%), methanol (100%), and water.
3. Film grids with any plastic; for example, butvar, parlodion, or formvar.
4. Coat filmed grids with carbon.

The grids can be stored or used immediately. In this laboratory better particulate adsorption has been achieved with freshly prepared grids.

3.2 Adsorption of particles to grids

There are many methodologies whereby particulate specimens can be adsorbed to the grid-substrate. The simplest method involves air-drying the sample on to the grid. Others involve variations on this theme; for example, samples containing excess salts can be sandwiched between a cube of 2% agar and a grid whereby the salts diffuse into the agar and the sample adsorbs to the grid. A popular variation of this method involves the deposition of plastic on to the dried agar surface, the edges are cut with a sharp razor blade and the plastic film (with sample) floated off in a dish of stain; this is called the pseudoreplica technique. Irrespective of the mode of adsorption, the particles will remain firmly attached to the carbon-filmed substrate despite repeated and prolonged washing steps incurred during IGS. If, however, the specimens are removed through washing, then a light fixation step (0.1% glutaraldehyde in appropriate buffer for 2 min or 2% (*para-*)formaldehyde for 15 min) may be used after adsorption, provided that the initial experiments indicate that this will not inactivate the antigens.

3.2.1 Conventional procedures

Conventional NCIEM involves the labelling of particles adsorbed to the grid substrate (as described above). Procedures do exist which involve labelling of particulate matter in suspension; however, these methods require substantially more time, greater volumes of antigen, and overall produce a higher background (that is, a poor signal-to-noise ratio). Due to the concise nature of this chapter only the procedures which produce good signal-to-noise ratios will be outlined.

Protocol 2 outlines the procedure for indirect IGS, an example of the results is illustrated in *Figure 2*. If the direct procedure is used then steps 7–8 should be omitted; however, if bridged-antibody procedures are used, then step 5 should be followed by step 4 and then the procedure repeated as listed.

Protocol 2. Immunolabelling schedule

1. Adsorb particles to grid.
2. Wash in buffer; for example PBS (×3).[a]
3. Incubatc in normal 'spccics' scrum (1%) or new bovine fetal calf serum in PBS (15–60 min) at RT.[b]

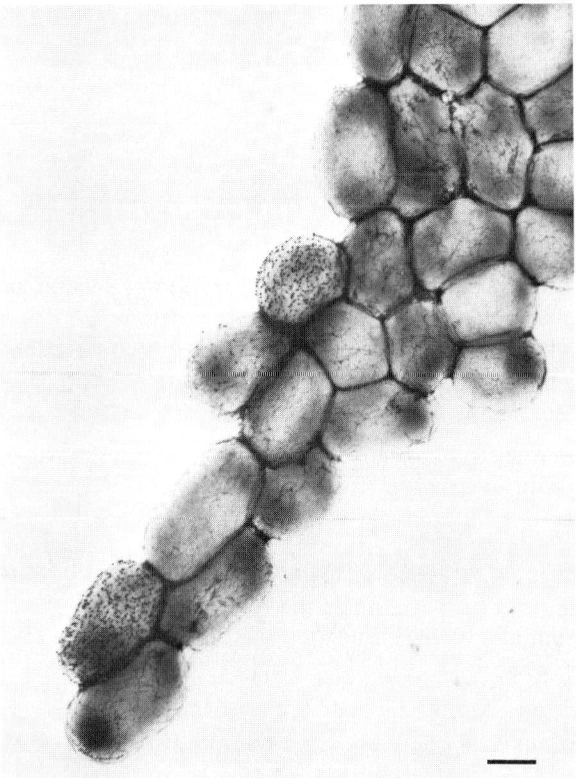

Figure 2. NCIEM of a mixed bacteria population. The labelled (protein A–gold) bacteria represent specific isolates of *Pasteurella multocida*. Bar represents 400 nm.

Protocol 2. *Continued*

4. Rinse in buffer containing blocker (10 min) – incubation buffer.
5. Incubate with antibody diluted in buffer (may contain blocker). For MAbs begin with dilutions of 1:10, 20, 40, and 80; for polyclonal antisera 1:50, 100, 200, and 400 (15–120 min RT) may be appropriate.
6. Rinse in incubation buffer (×6). (If step 5 involved gold labelled antibody then go to step 8).
7. If an indirect method is used then incubate for 15–20 min at RT in appropriately diluted gold complex (note that special blockers, described above, may be required).
8. Rinse in buffer (×6).
9. Fix in 2.5% buffered glutaraldehyde (5 min).
10. Rinse in distilled water (2 × 1 min).
11. Contrast with appropriate negative stain (for example, 2% phosphotungstic acid (PTA) or 1% ammonium molybdate pH 6.8).

[a] Rinses may be in the form of transferring grids from one pool to another, exchanging fluid in the pool via micro-pipettes, dipping in beakers of buffer, or sprays/squirts from wash bottles.
[b] Include this step if non-specific labelling is a problem; cannot be used with protein A.

3.2.2 Grid-cell-culture technique

Other popular methods of NCIEM involve those described by Hyatt *et al.* (5) and Hyatt and Eaton (6). The methodology known as the grid-cell-culture technique is outlined below. Only the preparation of seeded grids (grids on which cells are adsorbed and grow) and cytoskeletons are described in *Protocols 3* and *4*. The immunolabelling is identical to that described in *Protocol 2*. For the concentrations of glutaraldehyde used in these procedures it should be noted that quenching of free aldehydes is not required.

Protocol 3. Seeding of grids

1. Prepare gold grids as described (*Protocol 1*).
2. Wash grids in 0.1% NP40 followed with appropriate cell maintenance media. (The detergent facilitates the removal of air bubbles which would normally remain on the grids when placed in solution.) Place the grids in a 35 mm Petri dish.
3. Seed cells at an absolute level or 0.1×10^6 in a 35 mm plastic Petri dish. Note, if too many cells are attached the film will break upon irradiation.
4. If the cultures are to be used for virus infection, then cells are allowed to settle and grow overnight before a 1 h virus adsorption step at 37°C.

Protocol 4. Preparation of cytoskeletons

1. 18–24 h post seeding or infection, transfer grids to pools containing either 0.1% buffered glutaraldehyde or 2% (*para-*)formaldehyde and 1% NP40 (2 min and 15 min respectively). The fixative is required to retain both cells on the grid and components of the cytoskeleton such as microtubules.
2. Rinse in buffer until all detergent is removed at which time the pool will lose its flattened appearance.
3. Perform IGS.
4. Post-fix specimens in 2.5% buffered glutaraldehyde (10 min) followed with 1% osmium tetroxide—may be aqueous (10 min).
5. Wash in distilled water and dehydrate (70–100%) ethanol (5 min each).
6. Critical point dry from carbon dioxide.
7. Carbon coating may be required if the specimen is unstable under the electron beam.

3.3 Double labelling

NCIEM involves the adsorption of particles on to a support film. Multiple labelling (*Figure 3*) can be performed by using colloidal gold of different sizes which are complexed to a variety of ligands but is restricted to one side of the grid. A popular procedure involves the use of protein A complexed to different-sized (monodispersive) gold particles. This is outlined in *Protocol 5*.

Protocol 5. Double labelling, protein A–gold

1–5. As *Protocol 2*.

6. Rinse in incubation buffer (×6). (If step 5 involved gold labelled antibody then go to step 10.)
7. Protein A–gold (small) (1:20–1:100), 15–120 min RT.
8. Wash in incubation buffer (6 × 3 min).
9. Free protein A (20–100 μg/ml), (15–30 min RT).
10. Wash in incubation buffer (6 × 3 min).
11. Secondary antibody (refer to step 5, *Protocol 2*).
12. Wash in incubation buffer (6 × 3 min).
13. Protein A–gold (large) (1:20–1:100), 15–120 min RT.
14. Steps 8–11 in *Protocol 2*.

Whilst many workers have enjoyed success with this method some have reported problems with cross-contamination (1, 7, 8). If this procedure is

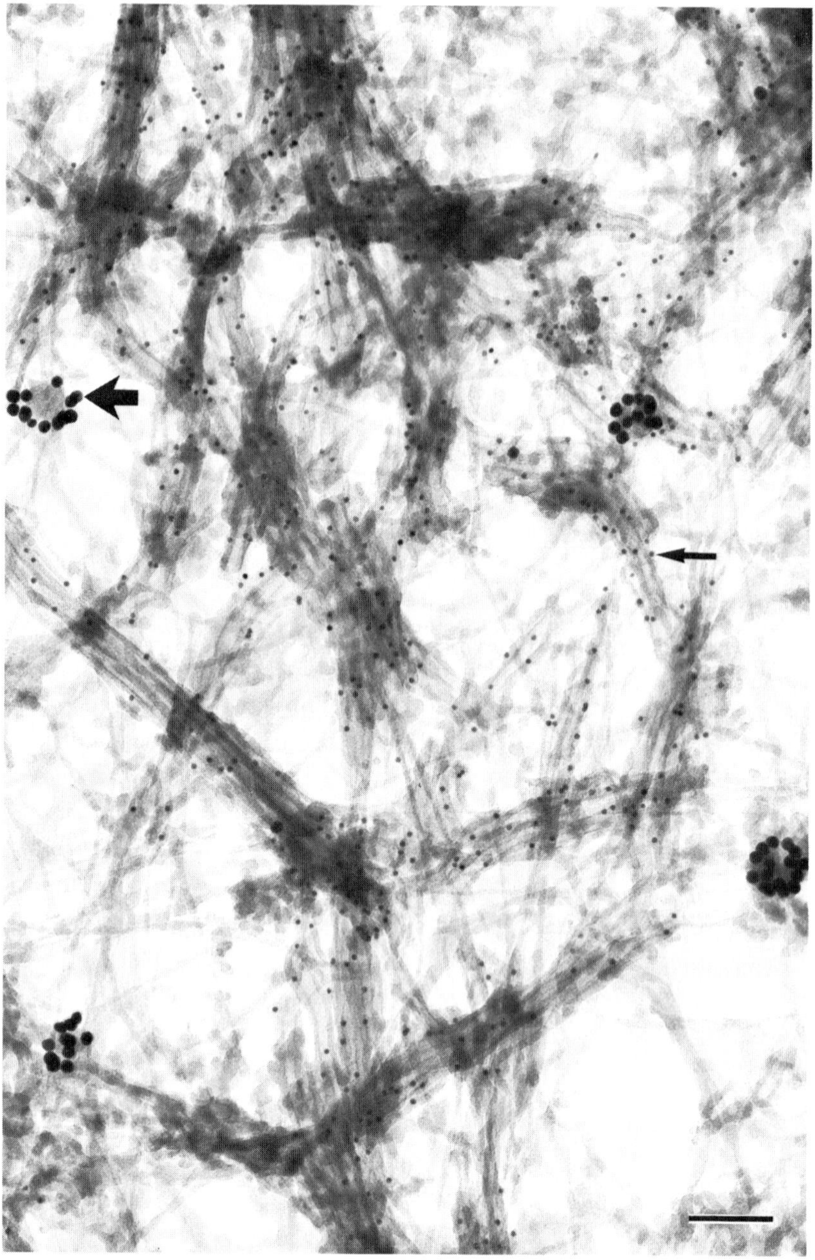

Figure 3. Double-labelling of bluetongue virus-infected cytoskeleton. The cytoskeleton is a whole mount; small probes (small arrow) indicate the presence of virus-tubule specific protein (NS1) and the large probes (large arrow) a virus specific protein (VP2). Labelling procedure, indirect [MAb(NSI) and protein A–gold] followed with direct labelling [gold complexed MAb (VP2)]. Bar represents 100 nm.

attempted then it is recommended that the first probe in the protocol should be the smaller of the two. Furthermore, it is recommended that the system be assessed by the addition of the larger protein A–gold complex in the absence of a secondary antibody. If cross-contamination occurs then the concentration and time of incubation of free protein A should be varied. Failure to abolish the above 'contamination' will indicate the inappropriateness of this labelling system to the specific application and other protocols should be attempted.

Variations of the above protocol can be attempted. The obvious variation involves the use of silver-enhancement of the smaller probe. Theoretically, the physical enhancement will occlude the reactive regions of the primary antibody and protein A–gold. Silver-enhancement can be performed by the method described by Danscher (9), or with the reagents supplied with commercial kits such as inteSEM (Amersham) or Aurion R-Gent (Aurion).

Double labelling can also be performed with antibody followed by anti-species gold complexes. To avoid similar problems to that described above the primary and secondary antibodies should be raised in different species which do not have a high degree of cross-reactivity. The procedure would be similar to that of *Protocol 5*. The ideal multiple labelling systems would involve direct gold complexes. The protocol would be similar to that of *Protocol 5* except the protein A–gold and associated steps would be omitted. Another multiple labelling system was developed by Wang and Larsson (10) which permits the use of antibodies of differing specificity (that is, to different epitopes) but raised in the same species. Briefly, the primary antibodies are saturated with gold labelled anti-IgG antibodies, the grids are then exposed to formaldehyde vapour thereby denaturing any free anti-IgG binding sites. The second (or third) labelling step is then permitted without the threat of cross-labelling. This procedure can also be applied to plastic (10) or cryosections (11). In any immunolabelling procedure the quality of the results are only as good as the controls, therefore it is important to test each relevant step particularly in multiple IGS.

3.4 Negative staining

Commonly used stains (see also Chapter 9) include uranyl acetate (UA, 1% pH not adjusted) phosphotungstic acid (2–4%, pH 6.5–6.8 adjusted with 1 M KOH), ammonium molybdate (1–4%, pH 6–8 adjusted with 1 M NaOH or NH_4OH) and sodium silicotungstate (1–4%, pH 6–8 adjusted with 1 M NaOH). If deep pooling of stain occurs around the specimen then shorter staining times or lower concentrations of stains should be used. The stains should be made up in distilled water, filtered prior to use and applied to the samples from 10 sec to 3 min. If fine detail is required then polyvinyl alcohol and uranyl acetate may be used; briefly, samples are stained in 1% UA (5 min) and a mixture of 1% PVA and 0.1% UA (5–20 min) (Tokuyasu, personal communication).

4. Immunogold staining (IGS) of plastic sections

4.1 Pre-embedding IGS

4.1.1 Non-extraction techniques

In non-extraction techniques, specimens can be maintained in buffer or lightly fixed and gold labelled prior to dehydration, resin infiltration and polymerization. Such procedures are generally limited to surface associated antigens in tissue culture or single-cell isolates. The method has the advantage that entire surface antigens are available for antibody interactions without the possible adverse effect of solvents and resins. If single cells are used then care should be taken in their preparation. For example, if the cells are pelleted into agar or gelatin then the concentration of this matrix can mask the antigenic sites (J. Leunissen, personal communication). High concentrations of gelatin for example (10–20%), whilst compatible with intracellular antigens can prevent the optimal interactions of antibodies with cell surface antigens; thus, for the detection of surface-associated antigens lower concentrations (2–5%) are recommended. The steps involved are outlined in *Protocol 6*.

Protocol 6. Pre-embedding IGS

1. Fix cells, for example 0.1–0.5% buffered glutaraldehyde (5–60 min).
2. Wash in buffer (6 × 3 min).
3. Wash in 50 mM glycine (15–60 min).
4. Label cells as per *Protocol 2* (steps 2–8).
5. Fix in 2.5% buffered glutaraldehyde.
6. Wash in buffer (3 × 20 min).
7. Post-fix (for example, 1% buffered osmium tetroxide) 1 h.
8. Proceed as per conventional resin procedures.

Note:
If gelatin is included, the subsequent fixation is required for stabilizing the gelatin. The gelatin may also give the extracellular background a textured appearance.

4.1.2 Extraction techniques

Procedures exist whereby penetration of cells with antibody and/or gold probes have been attempted. In general, these attempts have resulted in labelling of the upper few microns of the sample. Permeablization of biological samples has been achieved with detergents such as NP40, Triton X-100, and saponin (0.02–4%). These detergents do, however, have a destructive effect on membranes and effectively extract cells, thus at higher concentrations the cytoplasmic ground substance and organelles are eluted from the sample. Another problem associated with rendering samples permeable, even with very low concentrations of detergents, is that there is the risk of losing or redistributing diffusable antigens. Some workers (12) overcome this problem by using vibratome sections derived from freeze–thawed cryoprotected specimens. Another alternative is to use 1 nm gold probes. It is claimed that these probes can penetrate sections (non-plastic) and even whole cells (refer to Section 5); however, the suitability of this latter approach to pre-embedding IGS is yet to be substantiated in the scientific literature. Overall, if extraction techniques are to be used then they should be restricted to work on the insoluble components of the cell. If IGS is to be attempted on the soluble intracellular components of non-extracted cells at a depth of, say, greater than 3 nm, then other procedures involving cryoultramicrotomy or thin-section fracturing labelling or post-embedding IGS should be attempted.

4.2 Post-embedding IGS

Resins which are routinely used in electron-microscopy laboratories, namely the epoxy resins, polymerize into irreversible three-dimensional cross-linked solids. These resins, however, are generally ignored in IGS in favour for the cross-linked acrylic resins such as L. R. White, L. R. Gold, and the Lowicryls, K4M, K11M, HM20, and HM23 (also see Chapter 2). It should be noted that before proceeding to the acrylic resins the viability of antigens should be tested in a range of fixatives and resins; for example, in this laboratory some viral antigens have survived conventional fixation in glutaraldehyde, osmium tetroxide, and embedding and polymerization in Epon 812 (60°C, 48 h). The resultant sections, however, had to be etched with hydrogen peroxide (13) or treated with sodium metaperiodate (14) if successful labelling was to occur. The procedure is as given in *Protocol 7*.

Protocol 7. IGS of epoxy sections

1. Cut conventional sections and mount on coated or uncoated nickle or gold grids.
2. Treat with 10% H_2O_2 (15 min) or saturated sodium metaperiodate (30–60 min).
3. Rinse in buffer (6 × 3 min).
4. Continue as described in *Protocol 2* or *5*.

If the conventional resins prove to be unsuitable one of the acrylic resins should be used. The selection of the resin depends primarily upon which protocol is used, susceptibility of antigens to fixatives, extractive effects of alcohols, and simplicity of use in the work-place.

The Lowicryls which do not become very viscous at low temperatures (for example, −20 to −30°C) can be used for the progressive lowering temperatures (PLT) and freeze-substitution techniques. Although the success of freeze-substitution in IGS is still being assessed the PLT method is, at present, considered the superior method for the gold-labelling of plastic sections. Typical results are illustrated in *Figure 4*. The protocols for PLT and freeze-substitution are given below.

Protocol 8. PLT

1. Fix specimens (for example steps 1–5 protocol 6).
2. Cut specimens into small blocks (minimum of 1 mm^3).
3. Rinse in buffer (3 × 5 min).
4. 30% ethanol[a] 30 min (0, 0, 0, 0,°C).
5. 50% ethanol (1 h) (−20, −20, −20, −20°C).
6. 70% ethanol (1 h) (−35, −50, −50, −50°C).
7. 95% ethanol (1 h) (−35, −50, −60, −60°C).
8. 100% ethanol (1 h) (−35, −50, −60, −80°C).
9. 100% ethanol (1 h) (−35, −50, −60, −80°C).
10. Ethanol: Lowicryl (1:1 u/v) (−35, −50, −60, −80°C).
11. Ethanol: Lowicryl (1:2 v/v) (−35, −50, −60, −80°C).
12. Pure resin (1 h) (−35, −50, −60, −80°C).
13. Pure resin (1 h, or overnight) (−35, −50, −60, −80°C).
14. Fresh change in resin and polymerize by indirect UV lights (for example, two 15-W tubes, at approximately 30–40 cm above the sample (24 h).

Note: temperatures are for K4M, HM20, K11M, and HM23 respectively.

[a] Acetone has a freezing point of −27°C.

Protocol 9. Freeze-substitution (rapid) (see also Chapter 3)

1. Cryoprotect[a] samples in 0.6 M sucrose.
2. Cryofix by metal mirror or secondary cryogen (for example, liquid propane).
3. Ensure samples are small (minimum 1 mm^3).

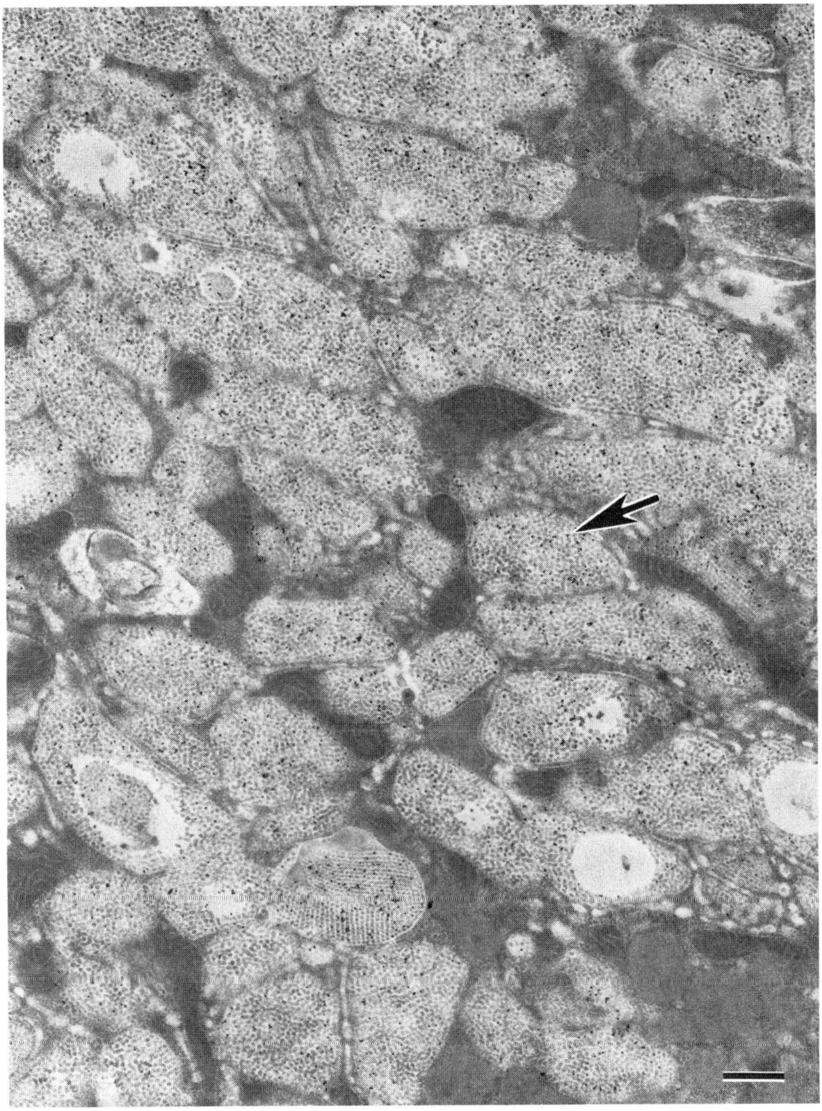

Figure 4. Post-embedding IGS. Specimen was prepared by PLT procedure. Labelling (direct) represents presence of RNA containing structures (arrow) within the distended cisternae of the Malpighian tubules from the common Australian field cricket. Bar represents 400 nm.

Protocol 9. *Continued*

4. Place in substitution medium; for example, acetone, methanol, or ethanol, or methanol +0.5% uranyl acetate (−80°C, 5 h).
5. Raise temperature to (−25° to −35°C) over 5–6 h period or to specific polymerization temperature.
6. After an additional 7 h, replace substitution media with ethanol: Lowicryl (1:2 v/v).
7. Proceed as per *Protocol 8*, Section 11.

[a] Cryo-protection will minimize ice-crystal damage.

The major limitation associated with PLT and freeze-substitution is the availability of cold-temperature rooms, refrigerators, or specialized equipment such as the Reichart CS auto. In addition, care must be taken to exclude all oxygen and water from the polymerization mixtures; osmicated and large samples can also prevent satisfactory polymerization. In the absence of sophisticated equipment the above conditions for successful polymerization can be achieved as follows. Replace the base of a plastic desicator with an aluminium well. Place ethanol (100%) in the well and a dessicant on the surrounds. Place the small gelatin capsules in wire baskets and ensure that they are in contact with the medium; heat generated during polymerization is withdrawn from the sample by the cold ethanol. Alternatively, the capsules can be placed directly on the base of the aluminium well (in the absence of alcohol). Replace the air with dry nitrogen gas (negative to ambient pressure) and put in a coldroom or freezer after placing a UV source above the sample.

LR White unlike the Lowicryls becomes viscous at low temperatures, but has the advantage of being miscible with water. In general, samples polymerized in LR White are subjected to comparatively high concentrations of fixatives (at low fixative concentrations ultrastructural preservation is better in Lowicryl sections). The procedure is thus used for routine cytochemical and immunocytochemical work provided the antigens can survive these conditions. The viscous nature of LR White at low temperatures translates to poor infiltration of specimens at low temperatures, the resin is therefore suitable for protocols incorporating heat or catalytic polymerization. The procedure for heat polymerization is given in *Protocol 10*.

Protocol 10. LR White (routine)

1. Fix tissue (for example, steps 1–5, *Protocol 5*).
2. Transfer to 70% ethanol (3 × 1 h) 2 (vol.) ethanol: 1 vol. (LR White) (1 h).
3. Pure LR White (2 × 1 h); use rotary device, and freshly purchased stock.

4. Place specimens in fresh LR White within small, tightly capped ('O' gauge) gelatin capsules.
5. Polymerize at 50°C for 24 h.

Catalytic polymerization can also be performed at room temperature or in the cold. These procedures are reviewed by Newman (15).

4.2.2 Labelling

i. Basic method

The labelling protocol for post-embedding IGS is similar to that described above. Briefly, the sections should be mounted on nickel or gold grids. The grids may be filmed, carbon-coated, and rinsed briefly in 20% acetic acid and 90% alcohol so as to enhance section adhesion. If either no or non-specific labelling occurs then attention should be focused on the areas discussed in Section 2.

ii. Multiple labelling

Multiple labelling on plastic sections can be performed on the one or both sides of the section. If a single section face is to be used any of the IGS protocols discussed previously will be suitable (Section 3.3.1). Unlike IGS of particulate material and cryosections, plastic sections can be labelled on both sides/faces; this is advantageous when cross-reactivity of immunoreagents is a problem. The double-side technique can incorporate a broad range of gold labels including protein A–gold. The labelling protocol is similar to that described in *Protocol 5* with the exceptions that (a) it does not matter which sized gold probe is used first, (b) no free protein A–gold is required, and (c) the section must be dried after rinsing in buffer and water. Care must be taken to ensure that both sides are not inadvertently contaminated.

5. Immunogold staining (IGS) of cryosections

It is not within the scope of this chapter to describe the preparation, sectioning, embedding, and staining of cryosections. (See Chapter 3 and refer to Leunissen and Verkleij, ref. 16.) The method for the gold labelling of cryosections is similar to that described within the context of this chapter. One aspect of IGS on cryosections which is frequently misunderstood is the permeability of the sections to the labelling reagents. Contrary to popular belief, labelling occurs only on the section surface and within cracks or fractures of the section, that is the sections are not permeable to immunoreagents. Recently, ultra-small gold particles ($\leqslant$1 nm) have become available, these have the potential to penetrate cryosections and provide *in toto*-IGS; this is an important feature when antigens of low antigenic mass are targeted. The probes require silver-enhancement because they are not readily detected in a conventional transmission electron microscope. The IGS schedule is similar to that described for conventional probes except that particular attention may have to be paid to background labelling.

6. Thin-section fracture-labelling

Cryotechniques involved in IGS extend to freeze-etching and fracture-label techniques, the latter of which were developed by Pinto da Silva (17–19). The simplest and perhaps the most useful of these techniques, in a routine electron microscopical laboratory, is the fracture-label technique. Details of other techniques and variations thereof are discussed elsewhere (20, 21). The fracture-label or thin-section fracture-label technique permits components of membranes and cytoplasm, including surfaces and internal components of organelles to be investigated by exposing split membranes and cross-fractured profiles of plasma, intracellular membranes, and cytoplasm to immunoreagents. This technique has been used successfully in this laboratory (*Figure 5*) and the general procedure described by Pinto da Silva (17, 18) is given in *Protocol 11*.

Protocol 11. Thin-section fracture-label

1. Fix biological sample (1–3% buffered glutaraldehyde, 30 min, RT).
2. If isolated cells are used embed in 30% BSA, centrifuge (1000 *g*, 3 min); repeat.
3. Post-fix in 1% buffered glutaraldehyde. Overlay resultant gel with glutaraldehyde (final concentration, 1%). Leave for 30 min, RT.
4. Cut block into 2 mm^3 cubes (may be stored at 4°C).
5. Impregnate cubes with 30% buffered glycerol. Add first drops of glycerol slowly and mix gently.
6. Freeze cubes in a secondary cryogen such as Freon 22.
7. Transfer samples to liquid nitrogen and crush in a tissue grinder such as a manual or motor-driven Dounce homogenizer which contains liquid nitrogen.
8. Allow tube containing the fractured sample to warm to RT. When liquid nitrogen has evaporated add 0.5–1% glutaraldehyde in 30% glycerol buffer solution.
9. Upon thawing (which can be enhanced by placing the tube in a 37°C water bath) wash (×2) in 30% glycerol buffer.
10. Deglycerinate by drop-wise addition of 1 mM glycyl–glycine (in buffer) or other aldehyde quenching reagent.

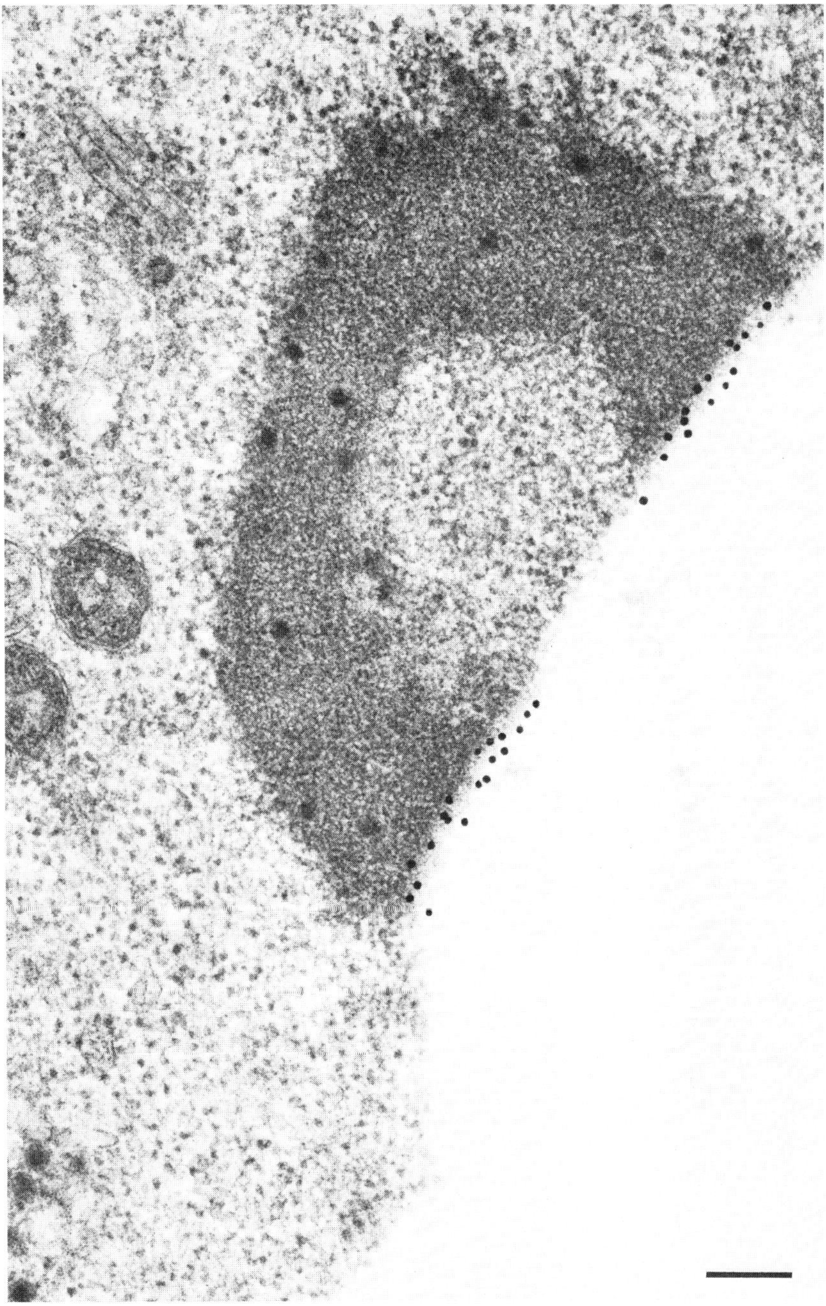

Figure 5. Thin-section fracture label of an intracellular virus induced inclusion body. Labelling (indirect) represents presence of a non-structural viral protein. Bar represents 200 nm.

Protocol 11. *Continued*

11. Label sample as described in other sections. (Note that centrifugation may be required between the labelling steps.)
12. Post-fix in 2.5% glutaraldehyde (buffered), (20 min, RT).
13. Wash in buffer (3 × 20 min).
14. Post-fix in 1% osmium tetroxide (buffered), (1 h, RT).
15. Proceed as per conventional plastic embedding protocol.

7. Scanning immunoelectron microscopy (SIEM)

SIEM has largely been restricted to large diameter markers (>20 nm). The reasons for this are that the smaller (5–15 nm) markers can only be detected by backscattered imaging and high-resolution scanning electron microscopes. It is generally not possible to image these probes by standard secondary electron imaging. Contributing factors to this problem include resolution, comparatively thick conductive coats, and the rough undulating surface of biological surfaces which obscure or mask the small and sometimes sparsely distributed gold markers. Whilst modern scanning electron microscopes can detect probes down to 15 nm by imaging backscattered electrons, these instruments are usually fitted with LaB_6 filaments and a backscattered electron detector capable of working at TV scan rates.

Large probes (20–40 nm), on the other hand, can be visualized with standard scanning electron microscopes, both in the secondary electron and backscattered electron image modes. Unfortunately, the use of the larger probes results in a decrease in labelling efficiency (6). The reasons for this are probably due to steric hindrance (that is, the ratio of binding sites to markers decrease—one marker physically blocks binding to other neighbouring sites), the inability of antibodies bound to the large markers to interact efficiently with the antigenic sites, and finally adverse electrostatic (repulsion) between the gold probes. In this laboratory we have overcome the above problems by silver-enhancing smaller (12 nm) gold probes bound to the surface of biological structures (*Figure 6*). This approach ensures the advantages incurred by the use of small gold probes (that is, increased labelling efficiency) while avoiding the problems frequently incurred by the use of the larger probes. A general method for SIEM using small gold probes and silver enhancement is given in *Protocol 12*.

Protocol 12. SIEM (silver-enhancement)

1. Rinse in a buffer of choice.
2. Fix in 0.1% buffered glutaraldehyde (10 min).

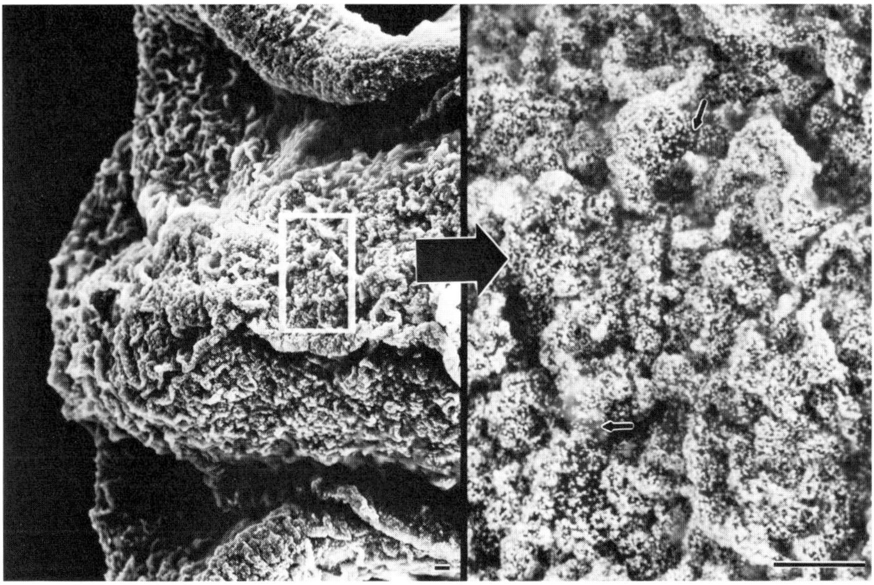

Figure 6. Scanning electron micrograph of the echinoderm 'crown of thorns' which has been gold labelled (indirect) and silver-enhanced. Small arrows, silver-enhanced gold particles. Bars represents 2 μm.

3. Rinse in buffer (3 × 5 min).
4. Quench aldehydes (especially if higher concentrations of glutaraldehyde are used).
5. Incubate • primary antibody, or
   • direct gold complex (37 °C, 1 h).
6. Wash in buffer (6 × 3 min).
7. Incubate on protein A–gold (37 °C, 1 h). (If direct method used, go to step 9.)
8. Wash in buffer (6 × 3 min).
9. Post-fix in 2.5% glutaraldehyde (20–40 min, RT).
10. Wash in buffer (3 × 15 min).
11. Wash thoroughly in double-distilled water.
12. Silver-enhance[a] (for example, Amersham intense M, 2 × 10 min).
13. Wash well in distilled water.
14. Dehydrate and critical point dry.
15. Coat with carbon or aluminium.

[a] Other silver-enhancement reagents/techniques, for example that described by Danscher (9), can be used.

When the gold probes are silver-enhanced, extreme care must be taken to ensure minimum self-nucleation; thus the range of blockers, buffers, and incubation times and concentration of reagents should be fully explored for any contribution to non-specific silver nucleation. If larger gold probes are used then the silver-enhancement step and washes can be omitted from the protocol. Imaging of the labelled specimen should also be performed with care and patience. Generally, a short working distance (8–10 mm), a probe current in the range of $3–6 \times 10^{-10}$ A, and an accelerating voltage of 15 kV will provide satisfactory results. Although the silver-enhanced particles produce a strong secondary electron image, the putative probes should be checked via backscattered imaging as to remove any possibility of confusing probes with focal areas of charging.

8. Concluding remarks

A number of protocols are described in this chapter which use colloidal gold as a marker for localizing intracellular and surface-associated antigens. It should be noted that for each specific application any one protocol may have to be varied according to the susceptibility of the antigen to different fixation regimes and efficient detection of the primary antibody. Attention should also be given to the information given in Section 2 of this chapter and an attempt made not to confuse background labelling with that incurred as a result of *bona fide* antigen–antibody interactions. In any immunocytochemical experiments controls must be included. The control experiments should be designed to produce information on (a) inherent 'background' labelling which may be due to interactions of the gold probe with the resin and/or components of the specimen, (b) the specificity of the antibody–antigen interactions and (c) for protein A or G immunolabelling, the specificity of the protein A– or G–antibody interactions. Finally, intense immunolabelling should not be perceived as the final goal. Optimum IGS occurs when non-specific labelling is minimized and the target can be concomitantly identified by a marker and ultrastructural detail. The presence of too many markers will obscure the accurate position and structure of the antigen.

Acknowledgements

I am indebted for the assistance of Ms S. Brookes, Mr T. Wise, and Ms S. Hengstberger, who are members of the electron microscopy group at this laboratory.

References

1. Hyatt, A. D. (1989). In *Colloidal gold: Principles, methods, and applications* (ed. M. A. Hayat), Vol. 2, pp. 19–31. Academic Press, London.

2. Roth, J., Taatjes, D. J., and Warhol, M. J. (1989). *Histochemistry*, **92,** 47.
3. Behnke, O., Ammitzboll, T., Jenssen, H., Klokker, M., Nilausen, K., Tranum-Jensen, J., and Olsson, L. (1986). *European Journal of Cell Biology*, **41,** 326.
4. Slot, J. W. and Geuze, H. J. (1985). *European Journal of Cell Biology*, **38,** 87.
5. Hyatt, A. D., Eaton, B. T., and Lunt, R. (1987). *Journal of Microscopy (Oxford)*, **145,** 97.
6. Hyatt, A. D. and Eaton, B. T. (1990). *Electron Microscopy Reviews*, **3,** 1.
7. Bendayan, M. (1982). *Journal of Histochemistry and Cytochemistry*, **30,** 81.
8. Bendayan, M. and Stephens, H. (1984). In *Immunolabelling for electron microscopy* (ed. J. M. Polak and I. M. Varndell), pp. 143–54. Elsevier, Amsterdam.
9. Danscher, G. (1981). *Histochemistry*, **71,** 81.
10. Wang, B.-L. and Larsson, L-I. (1985). *Histochemistry*, **83,** 47.
11. Bastholm, L., Nielsen, M. H., and Larsson, L-I. (1987). *Histochemistry*, **87,** 229.
12. Piekut, D. T. and Casey, S. M. (1983). *Journal of Histochemistry and Cytochemistry*, **31,** 669.
13. Baskin, D. G., Erlandsen, S. L., and Parsons, J. A. (1979). *Journal of Histochemistry and Cytochemistry*, **27,** 1290.
14. Bendayan, M. and Zollinger, M. (1983). *Journal of Histochemistry and Cytochemistry*, **31,** 101.
15. Newman, G. R. (1989). In *Colloidal gold: Principles, methods and applications* (ed. M. A. Hayat), Vol. 2, pp. 47–73. Academic Press, London.
16. Leunissen, J. L. M. and Verkleij, A. J. (1989). In *Immuno-gold labelling in cell biology* (ed. A. J. Verkleij and J. L. M. Leunissen), pp. 95–114. CRC Press, Boca Raton, Florida.
17. Pinto da Silva, P., Parkinson, C., and Dwyer, N. (1981). *Proceedings of the National Academy of Science, USA,* **78,** 343.
18. Pinto da Silva, P., Parkinson, C., and Dwyer, N. (1981). *Journal of Histochemistry and Cytochemistry,* **29,** 917.
19. Pinto da Silva, P., Kachar, B., Torrisi, M. R., Brown, C., and Parkison, C. (1981). *Science,* **213,** 230.
20. Pinto da Silva, P. (1989). In *Immuno-gold labelling in cell biology* (ed. A. J. Verkleij and J. L. M. Leunissen), pp. 179–97. CRC Press, Boca Raton, Florida.
21. Nermut, M. V. and Nicol, A. (1989). In *Colloidal gold: Principles, methods, and applications* (ed. M. A. Hayat), Vol. 1, pp. 350–73. Academic Press, London.

5

Electron microscopic autoradiographic techniques

GÉRARD MOREL

1. Introduction

At the electron microscopic level, autoradiography consists of the subcellular detection of radiation from a radioactive source in a nuclear emulsion over an ultra-thin section containing the radioactive isotope.

In autoradiography, the specimen itself is the source of the radiation which originates from radioactive material incorporated in it. The first step is the incorporation of this radioactive material. The second is the process of obtaining the specimen for electron microscopy. The third is the recording and visualization of radiation in nuclear emulsion. The last consists of demonstrating that these images reflect the distribution of radioactivity in the specimen.

The first autoradiograph was obtained over a century ago by Niepce de St. Victor (1). Thirty years later, Becquerel's experiment (2) confirmed these previous observations. Then, in 1898 the Curies demonstrated the existence of this radioactivity. Autoradiography is thus a technique which pre-dated the concept of radioactivity itself and contributed to its discovery.

The first biological applications of autoradiography were those of Lacassagne *et al.* (3, 4), who studied the polonium distributions. The development of this technique was very slow until after the Second World War and was limited by the small amount of interest shown by biologists in radioactive substances and the nature of photographic emulsions.

Advances in physics during and after the Second World War led to the production of nuclear emulsions (photographic emulsions with specialized characteristics) which recorded the tracks of charged particles with greater precision and sensitivity than before. At the same time, new radioactive isotopes became available, allowing new possibilities for biological systems, of which the first was the demonstration of the uptake of radioactive iodine by the thyroid gland (5), and then its distribution (6).

These results were obtained by placing the sectioned specimen in direct contact with a lantern plate. Belanger and Leblond (7) evolved a technique with liquid emulsion considerably increasing the resolution. Then, by 1955, Softes and Warren (8) had described the dipping technique, where the sections on the slides were dipped in melted nuclear emulsion. Several other authors (9, 10) attempted to improve the contact between emulsion and specimen by stripping the emulsion off its glass support and applying it directly to the specimen.

The electron microscope did not allow the use of liquid emulsion and stripping film techniques which produced emulsion layers a few microns thick over the surface of the specimen. The first autoradiographs made using electron microscopy were published in 1956 by Liquier-Milward (11). Since then, new techniques have been proposed, namely:

(a) the loop method, which was first proposed by Caro and Van Tubergen (12) and

(b) the dipping technique, developed by Saltpeter and Bachmann (14) using new nuclear emulsions. It is now possible to resolve the site of incorporation of radioactive material to within 500–700 Å in favourable circumstances.

Similarly, quantitation of signal is necessary to the study of the statistical significance of the autoradiograph, the sensitivity and the resolution of this method (15–19). Two different methods are now available: the probability circle method (20) and the cross-fire method (21).

Since the first application of the technique to biological material at the electron microscopic level reported by Liquier-Milward (11), the principle has not been modified.

2. Radioisotopes

Particles. Autoradiographs are produced by particles emitted during radioactive disintegrations. Gamma rays and X-rays are ineffective. Thus, only the emission of α particles, electrons (β^- particles) or positrons can be used. Auger electrons from electron capture reactions can also be utilized for biological material.

The main radioisotopes used are tritium (^3H), iodine-125 (^{125}I), and sulphur-35 (^{35}S).

Particle energies. The mean particle energy affects the efficiency of electron microscopic autoradiography (efficiency being defined as the number of silver grains formed for each radioactive disintegration in the specimen). The higher the β^- particle emission energy, the lower the efficiency of the recording process in very thin emulsion layers. β^- particle energies above 0.5 MeV are very difficult to detect. In other words, particle energy also affects the resolution of the autoradiographs, since it determines the amount of scattering in the emulsion and the maximum particle range.

Half-life. The half-life of the radioisotope is crucial in determining if it is suitable for electron microscopic autoradiography because it determines the specific radioactivity which can be attained in labelled molecules [specific activity is the amount of radioactivity in Curies (Ci) present in one mole or millimole of molecules]. The longer the half-life, the lower the number of nuclear transformations within a given time unit. However, if the half-life is too short, radioactivity will disappear during electron microscopic processing (labelling tissue, embedding, cutting ultra-thin sections, coating them with nuclear emulsion, and exposure). This half-life must be greater than 1 month.

3. Radiomolecules

The choice of a radioactive molecule depends on the particular experimental purpose:

- metabolism study, or
- binding study,

and involves three overlapping decisions. First, the molecule must be selected to be metabolically, physiologically, or pharmacologically appropriate taking account of the literature. Second, the biochemical form of the labelling molecule. Third, the cost (the price and the toxicity on the experiment must be evaluated).

(a) *Labelling*. Three different pathways are available to obtain radiomolecules:
   - substitution of one or more atoms of the stable isotope by a radioactive one
   - addition of a radioactive isotope to the molecule
   - incorporation of a radioactive isotope or radioactive molecule into the molecule during its synthesis

(b) *Purification*. After the labelling procedure, it is necessary:
   - to separate the labelling molecules from the excess of radioactive isotope, and
   - to check the integrity of the labelling molecule

(c) *Controls*. The first point is to determine the specific activity of the labelling molecule. The second is to check the biological activity of this labelled molecule. In some cases, addition or incorporation of a radioactive isotope modifies the structure of the molecule and its biological activity.

4. Administration of labelled molecule

4.1 Route of administration

The injection method may profoundly influence the distribution of labels in animal tissues (see reviews in refs 22 and 23). Briefly, some parts of the brain may best be labelled by injection into the cerebral ventricles; the intestine is especially well labelled by intraperitoneal injections, whilst the kidney can sometimes be conveniently labelled by intra-aortic injection (24–28). The state of alert is necessary to inject somatostain, or corticotropin-releasing hormone through the cannula, without anaesthesia (29), and growth hormone-releasing factor, with anaesthesia (ref 30, and Morel, unpublished data). The cannula is introduced two days previously into the jugular.

Invertebrates can be labelled by injection into the haemocoele or into the coelomic cavity. Labelling *in vitro* has also been successfully used (31).

4.2 The dose of radiomolecule injected

In electron microscopic autoradiography, a high radioactive isotope concentration in the tissue is necessary, since the mass of tissue in an ultra-thin section is minute. Time exposure depends of the concentration of labelled molecules. On average, these concentrations of radiomolecules are respectively up to 10 times greater, and 10–100 times greater, for electron microscopic autoradiography than those for the light microscopic and scintillation counting (32). For the labelling of animals *in vivo*, to study the biosynthesis process, radioactive precursors are generally administered at levels of 10–40 μCi/g body weight, while to study hormone binding ligands are injected at less than 1 μCi/g body-weight via intravenous injection (29) or into the third ventricle (28). *In vitro*, the concentration of the ligand is about 0.2 μCi/ml of culture medium (31). For more consideration of the ways of determining effective doses for work *in vivo* and *in vitro*; see also Williams, refs 23 and 32.

5. Preparation of labelled tissues and cells

5.1 The choice of protocol

The method of preparing a specimen in order to achieve an autoradiography, e.g. to produce a consistent and informative picture without any direct regard for the quantitative retention or extraction of particular tissue components, involves both necessities:

- to retain the specificity of the labelling, and
- to preserve the ultrastructural aspect

In this way, the preparation differs according to the experiment and the nature of the radiomolecule.

5.1.1 Amino acids, nucleosides, and sugars used as precursors

The macromolecules (proteins, nucleic acids, polysaccharides) are frequently assumed to be preserved in tissues fixed and embedded for electron microscopy. Also, most common fixing methods (formaldehyde, glutaraldehyde, osmium tetroxide, used alone or in association) coupled with hydrophile or epoxy resin embedding do indeed preserve a high proportion of these macromolecules (32).

5.1.2 Lipids used as precursors

(a) Neutral fats (e.g. cholesterol, triglycerides, diglycerides, monoglycerides, fatty acids) are not retained in their entirety by any of the commonly used embedding schedules (32–40).

(b) Phospholipids as precursors (e.g. phosphatidyl choline, phosphatidylethanolamine) are preserved by osmium tetroxide fixation and araldite embedding (33–35, 41).

(c) An alternative method using ultra-thin frozen sections. Tissues could be previously fixed (42) or unfixed (43) in order to visualize glycerol or steroids respectively.

5.1.3 Drugs, hormones, and transmitter substances

The particularity of these molecules is to be strongly bound to fixable macromolecules called binding sites or receptors. Most common fixing methods (formaldehyde, glutaraldehyde associated with osmium tetroxide) are used (29–31, 44–51). This step extracts about 20–25% of the radioactivity. This loss of radioactive material could correspond to unbound or denatured radioactive molecules (e.g. non-specific binding). Embedding processes in epoxy resin are generally used, but for native peptide or steroid hormones freezing is necessary (43–51).

5.1.4 Diffusible molecules

Three distinct approaches have been made to the problem of localizing diffusible low-molecular-weight molecules.

(a) Fixation with a well-chosen fixative which precipitates the molecule of interest and preserves it for subsequent dehydration, embedding and sectioning, followed by autoradiography (52).

(b) Freeze-drying or cryosubstitution of labelled tissues and embedding them in epoxy resin. The main advantage of this process is that these tissues differ little in appearance from the results of conventional fixing and embedding (53–56).

(c) Freezing, cryoultramicrotomy, and freeze-drying is the most desirable method which does not involve fixation, but nevertheless permits immobilization of all potentially diffusible substances. The emulsion has to be applied as a dry membrane by the loop technique (see Section 6.2.2). The main problems remain the poor visibility of ultrastructure, and variable ice-crystal damage. However, some electron microscopic autoradiographs have been obtained (57) using this approach.

5.2 The protocol (*Figure 1*)

The protocols for fixation and embedding, freezing, freeze-drying, cryosubstitution are currently described in Chapters 1, 2 and 3 and do not show any particular modification for autoradiographic applications.

The uptake of radioactivity by the studied tissue can be evaluated by counting before the electron microscopic process (*Figure 2*).

The loss of radioactivity must be taken into account at all steps of fixation and embedding.

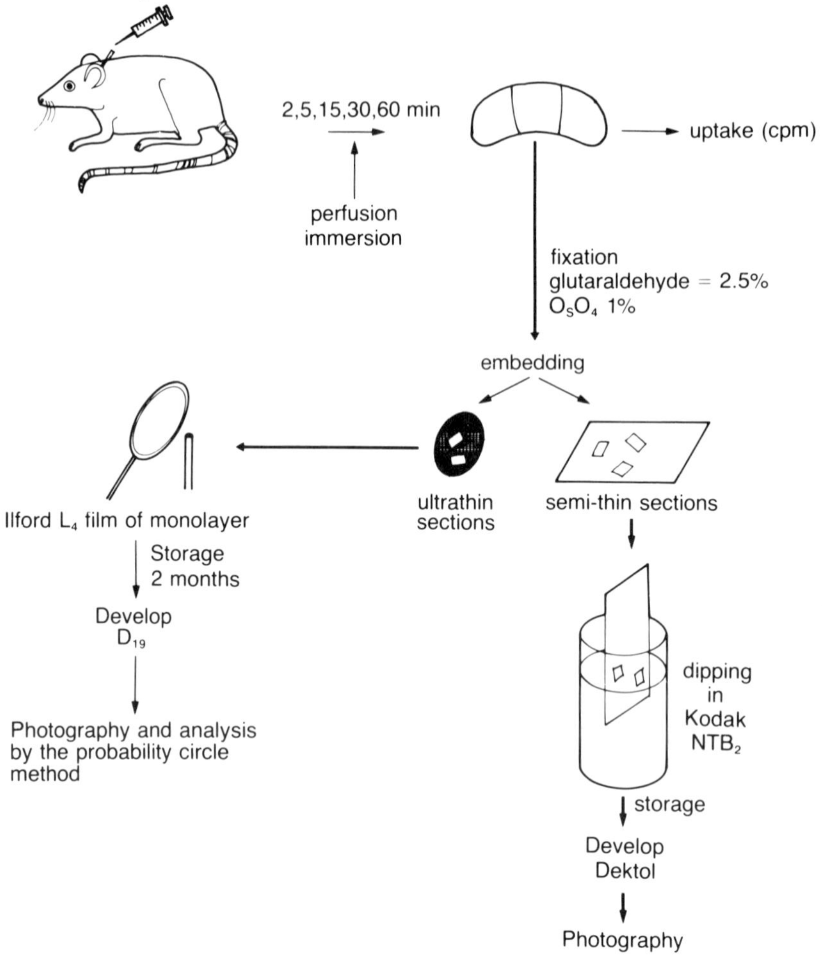

Figure 1. The principle of the electron microscopic autoradiographic method for drug uptake. After injection of drug via canula, the tissue is removed, uptake is estimated by counting (c.p.m.). After fixation and embedding, ultra-thin and semi-thin sections are coated: semi-thin sections are dipped into emulsion, ultra-thin sections are coated using the loop technique application. Autoradiographs are revealed before observation, photography and analysis.

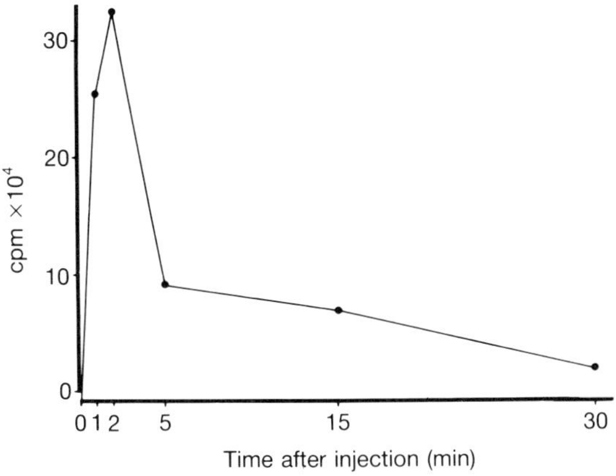

Figure 2. Time-course of the uptake of [^{125}I]-Atrial Natriuretic Peptide as evaluated by counting radioactivity in adrenal gland (80).

6. Preparation of autoradiographs

6.1 From semi-thin sections

Most electron microscopic autoradiographic experiments use semi-thin sections to check the specificity and the density of silver grain of autoradiographs.

No particular conditions are required for obtaining semi-thin sections between 0.5–1.0 μm thick. They are collected on very clean histological slides or on treated slides (gelatin, chrome alun, or poly-lysine) (see *Protocol 1*).

Protocol 1. Preparation of treated slides for semi-thin sections

1. Dip histological slides overnight in ethanol 95% containing HCl 1%.
2. Wash 2 × in running water.
3. Wash in distilled water 2 ×.
4. Dry in an oven to obtain clean histological slides, or Dip 5 min without drying in:
   - gelatin 1%–chrome alum 0.1% in distilled water or,
   - poly-lysine (mol. wt 70–150 000) 2.5 μg/ml distilled water
5. Dry and store in an oven (50°C).

Protocol 2. Procedure for preparing light microscopic autoradiographs from resin-embedding tissues

1. Section tissue at 1.0 μm thickness.
2. Expand section with solvent vapour while they are in the water bath.

Protocol 2. *Continued*

3. Transfer them to histological slides with a drop of distilled water.
4. Dry them by warming the slides on a hot plate at 40–50°C, Epoxy resin can be removed (see *Protocol 3*).
5. Apply nuclear emulsion (Amersham LM–1, Ilford K5, Kodak NTB2) diluted ≈ 1:1 (v/v) in distilled water by dipping (layer ≈ 3–5 μm) (see *Protocol 4*) or apply Kodak AR 10 stripping film.
6. Dry at room temp.
7. Expose the coated slides at 4°C in light-tight containers containing drying agent for approximately 2–30 days.
8. Develop the preparations (18°C in D19 or Dektol developer for 2 min).
9. Wash them briefly in water and then fix them in 30% sodium thiosulphate for 2 min or more.
10. Rinse carefully in water.
11. Stain them to view by brightfield or epifluorescence (see *Protocol 5*).
12. Mount with coverslide.

Protocol 3. Removal of epoxy resin from semi-thin sections

1. Prepare 50% methanolic sodium methoxide solution: Add 2 g of sodium metal to 110 ml methanol. After dissolution, add 100 ml benzene. It must be used immediately.
2. Plunge into the sodium methoxide solution for 3 min.
4. Wash in methanol–benzene (v/v) solution 5 min, then in acetone 5 min.
5. Air-dry.

Protocol 4. Liquid emulsion preparation

1. Check the dark-room conditions: safelight in agreement with nuclear emulsion, temperature (18–20°C), humidity (40–50%).
2. Check the dark-room equipment:
   - a water bath at 45°C
   - dipping jar half full of distilled water (up to a level showing final volume)
   - horizontal support to dry the histological slides
3. Place the jar of emulsion in the water bath and stand there for 30 min. Place the dipping jar in the bath.
4. Pour emulsion into the dipping jar to the required level.

5. Dip a clean slide into the emulsion, shake it gently and take it to the safelight to examine it for the presence of bubbles. If bubbles are present, gently tap the side of the jar and leave it for some minutes.
6. Dry another clean slide, withdrawing it slowly and carefully.
7. Drain the excess of emulsion on to tissue paper.
8. Leave the dipped slide on the horizontal support until it is dry, to judge appreciate the thickness and regularity of the emulsion.

Protocol 5. Staining of semi-thin sections

Autoradiographs from dipping in liquid emulsion can be easily stained with Toluidine Blue (or with several other dyes in particular circumstances) after development and fixation (staining before applying the emulsion is almost always unsuccessful due to positive chemographic effects).

1. Prepare 1% solution of Toluidine Blue in 1% sodium borax.
2. For semi-thin sections of tissue embedded in epoxy resin:
   (a) Apply a spot of dye to the region of the sections and warm on a plate at 60°C for 30 sec.
   (b) Differentiate the staining with 70% ethanol.
   (c) Air-dry the autoradiograph.
   (d) Mount with coverslide.
3. For semi-thin frozen sections:
   (a) Put the slide in a diluted solution of Toluidine Blue (1/5) for 5–10 min at RT.
   (b) Differentiate the staining with 95% ethanol.
   (c) Dehydrate quickly in 2 × 100% ethanol.
   (d) Mount with coverslide.

Protocol 6. Procedure for preparing light microscopic autoradiographs from semi-thin frozen section

1. Sections of tissue at 0.5–1 μm thickness are obtained with a cryoultra-microtome (specimen temperature: between −60 and −90°C, knife temperature: between −40° and −60°C) (58).
2. Transfer them from the knife to a histological slide with a drop of saturated sucrose.
3. Dry them for 1 h at RT.

Protocol 6. *Continued*

4. Wash sucrose with distilled water.
5. Air-dry.
6. Apply nuclear emulsion, expose, reveal and stain as described in *Protocol 2*.

6.1.3 Utilization

The autoradiographs from semi-thin sections are used:

(a) To look at the density of label in order to judge the period of exposure of ultra-thin sections (for example, if the signal from ^{125}I-hormone binding is abundant in target cells after an exposure of 15 days, the development of ultra-thin sections can be begun after 2 months of exposure).

(b) To characterize the specificity of labelling when an excess of unlabelled ligand was injected simultaneously with a radioactive ligand (see Section 7).

(c) To indicate the localization of labelled cells within heterogeneous tissue.

6.2 From ultra-thin sections

Experiments designed to obtain results by the autoradiographic technique at the ultrastructural level take a long time. Each step must be carried out with the maximum of attention and care, because mistakes are revealed only after some considerable time. Moreover, the number of grids must be sufficient to develop at different times and to quantify the silver grains.

6.2.1 Sections

The quality, area, and thickness of the section are important parameters. The quality since knife-marks and noise produce artefacts (59). The section area must be as large as possible to quantify a large number of silver grains in a similar condition. The thickness of the section is the most important parameter. Sections of gold interference colour (about 100 nm thick) will yield about 1.5 to 2 times the density of the grains given by a silver section (about 50 nm thick) depending upon the developer and the uniformity and nature of the radioactivity (32, 60). Thus, using thicker sections, more silver grains can be obtained or exposure time shortened, but with a loss of electron optical resolution and image quality and also of autoradiographic resolution.

The sections are collected and placed on a support appropriate to the particular method of emulsion application: on grids by the loop technique and on glass slide using a Heuse for dipping in liquid emulsion. Ultra-thin sections from epoxy-resin embedded tissue are put on coated (see *Protocol 7*) or non-coated grids but ultra-thin frozen sections are put on coated carboned grids (400 mesh) (see *Protocol 8*).

Protocol 7. Preparation of coated grids

Two membrane supports are used if the emulsion is applied by the loop method: the collodion or the formvar.

1. Formvar-coated grids (61).
   (a) The following equipment is required:
   - a histological glass slide
   - a glass rod
   - a tall beaker
   - a dropper
   - an acetate film
   - dichloro 1-2 ethane
   - 0,5% formvar solution in dichloro 1–2 ethane
   - 200 mesh nickel grids

   (b) Put some drops of the formvar solution on the end of the glass slide cleaned with dicholoro 1–2 ethane.
   (c) Spread the drops over the glass slide with the glass rod.
   (d) Air-dry to produce the membrane.
   (e) Remove the membrane by introducing the slide at a 45-degree angle into the water trough. The formvar film floats to the surface of the water.
   (f) Put the grids on the formvar film.
   (g) Put the acetate film over the formvar film and grids and remove.
   (h) Dry.

2. Collodion-coated grids
   (a) The following equipment is needed:
   - a tall bcakcr
   - a dropper
   - an acetate film
   - 200-mesh nickel grids 1% collodion solution in isoamyl acetate

   (b) Put down 1 drop of collodion solution over the water beaker. The collodion film floats to the surface of the water.
   (c) Air-dry.
   (d) Put the grids on the collodion film.
   (e) Put the acetate film over the collodion film and grids and take out.
   (f) Dry.

3. Deposit by vacuum. Evaporation a ~10 nm layer of carbon.

4. The thickness of the formvar or Collodion film can be reduced by floating grids on the solvent.

Protocol 8. Preparation of ultra-thin frozen tissue (51, 62)

After fixation, cryoprotection, and freezing (62) the ultra-thin frozen sections are performed following the dry sectioning method according to Tokuyasu (63). Using an cryoultramicrotome (specimen temperature: −120°C, knife temperature: −90°C). (Chapter 3, Section 2.)

1. Put the drop of saturated sucrose over the knife of the cryoultramicrotome.
2. Transfer the drop and sections on to the carbon-coated grid film.
3. Wash in distilled water.
4. Air-dry for 1 h or more.

Protocol 9. Emulsion application by the loop technique (4, 5, 32) (*Figure 3*)

1. Check the conditions in the dark-room, which must be small, completely light-tight and have a totally effective light trap on the door, safelight in agreement with nuclear emulsion, temperature 35–45°C, humidity ≈100%.
2. Check the dark-room equipment:
   - water bath (45°C)
   - jar to prepare diluted emulsion
   - wire loop of tungsten or platinum or nichrome

   The loop is obtained by winding the wire around two haemolytic tubes to form an ellipse and fusing the ends into a support.
   - support with a line of poles 3 mm in diameter

   The poles are 2 cm apart. The end of each pole is polished and cleaned before putting it on the grid.
   - small movable safe light
   - small boxes containing a drying agent wrapped in paper tissue
   - histological slide with a short section, cut from one of its ends, stuck to its surface, and a piece of double-sided adhesive tape stuck to the top of the section
3. Preparation of diluted nuclear emulsion (Ilford L4 or Amersham EM1). The emulsion is prepared the day before the coating session. The concentration of emulsion must be checked for each new batch.
   (a) Place the jar of emulsion in the water bath and allow it to stand there until it melts. Place the sterilized distilled water in the bath.
   (b) Dilute the emulsion with sterilized distilled water (for Ilford L4: 2 vol. emulsion add 1 vol. water).

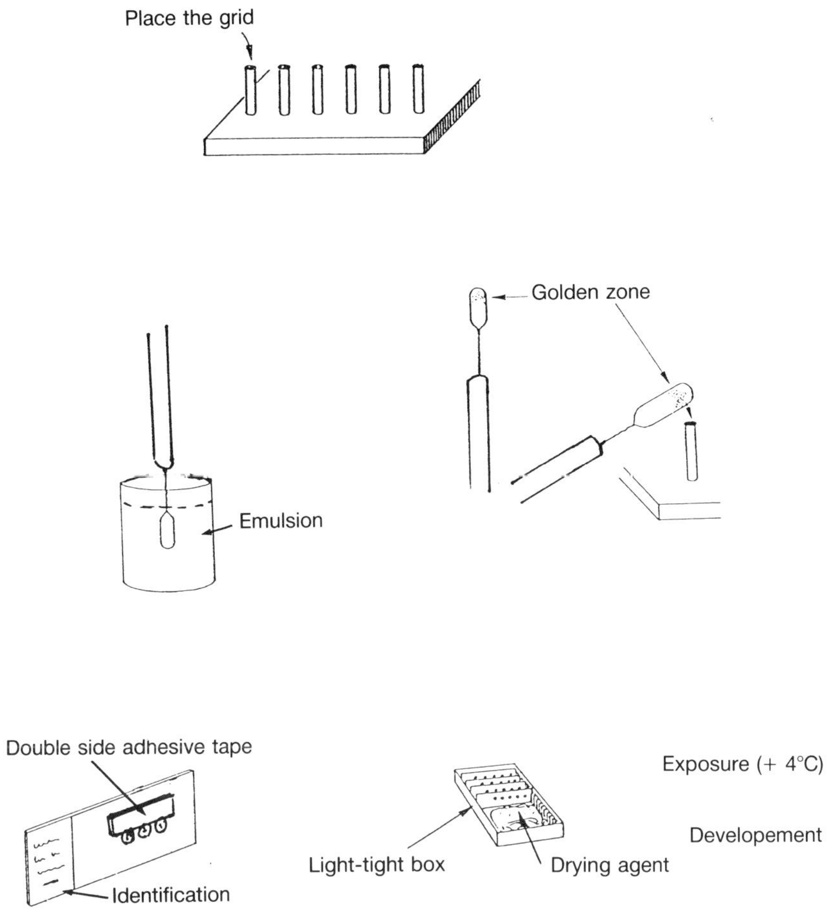

Figure 3. Emulsion application by the loop technique.

(c) Mix without bubbles for 30 min.

(d) Follow the instructions given below (steps 4–9) to obtain a monolayer film, and check it with an electron microscope[a] (32, 65). If the film is too dense, add water. Mix without bubbles for 30 min. Check the film. If the film is not quite dense enough, add emulsion, mix, and check until a monolayer dense film is obtained.

(e) Store the emulsion at 4°C in dark conditions.

4. Put a grid over the pole.

5. Dip the whole loop in the diluted emulsion.

6. Withdraw it carefully, edge-first, bearing a thin membrane of emulsion.

7. Rotate the loop vertically in front of the red lamp to see if the emulsion

Protocol 9. *Continued*

still flows when the membrane no longer moves, i.e. it is a firm gelled structure, the emulsion is ready to use.[b]

8. Apply the film over the grids. The best area is generally at the upper third of the loop.

9. Blow over the film to close contact between film and grid.

10. Replace the grid and attach it by an edge on to adhesive tape over the histological slide (3–4 grids per slide).

11. Put the specimens away in a light-tight box with tape and store them at 4°C to expose.

[a] Without treatment the AgBr crystals are electron-dense.

[b] If the film emulsion is applied ungelled to a collodion-coated grid AgBr crystals crowd near the grid bar (32).

Protocol 10. Emulsion application by dipping in liquid emulsion (*Figure 4*)

1. The following equipment is needed:
- clean histological glass slide (see *Protocol 1*)
- 1% collodion solution in isoamyl acetate or 0.5% formvar solution in dichloro 1–2 ethane
- a diamond pen
- 200 mesh nickel grids
- a completely light-tight dark-room (with safelight corresponding to the nuclear emulsion; temperature 20°C; humidity 20–40%)
- a water bath
- a jar to prepare diluted emulsion
- small boxes containing drying agent wrapped in paper tissue
- adhesive tape

2. Prepare the slide:

(a) Make 2 crosses 2.5 cm from the bottom and 0.5 cm from the edge of the back of the slide.

(b) Put the slide into collodion or formvar solution

(c) Drain each briefly.

(d) Dry at a shallow angle to the vertical.

(e) Cover the drying slides with an inverted jar, allowing a crack for air, but avoiding dust.

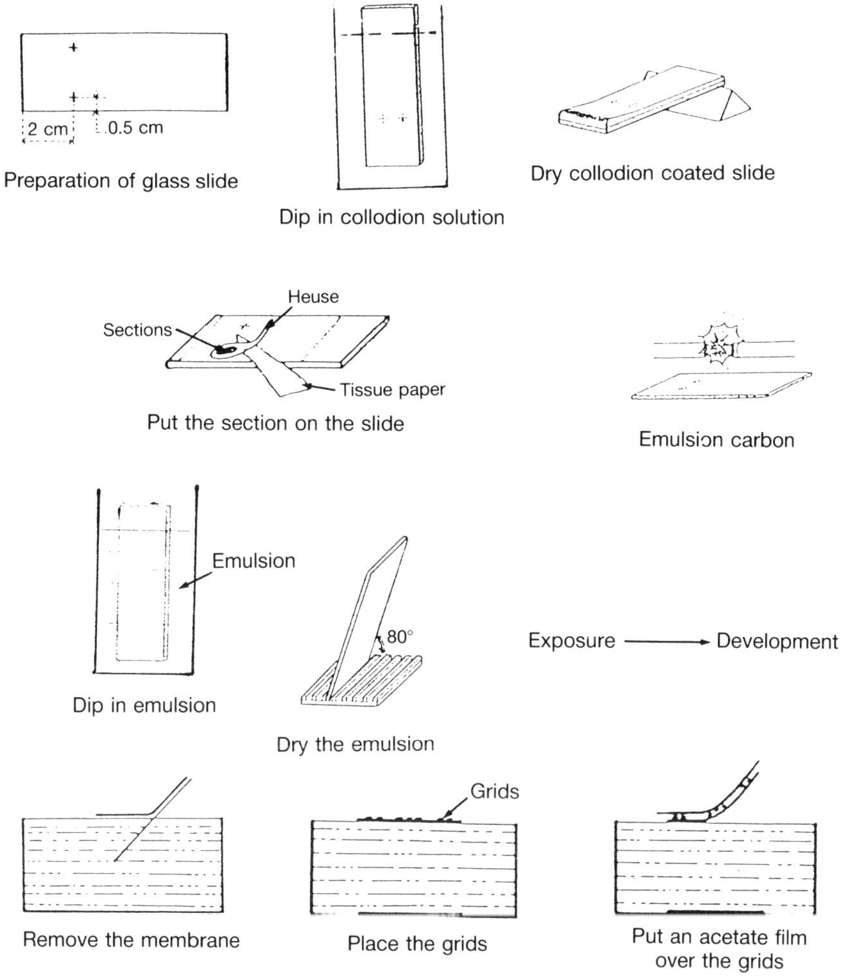

Figure 4. Emulsion application by dipping in liquid emulsion.

(f) After the slides are thoroughly dry, redip the lower end of each to a depth of about 0.5 cm then dry them as before.

3. With a Heuse, place ultra-thin sections on the front of the slide over the cross and dry with a filter paper. Do not move the Heuse during this step so as not to damage the film.

4. Counter-stain with lead citrate or uranyl acetate and lead citrate.

5. Evaporate on a 10–20 nm layer of carbon.

6. Dip the slide in correctly diluted liquid emulsion.[a]

7. Dry the lower end of the slide with a tissue paper.

Protocol 10. *Continued*

8. Air-dry the slide vertically for several hours.
9. Put the specimens away in a light-tight box with adhesive tape and place them, at 4°C to expose.

[a] As for the loop technique, it is necessary to check the layer of emulsion at the electron microscopic level then take the membrane off the slide and put the grid over it at the place where the sections are (see Protocol 7).

6.2.2 Exposure

Exposure times are usually long, 1–12 months according to the radioisotope used and the experiments. It is possible to gauge time of exposure from semi-thin preparations (32).

The exposure temperature is not critical. Some authors expose preparations at RT (≈20°C), but most use 4°C. The lower temperature would be expected to reduce chemographic effects.

To prevent damage, drying agent should always be included. The boxes must be carefully light-proofed with tape and preferably placed inside a second sealed container.

6.2.3 Development

The purpose of the development process is to cause the precipitation of enough silver metal on to the latent image so that it is visible in electron microscopy (32, 66). The latent image is composed of a few atoms. This small nucleus is used as a site where silver precipitates out from the solution. This results in the structure that is called a 'grain'. The size of the grain which is formed in this way depends to some extent on the size of the original latent image, and also, if development is extended for too long, even crystals unexposed to the radiation will result in grains. The development conditions need close control to achieve grain formation from as many exposed crystals as possible, while at the same time generating no grains from unexposed crystals.

In light microscopic autoradiography two developers are used: Dektol and D19. Silver grains are bigger with Dektol than with D19.

In electron microscopic autoradiography two standard developers are very widely used: D19 and Microdol X. In practice, this means that they are highly effective when used to develop Ilford L4 and Amersham EM1. The choice between Microdol X and D19 must be based on two main factors.

(a) D19 develops more of the latent images than Microdol X (67).

(b) D19 has a higher pH and may cause more contamination on the sections than Microdol X (32). The fine grain developers with or without gold latensification (see *Protocol 11*), paraphenylenediamine and the phenidon developers all have to be made up from the individual chemicals and do not keep (*Protocol 11*). The lesser reproducibility of treatment with these fine grain developers than with D19 or Microdol X is compensated for by a slight increase in the resolving power of the autoradiographic technique by reducing the photographic error (32). However, sometimes it is difficult to recognize what constitutes a single silver grain since several small silver granules may be developed from one bromide crystal (68). Moreover, the result is slightly variable, sometimes giving a round grain and sometimes several more irregular pieces of silver.

Protocol 11. Special development process: fine grain developers

Likewise the large size and irregular shape of silver grains after chemical development are not compatible with the size of cell components. In this respect other development processes have been introduced.

1. Paraphenylenediamine development[a] (64)

(a) Prepare a solution containing: sodium sulphite 0.1 M; paraphenylenediamine 0.01 M.
Dissolve the sodium sulphite at 50°C, then add the paraphenylenediamine.

(b) Filter and use while it is still fresh.

(c) Develop at 20°C for 1 or 2 min.

(d) Rinse briefly in 3% acetic acid and then in distilled water.

(e) Fix in 30% sodium thiosulphate for 2 min.

(f) Wash in distilled water.

2. Phenidon development (69)

(a) dissolve in 200 ml of distilled water, in the following order

• Ascorbic acid	6g
• Phenidon (1-ϕ-3 pyrazolidone)	1g
• KBr	2.4g
• K_2CO_3	5.2g
• Na_2SO_3	40g
• NaCNS	24g

(b) Make the solution up to 300 ml. At this stage the solution must be clear.[b]

(c) Filter and use while fresh.

(d) Develop at 17°C, 1 min.

(e) Rinse in 2% acetic acid stop bath.

(f) Fix in 30% sodium thiosulphate for 2 min.

3. Gold latensification (14, 65, 69–70).
The loss of sensitivity incurred when using fine grain development can be

Protocol 11. *Continued*

offset by prior treatment of the emulsion with gold thiocyanate. This treatment results in a three- to fourfold increase in sensitivity.

(a) Prepare 2% HAu Cl_4.
(b) Mix 1 ml of 2% HAu Cl_4 and 99 ml of distilled water, adjust the pH to 7.0 using NaOH.
(c) Add 250 mg of KCNS and 300 mg KBr, bringing to 500 ml with sterile distilled water.
(d) Soak the autoradiographs in this latensifying solution at 20°C for 5 min.
(e) Rinse them briefly in distilled water.
(f) Develop.

[a] The developer is unstable and the results are not as reproducible as with the standard developers. The relative efficiency is also rather low (32).
[b] Turbidity in the solution indicates poor dissolution of the reagents and the procedure must be carried out again.

6.2.4 Recovery of specimens from their supports

i. Grids on slide

Grids may be easily removed from slide after the final washing if the double-sided tape is not completely dry.

ii. Ultra-thin sections on membrane-coated slide

Collodion-coated slides:

(a) Scratch round the thick bottom edge of collodion on the slide with a razor blade.
(b) Remove the membrane (see *Protocol 7*).
(c) Place a 200-mesh grid over each group of sections on collodion film.
(d) Put an acetate film over the grids and take out.
(e) Dry in the oven at ≈40°C for 1–2 h.
(f) Detach the grids from the acetate film.[a]
(g) Dry the grids.

Formvar-coated slides

(a) Scratch around each ribbon of sections, ensuring that each one is completely circumscribed.
(b) Introduce a little hydrofluoric acid[b] into the scratch and leave the preparation until the carbon layer begins to lift.
(c) Repeat until the layer bearing the sections is floating free.

(d) Transfer the floating layer to a watch-glass of distilled water.
(e) Rinse by transfer until it is free of acid.
(f) Pick the layer up on a 200 mesh grid.

[a] If necessary the collodion can be thinned by immersing the grids in isoamyl acetate for 3 min.
[b] 3% of hydrofluoric acid (dilute 40–48% solution hydrofluoric acid).

6.2.7 Staining

Both possibilities are described: staining before coating with emulsion and staining after development. The first of these (see *Protocol 10*) is done in the same way as the staining of ultra-thin sections for electron microscopy. The main disadvantage is α tracts resulting from uranyl which can be seen in autoradiographs. The second possibility of staining is after-development. In this way, lead citrate (71) and uranyl acetate are used with (see *Protocol 12*) or without gelatin removal.

Aqueous or ethanol uranyl acetate (2–5%) are used for 10–60 min at room temperature alone or followed by lead citrate for 3–10 min without gelatin removal. The main disadvantages of this treatment are the production of diffuse areas of increasing density due to staining of gelatin and deposition of precipitates. Lead stains also add to electron opacity. This problem increases with the thickness of the gelatin coating. With a monolayer of gelatin this problem is reduced (29, 72, 73).

Great care must be taken with the method of gelatin removal that loss or displacement of silver grains does not occur.

Protocol 12. Staining of ultra-thin section on grids after removal of gelatin (74, 75)

1. Float the grids, emulsion side down, on distilled water at 37°C for 30 min.
2. Float the grids on 0.5 N acetic acid at 37°C for 15 min.
3. Rinse each grid in distilled water.
4. Dry.
5. Float the grids on uranyl aceate in aqueous solution for 20–30 min.
6. Rinse the preparation well.
7. Float each grid on lead citrate stain for 1–2 min.
8. Rinse each grid again well.
9. Dry.

Ultra-thin frozen sections can be observed without staining if they are fixed by osmium tetroxide (51) or by uranyl acetate in a 2% aqueous solution.

6.3 Observation

6.3.1 Semi-thin sections

These sections are observed at light microscopic level without staining in phase contrast or darkfield or with staining in brightfield or epifluorescence. Micrographs are used as a control of the labelling specificity (*Figure 7*).

6.3.2 Ultra-thin sections

Observations at electron microscopic level is generally easy. The methods described above give reasonable contrast permitting satisfactory micrographs (*Figures 5–6*). Negatives should be obtained at a magnification that allows enlargements to be made at a final magnification of 10 000–30 000.

In order to take full advantage of the autoradiographic approach, it is necessary to prepare micrographs in a way that gives representative sampling of the silver grain distribution and of the micro-anatomy. For this purpose, it will be necessary to record at least 500 silver grains and to spread the observations over several thin sections from different tissue blocks when considering one experimental treatment.

For quantitation, a series of micrographs must be prepared systematically; e.g. one from the top left-hand corner of each grid square right through each of several grids. In the case of uptake by a single target cell type within heterogenous tissue at least 100 cells must be studied. This process continues until enough silver grains have accumulated. Each series of micrographs must be taken at a calibrated magnification. This process requires considerable patience since these preparations have lower pictorial quality than ordinary thin sections.

7. Controls

A number of essential controls are required to validate the autoradiographic detection of radiolabelled molecules. The first of these is the determination of background counts. The second is the inhibition of labelling. The third is the integrity of radiomolecules in tissue. The fourth is physiological or experi-

---

Figure 5. A: Electron microscopic autoradiograph of rat adrenal medulla 1 min after injection of [^{125}I]-Atrial Natriuretic Peptide. Tissue was embedded in expoxy resin after fixation (glutaraldehyde 2.5%, osmium tetroxide 1%). Adrenalin-containing cells (**a**) are distinguished from noradrenalin-containing cells (**na**) on the basis of relative morphology. Silver grains are localized only in adrenalin-containing cells. Bar: 1 μm (74).

B: Electron-microscopic autoradiograph from ultra-thin frozen section 15 min after injection of [^{125}I]-monoiodo-native gonadoliberine in rat adenohypophysis. Tissue was fixed (glutaraldehyde 2.5%, osmium tetroxide 1%) and frozen. Silver grains are located over nuclear matrix, nuclear membrane and recutary granules of gonadotrophs. Bar = 1 μm (51).

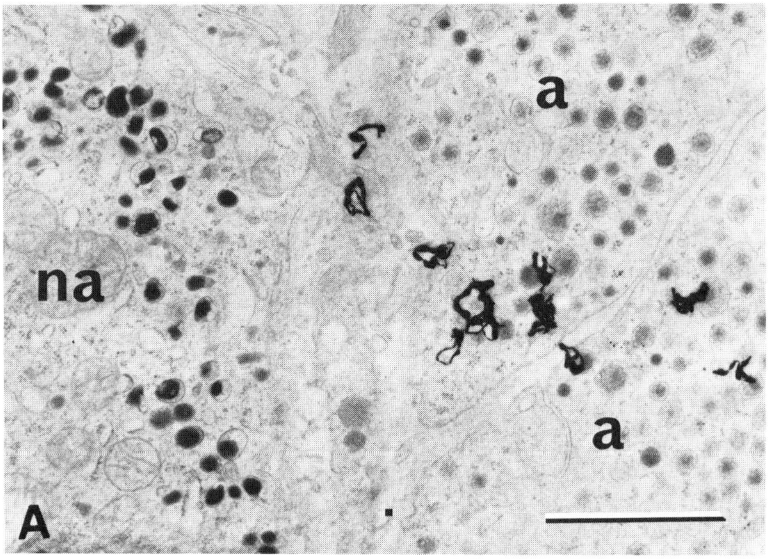
a
na
a
A

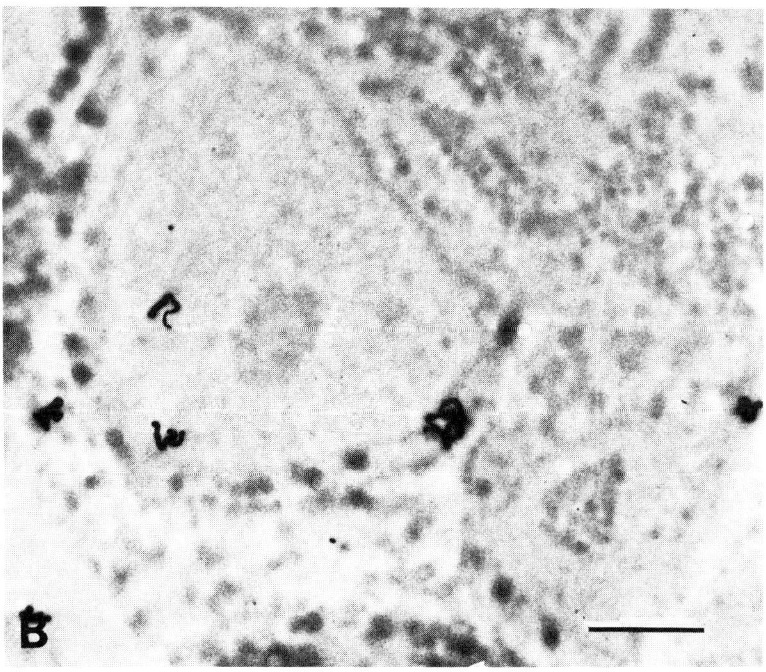
B

mental variation. Finally, the comparison of autoradiographic techniques with other methods and results described in the literature is desirable.

7.1 Determination of background

Some silver grains observed in autoradiographs do not result from disintegration of radioactive/isotopes. Determination of background count levels is achieved by taking a series of randomized micrographs of the support film around the tissue section. If the background grain frequency/unit area exceeds 5% of the grain frequency over labelled cells it is best to abandon this specimen (20).

The following possibilities should be examined (32)

(a) 'Fogging': developability of crystals due to long storage.
(b) 'Fog in the emulsion': if the background is high, this may originate from exposure to light during handling (insecure dark-room, working too close to a safelamp).
(c) 'Pressure': resulting from pressure during the drying of liquid emulsion.
(d) 'Fogging during exposure' may originate from storage near a source of radiation (usually γ radiation) or the possibility that the boxes used for exposure were not light-tight.
(e) 'Over-development' happens because of prolonged immersion in the developer, or it is caused by excessive developer temperature.

Some factors can induce background only over the tissue sections (66):

(a) Chemical effects of some tissue components induce latent image formation by direct chemical action (positive chemography). This phenomenon has been observed at light microscopic level, in fixed tissue and frozen tissue. Mulvaney (77) has reported one experiment in which certain lysosomes exhibited the effect.

The interposition of the carbon layer between the sections and the emulsion is undoubtedly helpful in reducing the incidence of the effect. Positive chemography can be tested for by preparing autoradiographs of identical unlabelled tissues. Chemographic grains are particularly large and complex. Copper grids exhibit the effect if not coated with a support film.

(b) Eradication of latent image by chemical action (negative chemography) has been oberved frequently in light microscopic experiments (78–79). This effect has rarely been seen in electron microscopic autoradiographs except for myelin if a layer of carbon is not interposed.

Figure 6. A: Electron microscopic autoradiograph of rat adenohypophysis 2 min after injection of [^{125}I] somatostatin-14 agonist. Tissue was embedded in Araldite after fixation (glutaraldehyde 2.5%, osmium tetroxide 1%). Silver grains are located at the periphery of somatotroph over the plasma membrane. Bar = 1 μm (75).

B: Electron microscopic autoradiograph of rat pancreas 30 min after injection of [^{125}I]-Atrial Natriuretic Peptide. Tissue was fixed and embedded; see **A**. Silver grains are localized over nucleus, zymogen granules, mitochondria, cytoplasm and basolateral membrane. Bar = 1 μm (76).

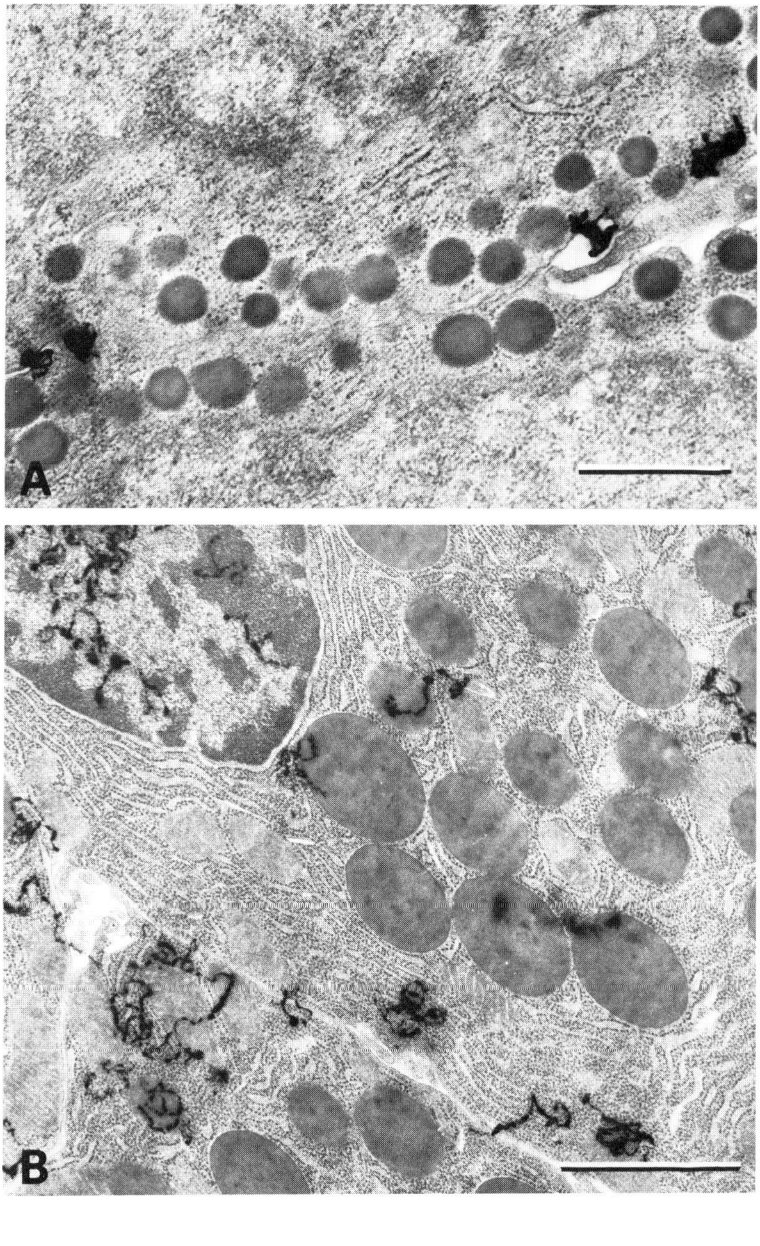
A
B

A test for negative chemography consists of light-fogging a few specimens, and these are then exposed with the others. Negative chemography will be seen as zones of decreased or zero grain density.

7.2 Diffusion of labelled molecules

The products may diffuse during the process. Thus the autoradiographs show high grain frequency over an empty embedding medium, but a much lower one over the support film. These molecules are soluble in the embedding resins (36): neutral lipids, triglycerides.

This problem can be avoided by changing or avoiding the embedding system.

7.3 Specificity of cell uptake

(a) Specificity is quantified by calculating the difference between the total d.p.m. taken up in all of the tissue following administration of the labelled molecule alone and the d.p.m. taken up in the tissue after co-administration of a labelled molecule and an excess (40–100-fold) of an unlabelled molecule. A decrease averaging 60–90% is obtained (29, 73).

(b) Specificity is visualized in semi-thin sections (*Figure 7*). An excess of unlabelled molecules drastically reduces the number of silver grains. Non-specific and background silver grains are always detected.

(c) *In vitro* specificity can be checked by decreasing the temperature of incubation (2–4°C). Add an inhibitor (e.g. for protein synthesis, puromycin at 2–5 μm/ml).

(d) Replacement of the labelled molecule (*Figure 8*): for hormone binding, the agonist or natural hormone is replaced by an antagonist or by another modified molecule without effect (51).

For protein synthesis, the natural precursor is replaced by a metabolically inert enantiomer (32).

7.4 Integrity of intracellular radiolabelled molecules

The detection, on a film, of a signal emitted by a radioactive atom does not permit the inference that the emission is coming from the molecule injected. It is evident that most molecules are subject to varying degrees of catabolic degradation following administration. In blood, one way to reduce this possibility is to inject the radioactive molecule into the jugular rather than the femoral vein in order to avoid hepatic portal circulation and first-past biodegradation. The plasma half-life of molecules depends on the molecule. For example, for hormones it varies between a few seconds and an hour.

In tissue, radioactivity quantified at the end of an experiment may not represent intact radiolabelled molecules. It is, therefore, necessary to ascertain whether the tissue under study contains a mixture of radiolabelled molecules, free isotopes, and free catabolite molecules containing the isotope, or

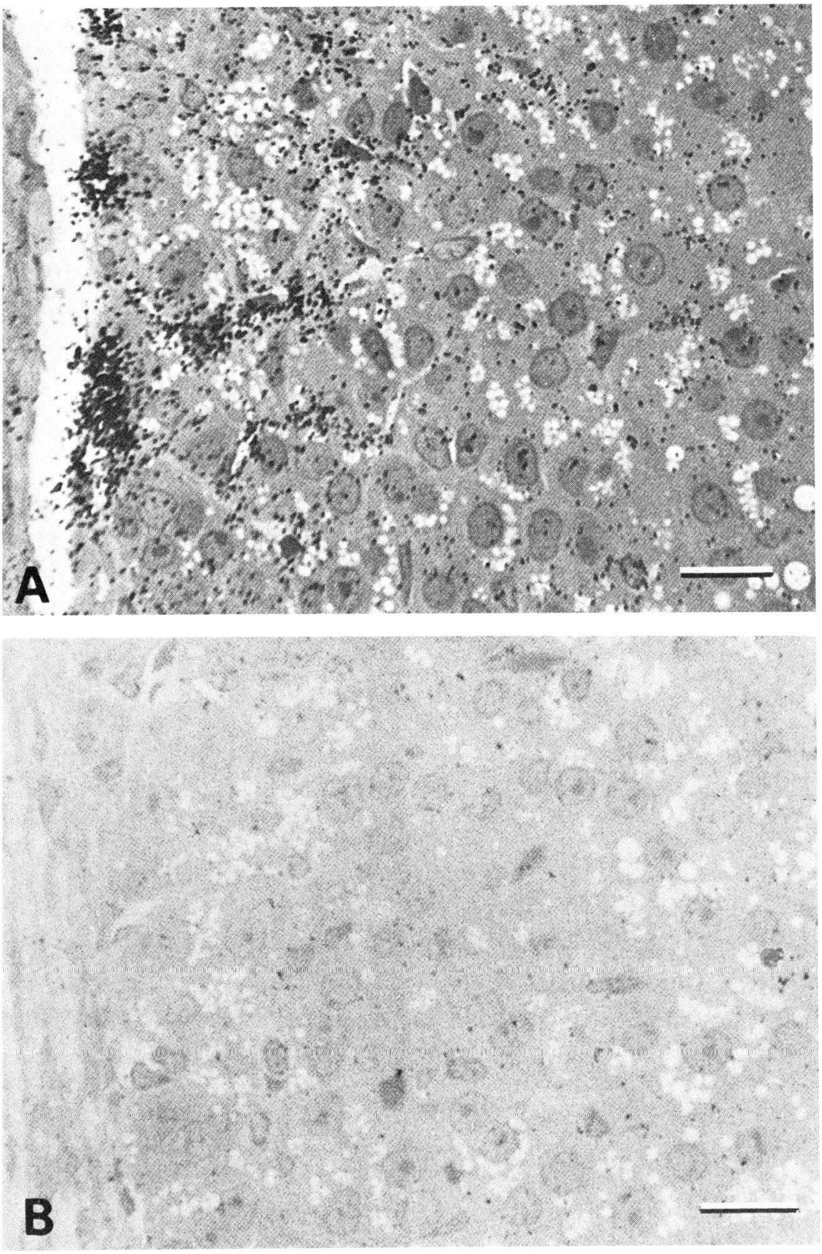

Figure 7. Light microscopic autoradiographs from semi-thin sections of glomerulosa and fasciculata cells of rat adrenal gland, 1 min after injection of [^{125}I]-Atrial Natriuretic Peptide alone (**A**) or in combination with a 70-fold excess of unlabelled Atrial Natriuretic Peptide (**B**). Tissue was fixed (glutaraldehyde 2.5%, osmium tetroxide 1%) and embedded in Araldite (**A**). Silver grains overlie the zona glomerulosa cells (**B**): a few silver grains accounting for background can be observed, showing the uptake specificity. Bar = 10 μm (80).

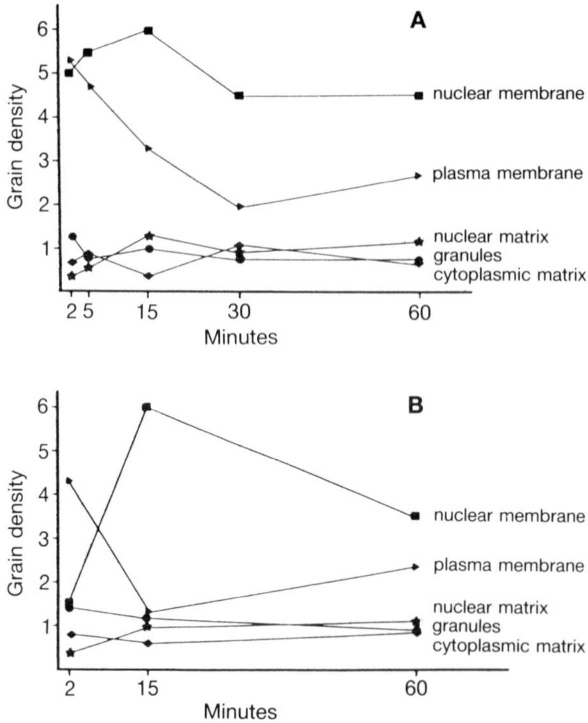

Figure 8. Grain densities (percentage of total grains/percentage of total area) as a function of time for cellular compartments significantly labelled by **A** [^{125}I]-monoiodo-native gonadoliberine, or **B** [^{125}I]-gonadoliberine agonist (74).

newly synthesized molecules into which the isotope was incorporated. It should be noted that radioactive breakdown products within a particular organ may originate as the result of intracellular degradation of intact labelled molecules, or the labelled product(s) may be taken up from the extracellular environment.

To test for the structural nature of a radiolabelled molecule a separation methodology can be used (50, 73). Generally, this procedure is used only at the longest experimental time interval, and any accumulation of degraded or synthesized products would have been expected to be maximal.

7.5 Physiological or experimental variations

The main control which is described in autoradiographic studies is the kinetics of uptaken radiolabelled molecules combined with quantitation. This aspect is always given in the literature.

Some other groups have compared *in vitro* studies (31, 72) and operated animals (70), with *in vivo* studies or with intact animals.

Pharmacological controls such as inhibition of protein synthesis, down regulation of binding sites, etc. Any modification of an autoradiographic experiment can be used as a control.

7.6 Comparison with other techniques

The main techniques of molecular detection in electron microscopy are immunocytology and *in situ* hybridization. In the field of endocrinology these three approaches are very interesting by providing confirmation of the presence of a hormone within its target cell(s) or its synthesis (73, 81, 82).

Autoradiography shows the presence of an exogenous or newly synthesized molecule if the radiolabelled molecule is a precursor. Immunocytology shows the presence of an endogenous molecule within a cell. *In situ* hybridization reveals messenger RNA which is responsible for the synthesis of one particular molecule.

8. Quantitation

The basis for quantitation is that the number of developed grains in the autoradiograph will be proportional to the radioactivity content of the tissue. The objective is to correlate the number of grains over a variety of structures to the relative proportion of radioactive molecules contained in these structures.

8.1 Efficiency

The efficiency is the percentage of disintegrations recorded. In electron microscopic autoradiographic technique it can be determined by experiment. Several groups (60, 83–86) have published estimates of efficiency for various systems.

Some elements can be evaluated:

- on account of the presence of an emulsion layer over one face of the section only and the fact that disintegrations do not have any preference, the efficiency is less than 50%
- for β^- particle energy, the efficiency values decrease with increasing energy
- for bromide crystal diameters, the efficiency increases with their size
- due to the nature of the developer, all latent images are not transformed into silver grains (20)
- according to the thickness of the sections, self-absorption is in the range of 40–100 nm (60)
- in fixation with a heavy metal (OsO_4–uranyl salt), self-absorption is less than 10% (87)
- the size of the radiation dose is inversely proportional to grain density

This efficiency can be determined by experiment. A standard radioactive source can be prepared from a tissue block, but it is not easy to make a totally satisfactory standard radioactive source. Commercial standards are now available for ^{3}H, ^{14}C, and ^{125}I.

Overall autoradiographic efficiencies for various isotopes using different developer-emulsion combinations are given by Williams (32).

8.2 Resolution

8.2.1 Light microscopic level

For light microscopic autoradiographs the silver grain dispersion around the radioactive source is larger than for electron microscopic autoradiography on account of the thickness of semi-thin sections, the multi-layers of emulsion, the developers used, which do not develop like latent images, and the size of the silver grains (92–94).

8.2.2 Electron microscopic level

A β^- particle leaves the source in the section and happens to pass through a silver bromide crystal displaced some distance from the source. The distance between the bromide crystal and the source is the geometric error thus produced. Development of the silver bromide crystal gives rise to a silver grain. The centre of the silver grain is displaced some distance from the centre of the bromide crystal. This displacement distance is the photographic error component. The total error is the addition of the geometric error and the photographic error. The factors which affect these errors are the thickness of sections, the emulsion and the angle of β^- emission, for the geometric error, and the size of bromide crystals and silver grains for the photographic error.

To summarize the data from Saltpeter (88–91) and Williams (20, 32), the image spread of a particular autoradiographic system may be described in either of two ways:

(a) The distance from a point radioactive source within which half of the silver grains fall: HR (half radius) value.

(b) The distance from a radioactive line within which half of the silver grains fall: HD (half distance) value.

These parameters are the 'point'- and 'line-resolution' values of the system which describe a physical characteristic of the system. This characteristic is an important factor in determining the extent to which radioactivity can be localized, but it is not the sole factor involved.

The ability of an autoradiographic system to distinguish between radioactivity in different parts of a complicated specimen is often a very complex matter involving the resolution, the subcellular components and indeed the relative degree of radioactive labelling of the different parts of the specimen.

It is essential to realize that the effect of image spread is to cause some of the silver grains to form outside the profile of this feature, causing grain formation. Grains formed outside the outline of the feature of origin are said to be 'cross-fired'. The proportion of cross-fired grains varies with the size and shape of the radioactive feature.

The relation between HD and HR is:

$$HR = \sqrt{3} \times HD$$

For the main isotopes used in biology the values are as given in *Table 1*.

Table 1. HD and HR values for useful isotopes

	HD (nm)	HR (nm)	
^{125}I	100–140	170–240	(90–91)
^{3}H	145–180	250–310	(17)
^{14}C–^{35}S	230	390	(88)

These values mean, for example for ^{125}I, that 50% of silver grains resulting from disintegrations in a point radioactive source are within a circle which is concentric upon this source and measures ≈170–240 nm. in diameter. This determination is the basis of the probability circle method of quantifying autoradiographs.

8.3 Statistical analysis

It is necessary to analyse autoradiographs since there exist background (see 7.1), low sensitivity of emulsion (see Section 8.1), and a difference of localization between radioactive sources and silver grains (see Section 8.2). These statistical analyses are mainly used for binding studies.

The following steps are carried out:

- measurement of background
- probability circle method
- cross-fire method

In all cases, an autoradiographic study is associated with a stereological study to give the grain density.

8.3.1 Measure of background

i. Principle

Take samples of area to count the silver grain number in each sample, in order to obtain two histograms (number of samples with 0, 1, 2, 3 . . . silver grains), the first without section and the second over section.

The two histograms must be analysed using a Poisson distribution.

ii. Method

The samples are observed at a magnification of 6000 to identify silver grains. Count 100–500 silver grains in the photographic field of electron microscopy. The area of this field is measured in order to obtain grain density (a number of silver grains/area of the photographic field). The signal/background ratio is obtained by dividing the two means.

In heterogeneous tissue, like the endocrine gland, it is possible to estimate the background by counting the silver grains in each cell population (100 cells). It is never the case that all of the cell populations in a given endocrine gland are simultaneously target cells for a hormone. In this way, there is always one or several cell population(s) which cannot bind the ligand. These cell populations can be used as samples for background.

For example, *Table 2* shows the number of silver grains in 100 of each adenohypophyseal cell type after injection of [^{125}I] Atrial Natriuretic Peptide (95).

Table 2. Percentage of labelled cells

Cell type	**Number of grains/cell**						
	0	1	2	3	4	5	≥6
Gonadotroph	38	35	11	8	1	–	7
Corticotroph	40	41	15	2	2	–	–
Lactotroph	43	39	11	2	3	1	1
Somatotroph	89	10	1	–	–		
Tyrotroph	88	12	–	–	–		

The experimental background was observed to be no more than 1 grain per target cell after concomitant injection of a radioactive ligand with an excess of unlabelled ligand (unpublished data).

8.3.2 Probability circle method

i. Principle

This method was described by Williams (96, 20), Salpeter (97) and Nadler (98). The aim is the determination of cell compartments which have specifically uptaken the radiolabelled molecule.

ii. Method

A common procedure expresses data in terms of percentage of total grains associated with any organelle. Such data can be converted into a percentage density by dividing the percentage of total grains by the percentage of total

Table 3. Relative percentage of area occupied by each cellular compartment in glomerulosa cells of rat adrenal gland (80)

Cellular compartment	Relative surface area (%)
Plasma membrane and cytoplasm near plasma membrane[a]	7.4
Golgi apparatus	1.5
Lysosomes (and multivesicular bodies)	1.1
Cytoplasmic matrix and rough endoplasmic reticulum	30.8
Mitochondria	23.6
Lipid droplets	12.0
Nucleus	23.6

[a] This cytoplasm area corresponds to an annulus equal in width to the radius of the probability circle

area occupied by that organelle.[a] If the distribution of grains over the entire tissue is purely random, the percentage density (percentage of grains/percentage area) should have a value of 1. Any value above 1 indicates a level of labelling above average and any value below 1 a level of labelling below average (89).

If resolution is considered in terms of the distance from a point source within which 50% of the developed grains fall, one obtains a marginal benefit in analysis (68).

A circle of given size (obtained by the probability of containing the source) is drawn around every grain centre.[b] A cell compartment (e.g. organelle) is considered to be associated with a grain if any part of the circle falls within that compartment.

Nadler (98) gives the details of this analysis. If only one organelle is located within the circle, the grain is classified as 'exclusive'. According to the method

[a] Determination of area occupied by cell organelles. This can be done by using a grid which is placed over the autoradiographs. The grid points are random relative to the fine structure of the cells (an intersection point/μ^2). For simple density analysis cell grid intersections lying over a cellular compartment are collected. Relative area is expressed as the percentage of intersections in one particular compartment/total intersections over the cell: the results are presented in *Table 3*.

[b] One procedure for locating this midpoint consists of placing over the developed grain a plastic mask with a series of concentric circles of increasing radii drawn around a central small hole. Once the smallest circle that can circumscribe the grain is fitted around it, the centre can be punched through the autoradiograph with a teasing needle. The punched centre then becomes the location of the grain for all further considerations. The size and shape of the developed grain no longer matter. Alternatively, one can use the midpoint of a line drawn between the two ends of a grain. Since the two ends of a grain are not always clearly seen in the general tangle of silver, the first method is preferable (89).

of Nadler (98), the 'shared grains' (there is more than one structure visible within the circle) were assigned on the basis of probability to only one structure. The corrected number of grains for each organelle is the sum of exclusive and adjusted shared grains and is expressed as a percentage of the total corrected number of grains.

After obtaining the corrected grain counts, the relative concentration of labelling in each organelle is calculated by dividing the percentage of silver grains over an organelle by the percentage of area occupied by this organelle (*Table 4*). Details of the statistical analysis are given in the work of Nadler (98).

A grain density greater than 1 is considered to represent significant labelling (91) (*Figure 9*).

If the density of silver grains is sufficient (≥400/experimental point) over the tissue, and the background is low, the correction proposed by Nadler (98) is not obligatory. In this case the location of a grain can be determined by its centre.

Table 4. Relative percentages of total grains counted, and grains densities[a] (GD), for cellular organelles of zona glomerulosa cells of rat adrenal gland 1–30 min after injection of [^{125}I] Atrial Natriuretic Peptide (80)

	Time after injection (min)									
	1		2		5		15		30	
	%	GD	%	GD	%	GD	%	GD	%	GD
Plasma membrane and cytoplasm near plasma membrane	23.6	3.1	16.1	2.2	15.4	2.0	6.6	0.9	9.4	1.2
Golgi apparatus	0.6	0.4	0.3	0.2	0.3	0.2	0.4	0.9	0.3	0.2
Lysosomes (and multivesicular bodies)	0.4	0.3	0.7	0.6	0.3	0.2	0.5	0.4	0.6	0.5
Cytoplasmic matrix and rough endoplasmic reticulum	23.1	0.7	33.5	1.0	28.3	0.9	27.6	0.9	33.9	1.1
Mitochondria	24.2	1.0	29.6	1.2	35.2	1.5	41.9	1.7	35.8	1.5
Lipid droplets	12.8	1.0	8.7	0.7	12.8	1.0	9.6	0.8	11.8	0.9
Nucleus	15.3	0.6	10.9	0.4	7.7	0.3	13.4	0.5	8.2	0.3
Total number of grains counted	454		561		597		418		457	

[a] GD: percentage of silver grain in one particular compartment/percentage of this cellular area (see *Table 3*).

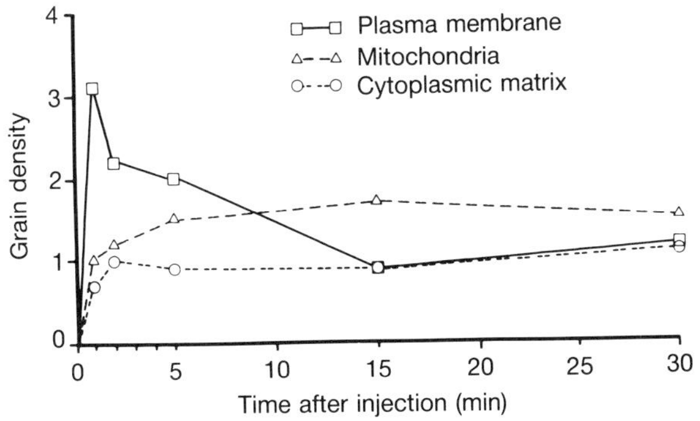

Figure 9. Grains densities (percentage of total grains/percentage of total area) as a function of time for cellular compartments significantly labelled by [^{125}I]-Atrial Natriuretic Peptide in glomerulosa cells of rat adrenal gland. Only grain densities >1 from *Table 4* are presented (80).

8.3.3 Cross-fire method

First described by Blackett and Parry (21, 99–101) as the 'hypothetical grain analysis method' and then by Salpeter *et al.* (91). The theoretical and practical aspects are considered by Williams and Downs (94, 102).

i. Principle

The method consists in the simulation of a hypothetical radioactive source distribution within the tissue, which produces hypothetical silver grains. A computer adjusts the distribution of hypothetical grains to that of the observed grains.

ii. Method

(a) Divide the tissue into compartments.

(b) Define the source compartments and the grain compartments. For statistical analysis the number of source compartments must be greater than the number of grain compartments.

(c) Make ×30 000 to ×50 000 magnified micrographs so as to obtain 500 to 1000 silver grains.

(d) Localize the silver grains. The grains are allocated to the compartment where is their centre is situated.

(e) Simulate a hypothetical radioactive source distribution and hypothetical silver grain distribution by placing a transparency over each autoradiograph. Each hypothetical grain is bound to a hypothetical radioactive source by two parameters: a directional angle for disintegration and a distance between grain and radioactive source.

(f) Construct the cross-fire matrix (table: the source compartments are given horizontally, the grain compartments vertically). At the same time take the localization of the observed silver grains from the autoradiographs.

iii. Calculation

The computer program analyses several equations (one for each grain compartment) with several unknowns (source density within each radioactive source compartment) (18, 94, 102–104).

iv. Results

This cross-fire method can be used only if the sources are localized in an area compartment; the cell membranes do not permit the use of hypothetical point radioactive sources (18).

Other methods can be used, with or without modifications. However, statistical analysis of autoradiographs is necessary to interpret the results in the light of background and resolution.

9. Applications of autoradiography

The synthetic table of Williams (32) showed that the particularly important applications of electron microscopic autoradiography have been those concerning the processes of macromolecular synthesis and secretion.

However, useful observations have also been made on nucleic acid synthesis, lipid metabolism, and polysaccharide synthesis.

Electron microscopic autoradiographic techniques are being used increasingly to study the binding sites of small molecules (hormones, peptides, drugs, toxins, and transmitter substances) (*Figures 5–6*) (29–31, 49–51, 73–76, 80, 81, 95).

The localization of truly diffusible substances has been made available by using the cryomethod (42, 43, 55, 56, 105–110).

The main developments of autoradiographic techniques are the *in situ* hybridization method (*Figure 10*) (111–119), the localization *in vitro* binding of hormones (120–124) and double staining using autoradiography and immunocytochemistry (*Figure 11*) (28, 125–127).

Figure 10. Electron microscopic autoradiographs of *in situ* hybridization with ^{35}S labelled growth hormone complementary DNA probe on ultra-thin frozen sections of rat pituitary fixed with 2% paraformaldehyde. Silver grains were observed in cells showing characteristic features of somatotrophs which synthesized growth hormone. Two populations of somatotrophs were distinguished: **A** shows a cell from the population with silver grains localized in a cytoplasmic matrix containing few secretory granules. **B** shows a somatotroph from the other population, with silver grains in both the nucleus and the cytoplasmic matrix containing a large number of secretory granules. Bar = 1 μm (115).

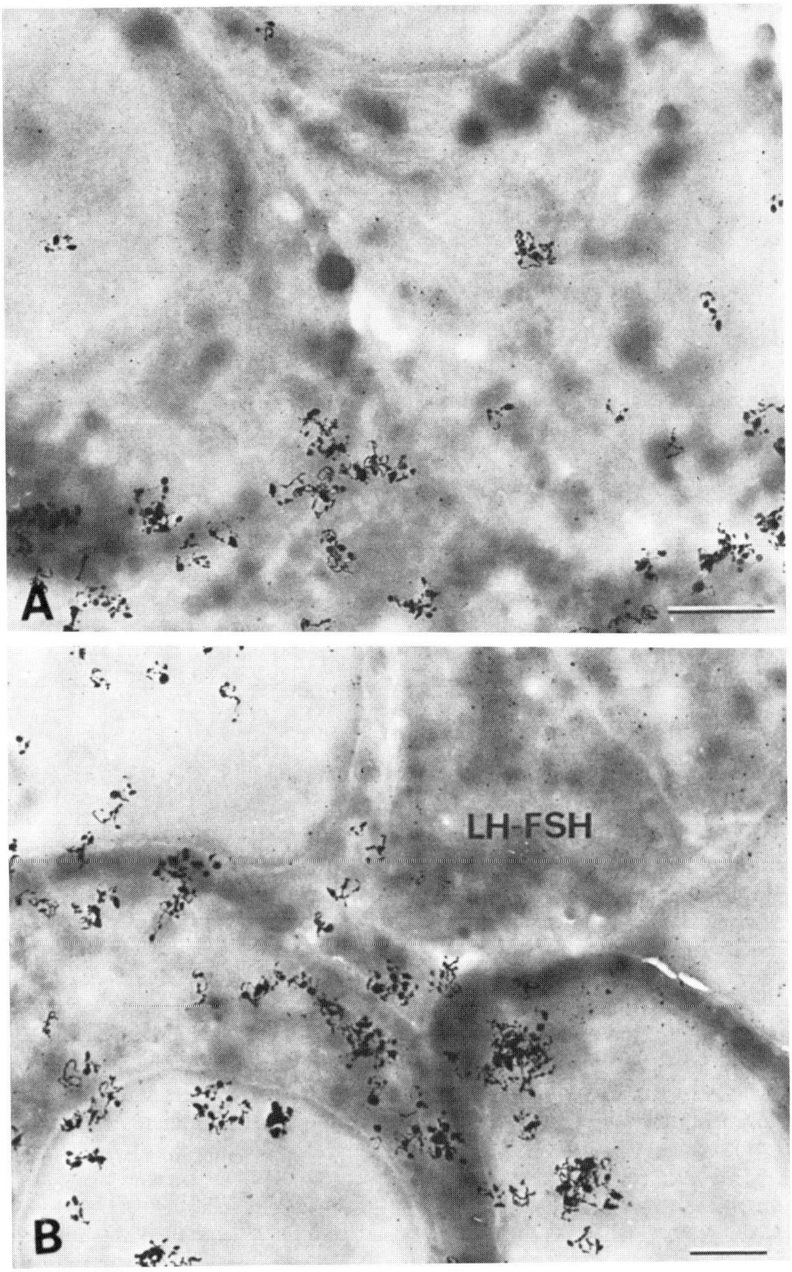
A
LH-FSH
B

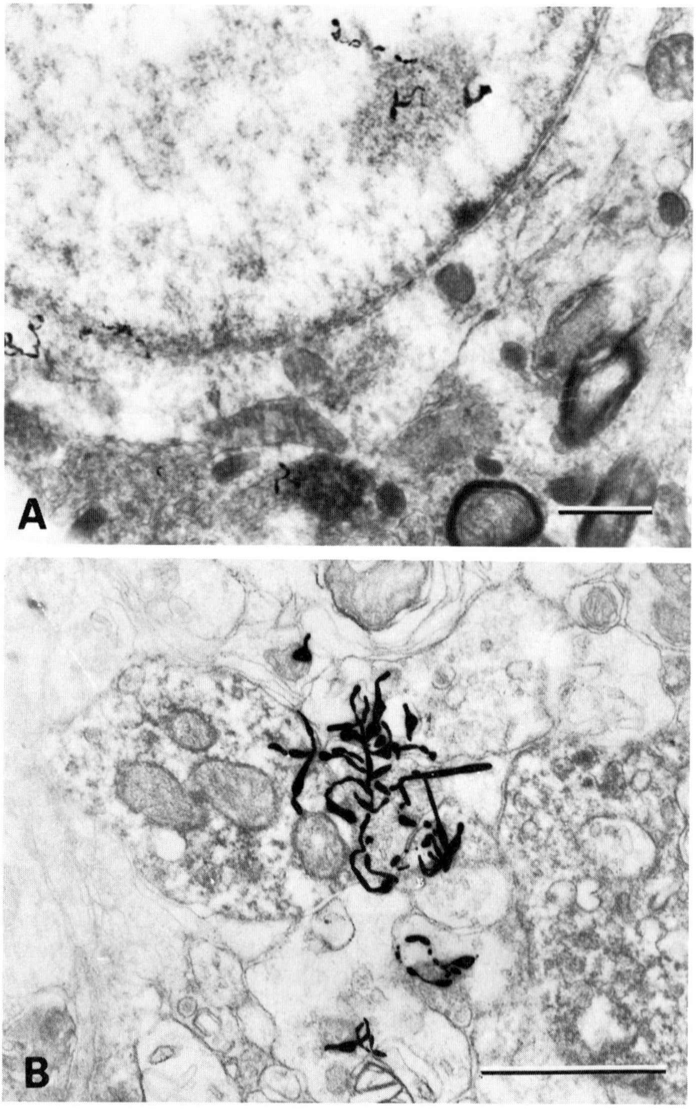

Figure 11. Electron micrographs showing association of autoradiography and immunocytochemistry.

A: The autoradiographic reaction indicates the uptake of ^{3}H dopamine 30 min after injection into the rat's third ventricle and fixation with a mixture of 4% paraformaldehyde and 0.2% glutaraldehyde. The immunostaining was performed on vibratome section with the pre-embedding staining technique using an anti-ACTH serum. Silver grains overlie the dopaminergic neuron. A POMC positive nerve ending is seen making contact with the dopaminergic neuron. Bar = 1 μm.

Acknowledgement

I would like to thank J. Doherty for corrections to the English of this manuscript, and M. Ficarelli for secretarial assistance.

References

1. Niepce de St. Vitor, N. (1867). *Comptes Rendus Hebdomadaires des Séances de l'Academie des Sciences, Paris*, **65,** 505.
2. Becquerel, H. (1896). *Comptes Rendus Hebdomadaires des Séances de l'Academie des Sciences, Paris*, **122,** 420, 501, 689, 1086.
3. Lacassagne, A. and Lattes, J. S. (1924). *Bulletin d'Histologie Apliquee et de Technique Microscopique*, **1,** 279.
4. Lacassagne, A., Lattes, J. S., and Lavedan, J. (1925). *Journal of Radiology and Electrology*, **9,** 1.
5. Hamilton, J. G., Soley, M. H., and Eichorn, K. B. (1940). *University of California (Berkeley) Publications in Pharmacology*, **1,** (28), 339.
6. Leblond, C. P. (1943). *Journal of Anatomy*, **77,** 149.
7. Belanger, L. F. and Leblond, C. P. (1946). *Endocrinology*, **39,** 8.
8. Joftes, D. L. and Warren, S. (1955). *Journal, Biological Photographic Association,* **23,** 145.
9. McDonald, A. M., Cobb, J., and Solomon, A. K. (1948). *Science,* **107,** 550.
10. Boyd, G. A. and Williams, A. I. (1948). *Proceedings of the Society for Experimental Biology and Medicine,* **69,** 225.
11. Liquier-Milward, J. (1956). *Nature,* **177,** 619.
12. Caro, J. J. and Skipper, S. J. (1954). *Science,* **119,** 141.
13. Caro, L. G. (1969). *Journal of Cell Biology*, **41,** 918.
14. Salpeter, M. M. and Bachmann, L. (1964). *Journal of Cell Biology*, **22,** 469.
15. Granboulan, P. (1965). In *Use of radioautography in investigating protein synthesis* (ed. C. P. Leblond and K. B. Warren), pp. 43–63. Academic Press, London.
16. Salpeter, M. M. and Bachmann, L. (1972). In *Principles and techniques of electron microscopy. Biological applications* (ed. M. A. Hayat), Vol. 2, pp. 219–78. Van Nostrand Reinhold, New York.
17. Salpeter, M. M., Bachmann, L., and Salpeter, E. E. (1969). *Journal of Cell Biology*, **41,** 1.

B: The autoradiographic technique shows the uptake of ^{3}H Serotonin into an ending in a rat suprachiasmatic nucleus. The immunostaining was performed using the pre-embedding method with an antivasoactive intestinal peptide serum. A radiolabelled varicosity is seen emerging from a thin unmyelinated axonal segment, making contact with two vasoactive intestinal peptide-immunoreactive dendrites without synaptic differentiation. Bar = 1 μm. This electron micrograph was kindly supplied by Dr O. Bosler.

18. Cau, P. (ed.) (1990). *Techniques en microscopie quantitative: stéréologie, autoradiographie et immunocytochimie quantitatives*. Editions INSERM, Paris.
19. Caro, L. G. (1962). *Journal of Cell Biology*, **15,** 189.
20. Williams, M. A. (ed.) (1977). *Quantitative methods in biology*. North Holland, Amsterdam.
21. Blackett, N. M. and Parry, D. M. (1973). *Journal of Cell Biology*, **57,** 9.
22. Waynforth, H. B. (1969). In *Techniques in protein biosynthesis* (ed. P. N. Campbell and J. R. Sargent), Vol. 2, p. 209. Academic Press, London and New York.
23. Williams, M. A. (1973). In *Techniques in protein biosynthesis*, Vol. 3 (ed. P. N. Campbell and J. R. Sargent), p. 125, Academic Press, London and New York.
24. Warshawsky, H., Leblond, C. P., and Droz, B. (1963). *Journal of Cell Biology*, **16,** 1.
25. Peterson, R. O. and Baserga, R. (1964). *Archives of Pathology*, **77,** 582.
26. Pelc, S. R. and Appleton, J. C. (1965). *Nature,* **205,** 1287.
27. Skourgaard, M. R. and Stewart, P. A. (1966). *Experimental Cell Research,* **45,** 158.
28. Morel, G. and Pelletier, G. (1986). *Peptides,* **7,** 1197.
29. Morel, G., Leroux, P., and Pelletier, G. (1985). *Endocrinology*, **116,** 1615.
30. Leroux, P., and Pelletier, G. (1984). *Endocrinology*, **114,** 14.
31. Morel, G., Pelletier, G., and Heisler, S. (1986). *Endocrinology,* **119,** 1972.
32. Williams, M. A. (1985). In *Autoradiography and immunocytochemistry. Practical methods in electron microscopy*, Vol. 6 (ed. A. M. Glauert), Part 1, pp. 77–208, North Holland, Amsterdam, New York and Oxford.
33. Korn, E. D. and Weisman, R. A. (1966). *Biochimica et Biophysica Acta,* **116,** 309.
34. Stein, O. and Stein, Y. (1967). *Journal of Cell Biology*, **33,** 319.
35. Stein, O. and Stein, Y. (1967). *Journal of Cell Biology*, **34,** 251.
36. Cope, G. H. and Williams, M. A. (1968). *Journal of the Royal Microscopical Society,* **88,** 259.
37. Saunders, D. R., Wilson, J., and Rubin, C. E. (1968). *Journal of Cell Biology*, **37,** 183.
38. Gautheron, C. and Chevalier, F. (1971). *Journal de Microscopie*, **10,** 99.
39. Carlier, H. and Bezard, J. (1975). *Journal of Cell Biology*, **65,** 383.
40. Hedley-Whyte, E. T. (1975). *Journal of Cell Biology*, **66,** 333.
41. Morgan, T. E. and Huber, G. L. (1967). *Journal of Cell Biology*, **32,** 757.
42. Boyenval, J. and Droz, B. (1976). *Journal de Microscopie et Biologie Cellulaire*, **27,** 129.
43. Magalhaes, M. M. (1976). *Journal de Microscopie et Biologie Cellulaire*, **27,** 231.
44. Wolfe, D. E., Potter, L. T., Richardson, K. C., and Axelrod, J. (1968). *Science,* **138,** 440.
45. Taxi, J. and Droz, B. (1966). *Comptes Rendus Hebdomadaires des Séances de l'Academie des Sciences*, *Paris*, **263,** 1326.
46. Williams, M. A. and Baba, W. I. (1967). *Journal of Endocrinology,* **39,** 543.
47. Orkland, P. M. and Kravitz, E. A. (1971). *Journal of Cell Biology*, **49,** 75.
48. Price, D. L., Stocks, A., Griffin, J. W., Young, A., and Peck, K. (1976). *Journal of Cell Biology*, **68,** 389.

49. Pelletier, G., Dube, D., Guy, J., Seguin, C., and Lefebvre, F. A. (1982). *Endocrinology,* **111,** 1068.
50. Duello, T. M., Nett, T. M., and Farquhar, M. G. (1983). *Endocrinology*, **112,** 1.
51. Morel, G., Dihl, F., Aubert, M. L., and Dubois, P. M. (1987). *Cell Tissue Research,* **248,** 541.
52. Mizuhira, V. and Futaesaku, Y. (1975). *Proceedings of the Royal Microscopical Society*, **10,** 158.
53. Stirling, C. E. and Kinter, W. B. (1967). *Journal of Cell Biology*, **35,** 585.
54. Benson, T. E., Burd, G. B., Greer, C. A., Landis, D. M., and Shepherd, G. M. (1985). *Brain Research*, **339,** 67.
55. Yaruwsky, P. J. and Boyne, A. F. (1989). *Journal of Neurocytology*, **18,** 121.
56. Jourdan, F. (1991). In *Techniques en microscopie électronique: cryométhode, immunocytologie, autoradiographie, hybridation in situ* (ed. G. Morel), pp. 377–413. Editions INSERM, Paris.
57. Baker, J. R. J. and Appleton, I. C. (1975). *Proceeding of the Royal Microscopical Society*, **10,** 157.
58. Morel, G. (1990). In *Techniques en microscopie électronique: cryométhode, immunocytologie, autoradiographie, hybridation in situ* (ed. G. Morel). (In press.) Editions INSERM, Paris.
59. Droz, B. (1976). *Journal de Microscopie et Biologie Cellulaire*, **27,** 133.
60. Vrensen, G. F. J. M. (1970). *Journal of Histochemistry and Cytochemistry,* **18,** 278.
61. Boumendil, J. (ed.) (1989). *Techniques en microscopie electronique: Principes et méthodes de préparation*. Editions INSERM, Paris.
62. Morel, G. (1991). In *Techniques en microscopie électronique: cryométhode, immunocytologie, autoradiographie, hybridation in situ* (ed. G. Morel), pp. 31–54. Editions INSERM, Paris.
63. Tokuyasu, K. T. (1973). *Journal of Cell Biology*, **57,** 551.
64. Caro, L. G. and Van Tubergen, R. P. (1962). *Journal of Cell Biology*, **15,** 173.
65. Bouteille, M. (1976). *Journal de Microscopie et Biologie Cellulaire*, **27,** 121.
66. Rogers, A. W. (ed.) (1979). *Techniques of autoradiography*. Elsevier/North-Holland Biomedical Press, Amsterdam, New York and Oxford.
67. Kopriwa, B. M. (1967). *Journal of Histochemistry and Cytochemistry*, **15,** 501.
68. Bachmann, L. and Salpeter, M. M. (1965). *Laboratory Investigations*, **14,** 1041.
69. James, T. H. (1948). *Journal of Colloid Science*, **3,** 447.
70. Lettré, H. and Paweletz, N. (1966). *Naturwissenschaften*, **53,** 268.
71. Reynolds, E. S. (1963). *Journal of Cell Biology,* **17,** 208.
72. Rosenweig, L. J. and Kanwar, Y. S. (1982). *Endocrinology*, **111,** 1817.
73. Morel, G. and Heisler, S. (1988). *Electron Microscopy Reviews*, **1,** 221.
74. Morel, G., Chabot, J.-G., Garcia-Caballero, T., Gossard, F., Dihl, F., Belles-Isles, M., and Heisler, S. (1988). *Endocrinology,* **123,** 149.
75. Morel, G. and Pelletier, G. (1986). *Experimental and Clinical Endocrinology* (Life Sci. Adv.), **5,** 31.
76. Chabot, J.-G., Morel, G., Belles-Isles, M., Jeandel, L., and Heisler, S. (1988). *American Journal of Physiology*, **254,** E 301.
77. Mulvaney, B. D. (1971). *Journal of Cell Biology,* **51,** 568.
78. Stevens, A. R. (1966). In *Methods in cell physiology* (ed. D. Prescott), Vol. 2, pp. 255. Academic Press, New York and London.

79. Rogers, A. W. and John, P. N. (1969). In *Autoradiography of diffusible substances* (ed. L. J. Roth and W. E. Stumf), p. 51. Academic Press, New York.
80. Morel, G., Mesguich, P., Chabot, J.-G., Belles-Isles, M., Jeandel, L., and Heisler, S. (1989). *Biologie Cellulaire*, **65,** 181.
81. Morel, G., Chabot, J.-G., Gossard, F., and Heisler, S. (1989). *Endocrinology,* **124,** 1703.
82. Le Guellec, D. and Frappart, L. (1990). In *Techniques en microscopie electronique: cryométhode, immunocytologie, autoradiographie, hybridation in situ* (ed. G. Morel), pp. 535–44. Editions INSERM, Paris.
83. Caro, L. G. and Schnös, M. (1965). *Science, New York*, **149,** 60.
84. Bachmann, L. and Salpeter, M. M. (1967). *Journal of Cell Biology,* **33,** 299.
85. Fertuck, H. C. and Salpeter, M. M. (1974). *Journal of Histochemistry and Cytochemistry,* **22,** 80.
86. Fertuck, H. C. and Salpeter, M. M. (1976). *Journal of Cell Biology,* **69,** 144.
87. Salpeter, M. M. (1973). *Journal of Histochemistry and Cytochemistry*, **21,** 623.
88. Salpeter, M. M. and Salpeter, E. E. (1971). *Journal of Cell Biology,* **50,** 324.
89. Salpeter, M. M. and McHenry, F. A. (1973). In *Advanced techniques in biological microscopy* (ed. J. K. Koehler), pp. 113–31, Springer-Verlag, New York.
90. Salpeter, M. M., Fertuck, H. C., and Salpeter, E. E. (1977). *Journal of Cell Biology*, **72,** 161.
91. Salpeter, M. M., McHenry, F. A., and Salpeter, E. E. (1978). *Journal of Cell Biology*, **76,** 127.
92. Salpeter, M. M., Budd, G. C., and Mattimoe, S. (1974). *Journal of Histochemistry and Cytochemistry*, **22,** 217.
93. Williams, M. A., Downs, A., and Junger, E. (1983). *Acta stereologica,* **2,** 249.
94. Downs, A. and Williams, M. A. (1984). *Journal of Microscopy*, **136,** 1.
95. Morel, G., Chabot, J.-G., Belles-Isles, M., and Heisler, S. (1988). *Molecular and Cellular Endocrinology,* **55,** 219.
96. Williams, M. A. (1969). In *Advances in optical and electron microscopy*, Vol. 3 (ed. R. Barer and V. E. Cosslett), pp. 219–72, Academic Press, London.
97. Salpeter, M. M. (1968). *Journal of Morphology*, **124,** 387.
98. Nadler, N. J. (1971). *Journal of Cell Biology*, **49,** 877.
99. Parry, D. M. (1976). *Journal de Microscopie et Biologie Cellulaire*, **27,** 185.
100. Parry, D. M. and Blackett, N. M. (1976). *Journal of Microscopy*, **106,** 117.
101. Blackett, N. M. and Parry, D. M. (1977). *Journal of Histochemistry and Cytochemistry*, **25,** 206.
102. Williams, M. A. and Downs, A. (1978). *Journal of Microscopy,* **114,** 157.
103. Land, D. R. and Salpeter, E. E. (1978). *Journal of Cell Biology*, **76,** 142.
104. Cau, P. (1988). *Journal of Microscopy,* **150,** 199.
105. Stumpf, W. E. and Roth, L. J. (1966). *Journal of Histochemistry and Cytochemistry,* **14,** 274.
106. Kennedy, C., Des Rosiers, M. H., Jemle, M. M., Reivich, M., Sharp, F. R., and Sokoloff, L. (1975). *Science,* **187,** 850.
107. Des Rosiers, M. H. and Descarries, M. H. (1978). *Comptes Rendus Hebdomadaires des Séances de l'Academie des Sciences, Paris,* **287,** 153.
108. Basinger, S. F., Gordon, W. C., and Lam, D. M. (1979). *Nature,* **280,** 682.
109. Jourdan, F., Duveau, A., Astic, C., and Holley, A. (1980). *Brain Research,* **188,** 139.

110. Yarowsky, P. J. and Boyne, A. F. (1989). *Journal of Neurocytology,* **18,** 121.
111. Jacob, J., Todd, K., Birnstiel, M. L., and Bird, A. (1971). *Biochimica et Biophysica Acta,* **228,** 701.
112. Croissant, O., Dauguet, C., Jeanteur, P., and Orth, G. (1972). *Comptes Rendus Hebdomadaires des Séances de l'Academie des Sciences, Paris*, **274,** 614.
113. Binder, M., Tourmente, S., Roth, J., Renaud, M., and Gehring, W. J. (1986). *Journal of Cell Biology*, **102,** 1646.
114. Morel, G., Dubois, P., and Gossard, F. (1986). *Comptes Rendus Hebdomadaires des Séances de l'Academie des Sciences, Paris*, **302,** 479.
115. Morel, G., Dihl, F., and Gossard, F. (1989). *Molecular and Cellular Endocrinology,* **65,** 81.
116. Puvion-Dutilleul, F., and Puvion, E. (1989). *European Journal of Cell Biology,* **49,** 99.
117. Guitteny, A. F. and Bloch, B. (1989). *Histochemistry,* **92,** 277.
118. Tong, Y., Zhao, H. F., Simard, J., Labrie, F., and Pelletier, G. (1989). *Journal of Histochemistry and Cytochemistry,* **37,** 567.
119. Trembleau, A., Calas, A. and Fevre-Montange, M. (1990). *Molecular Brain Research,* **8,** 37.
120. Hamel, E. and Beaudet, A. (1984). *Journal of Electron Microscopy Technique,* **1,** 317.
121. Stoeckel, M. E., Freund-Mercier, M. J., and Kremarik, P. (1991). In *Techniques en microscopie electronique: cryométhode, immunocytologie, radioautographie, hybridation in situ* (ed. G. Morel), pp. 329–38. Editions INSERM, Paris.
122. Hamel, E. and Beaudet, A. (1987). *Brain Research,* **401,** 239.
123. Beaudet, A. and Szigethy, E. (1989). In *Brain imaging: techniques and applications* (ed. Sharf and Lewis), pp. 186–208. Horwood, Chichester.
124. Dana, C., Vial, M., Leonard, K., Beauregard, A., Kitabgi, P., Vincent, J. P., Rostene, W., and Beaudet, A. (1989). *Journal of Neuroscience,* **9,** 2247.
125. Bosler, O. and Calas, A. (1982). *Brain Research Bulletin*, **9,** 151.
126. Bosler, O. and Beaudet, A. (1985). *Journal of Neurocytology,* **14,** 749.
127. Bosler, O., Beaudet, A., and Pickel, V. M. (1986). *Journal of Electron Microscopy Technique,* **4,** 21.

6

Electron microscopical enzyme histochemistry

CORNELIS J. F. VAN NOORDEN and
CAESAR E. HULSTAERT

1. Introduction

Good preservation of the ultrastructure of tissues and cells is a prerequisite for an adequate localization of enzyme activity at the electron microscopical level. For light microscopical purposes, tissue preservation is not as critical because the resolution of the microscope generally excludes enzyme localization within organelles or other cytoplasmic structures. Morphology is usually preserved by chemical fixation but for the localization of enzyme activity cells and tissues cannot be fixed in a similar way as for morphological ultrastructural investigations, because the majority of enzymes are inactivated rapidly by fixation. Exceptions are phosphatases and oxidases, and therefore these enzymes have been studied for years at the ultrastructural level. Today, a variety of methods is available for the light microscopical demonstration of a large number of enzymes (1) and improvements in tissue preservation and detection techniques have made possible the further development of electron cytochemical methods. The aim of this chapter is to describe the customary methods that are available for the ultrastructural localization of enzyme activity. The scope of this chapter only allows a brief description of these methods and the less well-known methods have had to be omitted. For an extensive review of electron cytochemical methods see Borgers and Verheyen (2).

Electron microscopical immunocytochemistry has developed rapidly since cryofixation was introduced, because it avoids denaturation of macromolecules such as proteins and thus of antigens. Although cryofixation appears to be theoretically ideal for enzyme histochemistry, it has not been successful for the ultrastructural localization of enzyme activities. A minimum amount of enzyme is required at a site to build up enough final reaction product to exceed the solubility product and to obtain a precipitate (3). Usually, ultra-thin cryosections are too thin to meet this requirement. Therefore, most enzyme reactions are performed after mild fixation on tissue preparations such as tissue chopper or vibratome sections, or cell suspensions. The precipitated reaction product is then examined in ultra-thin sections that are cut after dehydration and embedding of the tissue in a resin.

A number of criteria should be met in order to obtain reliable information on the localization of enzyme activity. Indeed, for all enzyme histochemical methods it should be possible to establish precision, reproducibility, specificity, and validity (4). A method is *precise* when the final reaction product is precipitated at the true *in vivo* subcellular site of the enzyme. A method is *reproducible* when it gives similar results irrespective of the place and time of its performance. A method is *specific* when the final reaction product is generated only by the enzyme studied. A *valid* method gives rise to formation of final reaction product proportionally with the *in vivo* activity of the enzyme. The latter requirement is usually not met in electron microscopy because tissue preservation leads to inactivation of at least some of the enzyme molecules. Inactivation may not occur to the same extent in all cellular compartments. The first three requirements are sometimes hard to fulfil, especially at the electron microscopical level, as discussed by Fahimi (5). Control reactions to test the *specificity* of reaction are essential for each enzyme histochemical method and therefore control reactions are given emphasis here for all the methods described.

2. Tissue pre-treatment

2.1 Fixation

Precise localization of enzyme activity can be determined only when the ultrastructural morphology is sufficiently intact. This requires fixation conditions that preserve enzyme activity on the one hand and cellular fine structure on the other. Moreover, fixation should be as uniform as possible to obtain *reproducible* results and therefore tissues have preferably to be fixed by perfusion (see *Protocols 1* and *2*). If perfusion fixation is impossible, immersion fixation should be performed with tissue blocks as small as possible (see *Protocol 3*). Individual cells can be fixed in suspension (see *Protocol 4*). Before incubation, the tissue is sliced into 10–40 μm sections to achieve an optimal penetration of the tissue by the compounds of the incubation medium. A wide variety of chemical cross-linking agents are available for fixation, but formaldehyde and glutaraldehyde are still the most commonly used fixatives for enzyme histochemical purposes. Impurities and contaminants of commercially available aldehydes may inactivate enzymes and interfere with *reproducibility*. Therefore, formaldehyde solutions should be freshly prepared from *para*formaldehyde powder, and only solutions of monomeric glutaraldehyde should be used (6). Purity of glutaraldehyde can be assessed by UV absorbance. Monomeric glutaraldehyde is commercially available.

Protocol 1. Perfusion fixation of whole animals[a] (7)

1. Anaesthetize animal with Nembutal.
2. Administer 0.45 mg sodium nitrite and 250 units of heparin in a 0.9% sodium chloride solution intravenously in tail vein to prevent contraction of blood vessels.
3. Open abdomen and cut the chest open at both sides of the sternum.
4. Remove pericardium from heart.
5. Insert needle of perfusion system in left ventricle.
6. Perfuse with 2.5% glutaraldehyde in 100 mM cacodylate buffer (pH 7.4; temp. of fixative 20°C). Flow rate of perfusion medium 2.5–15 ml/min.
7. Open right atrium to allow body fluids and fixative to leave the body.
8. Perfuse for 1–10 min depending on stiffening of the body.
9. Perfuse then for 5 min with 100 mM cacodylate buffer (pH 7.4).

[a] Volumes, flow rate, etc., are given for rats and mice. For larger or smaller animals these values need adjustment.

Protocol 2. Perfusion fixation of organs (8)

1. Perform perfusion with a pump with a flow rate of 13–15 ml/min at room temperature (RT).
2. Perfuse for 1 min with rinsing medium consisting of 2% polyvinylpyrrolidone (PVP, mol. wt 40 000), 75 mM sodium nitrite in 100 mM cacodylate buffer (pH 7.4). PVP compensates for the colloid–osmotic pressure; sodium nitrite prevents contraction of blood vessels.
3. Perfuse with 2% glutaraldehyde in 100 mM cacodylate buffer (pH 7.4) for 5 min.
4. Perfuse with 6.8% sucrose in 100 mM cacodylate buffer (pH 7.4).

Protocol 3. Fixation of individual cells (9)

1. Fix cells in suspension with 1–2% glutaraldehyde and 5% sucrose in 100 mM cacodylate buffer (pH 7.2) for 10–30 min at 4°C.
2. Centrifuge the cell suspension for 10 min at 4°C (50–2000 *g*, depending on the type of cell).

Protocol 3. *Continued*

3. Rinse cells overnight with 6.8% sucrose in 100 mM cacodylate buffer (pH 7.2) at 4°C.
4. Suspend cells in a 7.5% agar solution and let it solidify.

Protocol 4. Immersion fixation (9)

1. Cut tissue in blocks as small as possible.
2. Immerse tissue in the same fixative given in *Protocol 3* for 45 min at 4°C.
3. Rinse tissue blocks thrice in 100 mM cacodylate buffer (pH 7.2).
4. Immerse blocks in a 7.5% agar solution and let it solidify.

A large number of enzymes, such as many dehydrogenases and proteases, are completely inactivated by the above mentioned fixation methods. In general, glutaraldehyde is more deleterious than formaldehyde, but even formaldehyde may inactivate enzymes. For the demonstration of fixation-sensitive enzymes, 4% formaldehyde (freshly prepared from *para*formaldehyde) in 100 mM phosphate buffer (pH 7.6) containing 5% sucrose can be employed for perfusion fixation (10, 11). When the ultrastructural morphology is not satisfactory, a small amount of glutaraldehyde (e.g. 0.25–0.5%) may be added to the fixative. In this case, fixation is carried out for 5 min at 4°C.

For immersion fixation, small tissue fragments can be immersed in a solution of 0.1% glutaraldehyde in 100 mM cacodylate buffer (pH 7.4) for 20 min at 4°C (12). Subsequently, the tissue fragments are rinsed overnight in 100 mM cacodylate buffer (pH 7.4).

2.2 Sectioning of aldehyde-fixed tissue before incubation

Before incubation, 10–40 μm thick sections are made with a Smith-Farquhar tissue chopper, an Oxford Vibratome, or a TPI vibratome (Series 2000; Technical Products International). The sections are used immediately.

3. Demonstration methods

Electron cytochemical methods are based on the principle that conversion of a water-soluble substrate produces a water-insoluble electron dense or osmiophilic final reaction product.

3.1 Metal salt methods

Metal salt methods are used for the demonstration of activity of phosphatases, sulphatases, transferases, (cholin)esterases, and lipases. Cerium methods are also used for the demonstration of oxidase activity, because the final reaction product of oxidases, hydrogen peroxide, can be trapped by cerium to form a water-insoluble electron-dense compound. The methods are all based on the principle that during the enzyme reaction the substrate is converted by the enzyme and either the liberated group (for example phosphate or sulphate) or the compound generated (for example hydrogen peroxide) is captured by cations (for example lead or cerium) present in the incubation medium. As a result the water-insoluble electron-dense salt (for example lead phosphate) precipitates at the enzymic site.

3.1.1 Lead salt methods

The lead salt method developed by Gomori in 1941 (13) for the demonstration of acid phosphatase activity was the first enzyme histochemical method to be used for electron cytochemical investigations (2). Lead phosphate is easily recognized in the electron microscope but the method has some drawbacks (14). Non-specific lead precipitates and diffusion of intermediate reaction products often make it difficult to draw unequivocal conclusions on the precise localization of the final reaction product. Furthermore, lead ions are not easy to keep in solution in the incubation media, especially at a high pH, and they may also inhibit enzyme activity. In *Protocols 5–7*, the methods for acid phosphatase, aspartate aminotransferase, and ornithine carbamoyltransferase, respectively, are given.

Protocol 5. Lead salt method for acid phosphatase activity (15)

1. Dissolve 750 mg β-glycerophosphate in 12.5 ml acetate buffer (100 mM, pH 5.0).
2. Adjust the pH to 5.0 with 2 N hydrochloric acid and bring the solution to a volume of 25 ml.
3. Add this solution under continuous stirring to 225 ml of a solution of 300 mg lead nitrate in 50 mM acetate buffer (pH 5.0).
4. The final concentrations are 3.6 mM lead nitrate, 13.9 mM β-glycerophosphate, and 50 mM acetate buffer (pH 5.0).
5. Use immediately for incubation for 30 min at RT.
6. After incubation, rinse tissue preparations thrice (90 sec each) in 50 mM acetate buffer (pH 5.0) at RT.
7. Rinse thrice (5 min each) in 100 mM cacodylate buffer (pH 7.4) with 4% sucrose at RT.
8. Control incubations:
   - in the absence of β-glycerophosphate
   - in the presence of 2 mM sodium fluoride or 10 mM sodium tartrate

---

Protocol 6. Lead salt method for aspartate aminotransferase (16)

1. Prepare incubation medium consisting of 4 mM α-ketoglutarate, 20 mM L-aspartate, 6 mM lead nitrate, and 250 mM sucrose in 50 mM imidazole buffer (pH 7.5).
2. Incubate for 20 min at RT.
3. Rinse specimens for 5 min at RT in a solution of 20 mM L-aspartate in 50 mM imidazole buffer (pH 7.3).
4. Control incubations:
   - in the absence of either α-ketoglutarate or aspartate

Protocol 7. Lead salt method for ornithine carbamoyltransferase (17)

1. Dissolve 3 mg lithium carbamoyl phosphate and 5 mg L-ornithine in 4 ml of 50 mM Tris–maleate buffer (pH 7.2).
2. Dissolve 10 mg lead nitrate in 1 ml of 50 mM Tris–maleate buffer (pH 7.2).
3. Dissolve 0.8 g sucrose in 5 ml distilled water.
4. Add sucrose solution to the solution of carbamoyl phosphate and ornithine, and then add dropwise during continuous stirring the lead nitrate solution. Filter the slightly turbid incubation medium and use immediately.
5. Incubate for 10 min at RT. Longer incubations should be avoided because of unstability of carbamoyl phosphate and precipitation of lead phosphate.
6. Renew medium when necessary because of turbidity.
7. Rinse briefly in buffer (thrice).
8. Control incubations:
   (a) Replace L-ornithine by D-ornithine, L-glutamate, D,L-methionine, or D,L-alanine.
   (b) Pre-treat specimens before incubation with 5 mM *para*-chloromercuribenzoate and 8% sucrose in 50 mM Tris–maleate buffer (pH 7.2), and incubate in the presence of 1 mM *para*-chloromercuribenzoate to inhibit all enzyme activity.

3.1.2 Cerium salt methods

i. Phosphatases

Cerium was introduced in 1975 by Briggs *et al.* (18) as a capturing agent for hydrogen peroxide, the product of oxidase activity. In 1980, cerium was reported to be a suitable capturing agent for phosphate (19). It appeared that cerium was superior to lead as a trapping agent for phosphatase cytochemistry because of the high reproducibility of the methods, the fine granular form of the reaction product, the lack of non-specific precipitates and a less inhibitory effect on enzymes. The only serious disadvantage of cerium is its slow penetration into tissues. There are several ways to enhance penetration of cerium into tissues:

- pre-incubation of the specimens for 1 h at 37 °C in incubation media without substrate
- addition of 0.0001–0.0002% Triton X-100 to the incubation medium
- freezing of the specimens in melting nitrogen

In principle, all methods based on the metal salt principle can be performed with cerium, and they generally yield results that are superior to methods employing lead.

Incubations are usually carried out for 30 min at 37 °C, but the incubation period can be varied between 10 and 60 min, depending on the tissue and the enzyme. Non-specific precipitates such as cerium hydroxide may be formed, especially at alkaline pH; they are removed by washing specimens overnight at 4 °C in 6.8% sucrose in 100 mM cacodylate buffer (pH 6.0). The general procedure is given in *Protocol 8*. Examples of tissue incubated according to these methods are presented in *Figures 1* and *2*.

Protocol 8. Cerium salt method for phosphatases (8)

1. Pre-incubate sections for 1 h at 37 °C in medium[a] without substrate.
2. Incubate sections for 30 min at 37 °C in complete medium.[a]
   - acid phosphatase: 2 mM cerium chloride,[b] 1 mM sodium β-glycerophosphate, 100 mM acetate buffer (pH 5.0) (9)
   - alkaline phosphatase (see *Protocol 9*): 14 mM cerium chloride, 11 mM sodium citrate, 4 mM magnesium chloride, 10 mM *para*-nitrophenyl phosphate, 100 mM glycine/sodium hydroxide buffer (pH 9.3)
   - 5′-nucleotidase: 1 mM cerium chloride, 5 mM magnesium nitrate, 1.3 mM adenosine monophosphate, 70 mM Tris–maleate buffer (pH 7.2)
   - nucleoside di- and triphosphatases: 1 mM cerium chloride, 5 mM magnesium nitrate, 2.3 mM substrate (ATP, ADP, UTP, UDP, ITP), 70 mM Tris–maleate buffer (pH 7.2)
3. Rinse sections overnight at 4 °C in 6.8% sucrose in 100 mM cacodylate buffer (pH 6.0).
4. Control incubations:
   - for all enzymes: without substrate
   - for acid phosphatase: pre-incubation and incubation in the presence of 2 mM sodium fluoride or 10 mM sodium tartrate
   - for alkaline phosphatase: pre-incubation and incubation in the presence of 5 mM levamisole

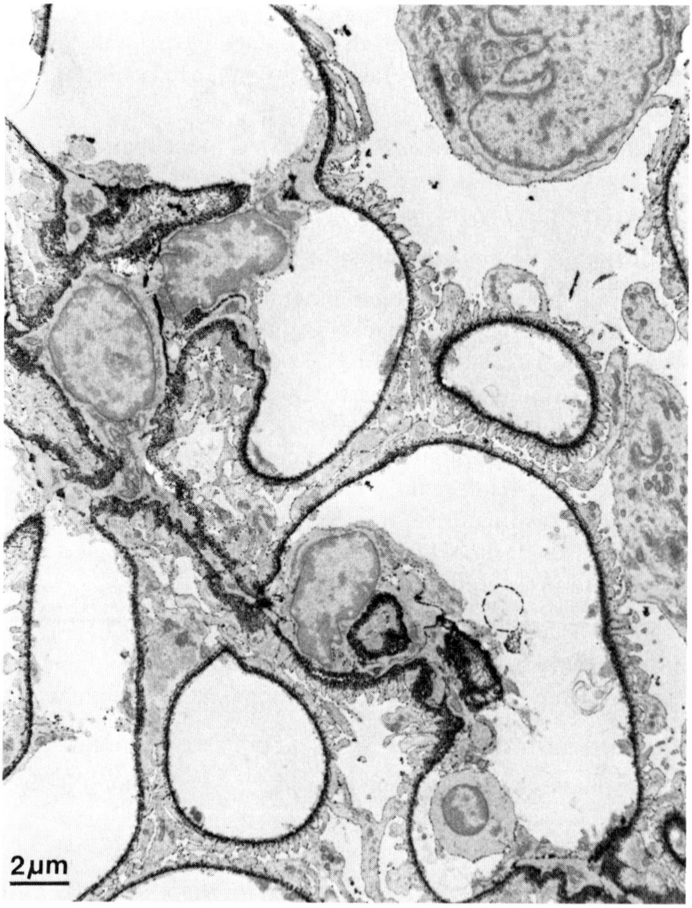

Figure 1. Nucleoside diphosphatase activity in rat kidney glomerulus as demonstrated with the cerium technique and with UDP as substrate. The reaction product in the basement membrane is very prominent.

Protocol 8. *Continued*

- for 5′-nucleotidase: pre-incubation and incubation in the presence of 5 mM levamisole. This should not affect the results
- for ATPase and ADPase: Pre-incubation and incubation in the presence of 1 mM vanadium pentoxide (K. Poelstra, personal communication)

[a] Use CO_2-free distilled water for the preparation of the incubation media to avoid formation of insoluble cerium carbonate. Prepare all media shortly before incubation.

[b] Use cerium chloride from Fluka AG (Buchs, Switzerland). Other cerium preparations can give inferior results.

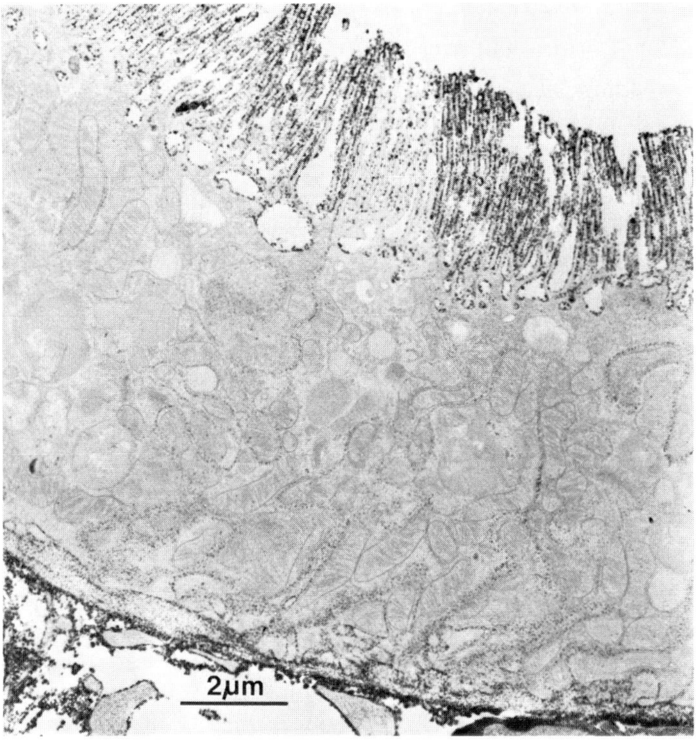

Figure 2. Nucleoside triphosphatase activity in rat kidney proximal tubular epithelial cells as demonstrated with the cerium technique and with ITP as substrate. Note the difference in activity along the microvilli between two neighbouring cells.

In some tissues such as rat kidney glomerulus, alkaline phosphatase is inactivated by glutaraldehyde fixation (20). In that case formaldehyde has to be used. The preparation of the incubation medium for this enzyme is given in *Protocol 9*.

Protocol 9. Preparation of alkaline phosphatase incubation medium (21)

1. Add to 6 ml 300 mM glycine/sodium hydroxide buffer (pH 9.3) 1 ml 110 mM sodium citrate, 1 ml 40 mM magnesium chloride, 1 ml 100 mM *para*-nitrophenyl phosphate, 1 ml 140 mM cerium chloride.
2. Adjust pH with 5 N sodium hydroxide to pH 9.3.
3. Stir for 10 min or warm slightly until the turbid medium becomes clear.

For the demonstration of glucose-6-phosphatase activity, a special procedure is developed to obtain reaction product throughout vibratome sections (see *Protocol 10*).

Protocol 10. Cerium salt method for glucose-6-phosphatase (22)

1. After fixation according to *Protocol 2* quench-freeze small blocks (5 × 5 × 5 mm) in melting nitrogen.
2. After thawing, cut vibratome sections.
3. Pre-incubate sections for 1 h at 4°C in complete incubation medium consisting of 1 mM cerium chloride (Fluka AG; see *Protocol 8*), 2 mM disodium glucose-6-phosphate and 80 mM Tris–maleate buffer (pH 6.5).
4. Incubate sections for 30 min at 37°C in complete incubation medium.
5. Rinse sections overnight at 4°C in 6.8% sucrose and 100 mM cacodylate buffer (pH 6.0).
6. Control incubations:
   - in the absence of substrate
   - in the presence of 100 mM citrate or 1 mM diethyl pyrocarbonate

ii. Oxidases

Another application of cerium in enzyme histochemistry is its employment for the demonstration of oxidase activity. One of the products of oxidases is hydrogen peroxide, and this can be captured by cerium ions to form a water-insoluble product, probably cerium perhydroxide. It has not yet been established whether this compound is actually formed, but a cerium-containing precipitate was found at the sites of enzyme activity by X-ray microanalysis. The precipitate is very fine and electron dense and non-specific deposits are usually not found. Therefore, the cerium-based method for the ultrastructural demonstration of oxidase activity is the method of choice. However, oxidases are mainly localized in peroxisomes, and catalase is a component of peroxisomes as well. The latter enzyme has to be inhibited during the oxidase reaction because hydrogen peroxide is used as a substrate by catalase. The most commonly used catalase inhibitors in oxidase incubation media are either sodium azide, or 3-amino-1,2,4-triazole. Usually, pre-incubations with the inhibitor are carried out. Rinsing of the specimens after incubation in 6.8% sucrose in 100 mM cacodylate buffer (pH 6.0) minimizes background staining (see *Protocol 11*). An example of a reaction for oxidase activity is shown in *Figure 3*.

Protocol 11. Cerium salt methods for oxidases (23)

1. Pre-incubate specimens for 30 min at 37°C without substrate in a medium containing 3 mM cerium chloride (Fluka AG; see *Protocol 8*), 100 mM sodium azide, in 100 mM Pipes buffer (pH 7.8).
2. Incubate specimens for 60 min at 37°C in the same medium to which 0.1–10 mM of the specific substrate is added: 0.8 mM NADH or NADPH for

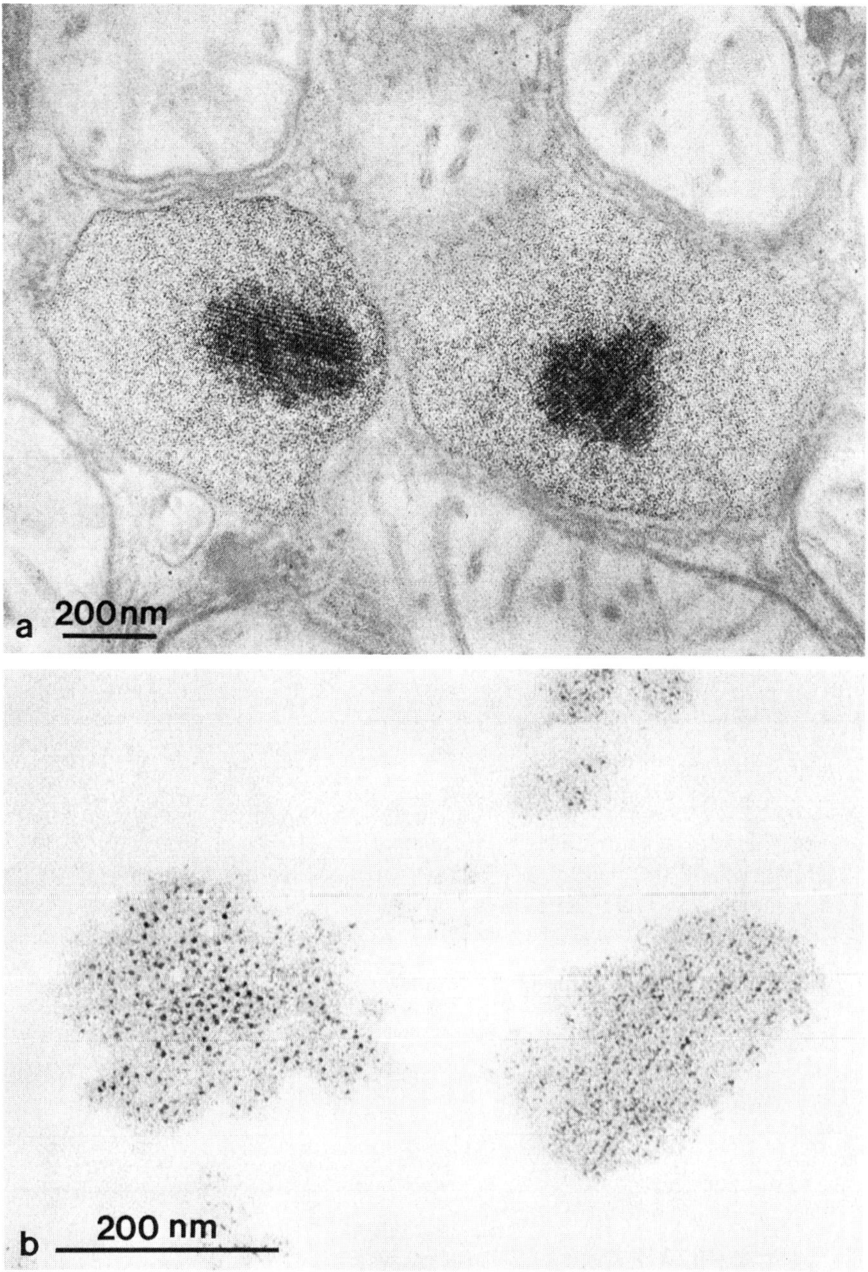

Figure 3. Oxidase activity in rat liver as demonstrated with the cerium method. (a) Urate oxidase in peroxisomes of a liver parenchymal cell. (b) Xanthine oxidase in isolated peroxisomal cores. (Courtesy of Dr S. Angermueller, Institute of Anatomy and Cell Biology, University of Heidelberg, Germany.)

Protocol 11. *Continued*

NAD(P)H oxidase, 10 mM D-proline for D-amino acid oxidase, 10 mM L-leucine for L-amino acid oxidase, 13 mM sodium glycolate or argininic acid for α-hydroxy acid dehydrogenase, 13 mM glycolate for glycolate oxidase, 2.5 mM tyramine hydrochloride for monoamine oxidase, 0.1 mM sodium urate for urate oxidase, 50 mM methanol for alcohol oxidase, 10 mM oxalate for polyamine oxidase, 0.1 mM hypoxanthine for xanthine oxidase.

3. Rinse the tissues thoroughly overnight at 4°C in 6.8% sucrose in 100 mM cacodylate buffer (pH 6.0).
4. Control incubations:

   For all oxidases:

   (a) Omit substrate from the incubation medium.

   (b) Bubble oxygen-free nitrogen through the medium because oxygen is one of the two substrates for oxidases.

   (c) Add 0.01% catalase to the incubation medium instead of azide or aminotriazole. This should abolish the reaction almost completely.

   For NAD(P)H oxidase:

   (a) Include 1 mM potassium cyanide in the medium. This should not affect the result.

   (b) Add 10 mM *para*chloromercuribenzene sulphonate to the pre-incubation and incubation medium.

   For D-amino acid oxidase:

   Add 100 mM D,L-β-hydroxybutyrate.

   For L-amino acid oxidase:

   Replace L-leucine with L-lysine or L-serine.

   For monoamine oxidase:

   Include 1 mM pargyline or 2 mM 2-phenethyl-hydrazine sulphate in the media.

   For urate oxidase:

   Add 2 mM 2,6,8-trichloro purine to the media.

   For xanthine oxidase:

   Add 1 mM allopurinol to the media.

3.1.3 Other metal salt methods

Before cerium was applied in phosphatase histochemistry, alternatives for the lead methods were developed such as the use of strontium for the localization of adenosine triphosphatase. Barium is used as a capturing agent for the localization of aryl sulphatase (see *Protocol 12*).

Protocol 12. Barium salt method for aryl sulphatase (24)

1. Prepare pre-incubation and incubation medium by adding 8 ml of a solution of 2.5% barium chloride to 12 ml 100 mM acetate buffer (pH 5.5).
2. Add 160 mg *p*-nitrocatechol sulphate to the incubation medium under continuous stirring.[a]
3. Pre-incubate in medium without substrate for 20 min at RT.
4. Incubate with medium containing substrate for 20 min at 37 °C.
5. Rinse thrice in 100 mM cacodylate buffer (pH 7.3) with addition of 6.8% sucrose for at least 2 h at RT.
6. Control incubations:
   - in the absence of substrate
   - in the presence of 10 mM sodium fluoride

[a] Final concentrations are 26 mM *p*-nitrocatechol sulphate, 96 mM $BaCl_2$ and 60 mM acetate buffer.

3.2 Diaminobenzidine methods

Oxidation of diaminobenzidine (DAB) leads to polymerization into DAB-brown. The monomer is faintly brown and water-soluble, whereas the polymer is brown, osmiophilic, and water-insoluble. Cytochrome *c* oxidase, catalase, and peroxidase activities can be demonstrated with DAB methods. These methods have become very popular, especially because peroxidase is frequently used as a label in immunohistochemistry. The methods described here allow discrimination between the activity of each of the three enzymes based on short incubation periods and composition of the media (see *Protocols 13–15*) (25). An example of a peroxidase reaction is presented in *Figure 4*.

Protocol 13. DAB method for cytochrome *c* oxidase (25)

1. Incubate tissue specimens in an incubation medium consisting of 2.5 mM DAB and 0.15% cytochrome *c* in 50 mM phosphate buffer (pH 7.2) for 5 min at RT.
2. Rinse thrice (10 min each) in 50 mM phosphate buffer (pH 7.2) at 4°C.
3. Post-fix specimens in a 2% aqueous solution of osmium tetroxide for 90 min at 4°C.

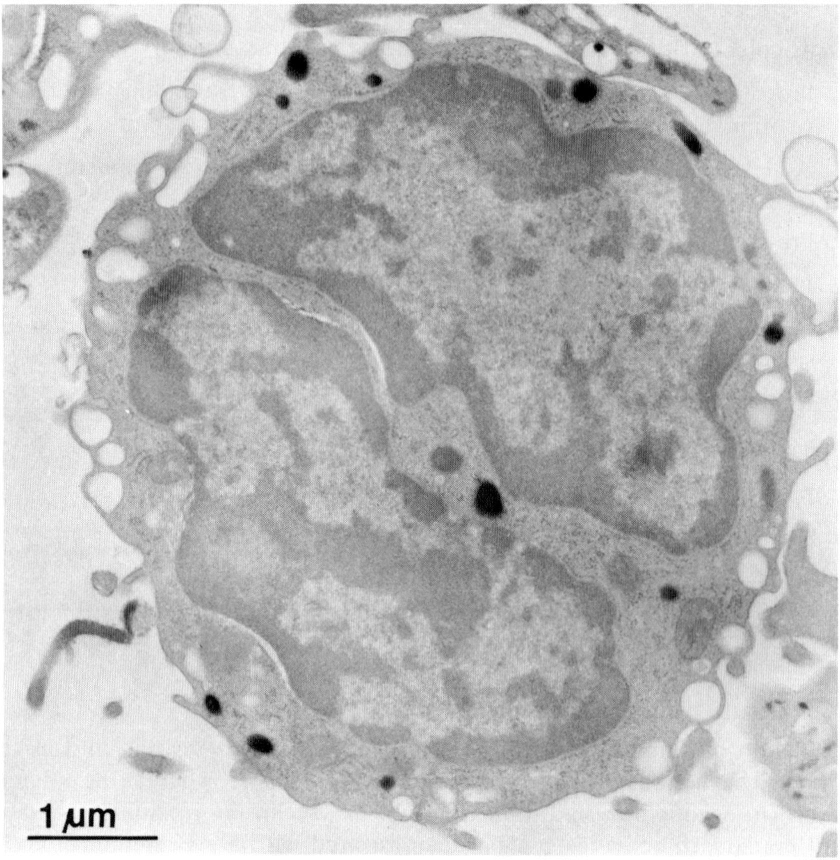

Figure 4. Peroxidase activity in a mouse monocyte as demonstrated with the diaminobenzidine method. (Courtesy of Dr H. K. Koerten, Laboratory of Electron Microscopy, University of Leiden, The Netherlands.)

Protocol 13. *Continued*

4. Control incubations:

(a) In the absence of cytochrome *c*.

(b) Pre-incubate for 10 min in a solution of either 5 mM potassium cyanide or 5 mM sodium azide and add cyanide or azide to the incubation medium as well.

Protocol 14. DAB method for catalase (25)

1. Incubate in medium containing 0.15% hydrogen peroxide[a] and 5 mM DAB in 100 mM glycine–NaOH buffer (pH 10.5) for 5 min at RT.

2. Rinse thrice (10 min each) in 100 mM glycine–sodium hydroxide buffer (pH 10.5) at 4°C.
3. Post-fix specimens in a 2% aqueous solution of osmium tetroxide for 90 min at 4°C.
4. Control incubations:
   (a) Omit hydrogen peroxide from the medium.
   (b) Pre-incubate for 10 min in the presence of 100 mM azide or 50 mM 3-amino-1,2,4-triazole and add either azide or triazole to the incubation medium as well.

[a] The high concentration of hydrogen peroxide (0.15%) inhibits peroxidase activity.

Protocol 15. DAB method for peroxidase

1. Incubate for 5 min at RT in a medium consisting of 0.02% hydrogen peroxide and 2.5 mM DAB dissolved in 100 mM cacodylate buffer (pH 6.5).
2. Rinse thrice (10 min each) in 100 mM cacodylate buffer (pH 6.5) at 4°C.
3. Post-fix in an aqueous solution of 2% osmium tetroxide for 90 min at 4°C.
4. Control incubations:
   (a) In the absence of hydrogen peroxide.
   (b) Pre-incubate for 20 min at RT in a solution of 0.15% hydrogen peroxide to inactivate all peroxidase activity.

3.3 Tetrazolium salt methods

Tetrazolium salts are used for the demonstration of the activity of dehydrogenases, reductases and oxidases. Tetrazolium salts are colourless or yellow and reasonably soluble in water; they can be reduced easily by the uptake of electrons to their formazans which are water-insoluble and strongly coloured. The reaction chain consists of electron transfers, with the first step that of the enzyme transferring electrons from one substrate to the other (for example lactate dehydrogenase transfers electrons from lactate to NAD thus generating pyruvate and NADH). In all tetrazolium salt reactions, only physiological substrates are used and in principle every dehydrogenase and oxidase can be demonstrated with tetrazolium salts. As cerium methods are superior for the ultrastructural localization of oxidase activity (23), tetrazolium salt methods for oxidases are not presented here.

A large number of methods have been described for the light microscopical detection of enzyme activity using tetrazolium salts, but for electron microscopy the number is limited. This is due to the fact that most formazans are not very electron-dense or osmiophilic. Exceptions are some derivatives of nitro blue tetrazolium (nitro BT) such as distyryl-nitro BT and the recently developed BPST [2-(2-benzothiazolyl)-3-(4-phthalhydrazidyl)-5-styryl-tetrazolium] which are osmiophilic (28). However, both the tetrazolium and formazan forms of the nitro BT-derivatives are osmiophilic. Therefore, care has to be taken because binding of the tetrazolium salt to tissue components will lead to false-positive results. In contrast, BPST itself is not electron-dense or osmiophilic, but its formazan does bind osmium. On the other hand, post-fixation with osmium tetroxide may also lead to false-negative results because it can reoxidize formazan to the tetrazolium salt which then may diffuse out of the specimen. Tetranitro BT is the most frequently employed tetrazolium salt for electron cytochemistry but its formazan provides a rather low contrast (26). Deimann *et al.* (10) used a scanning transmission electron microscope to detect tetranitro BT-formazan generated by lactate dehydrogenase activity in ultrathin sections without osmification. This approach ensures the visualization of the reaction product with sufficient contrast.

BPST does not permeate into tissues very easily, unlike most other tetrazolium salts and this may lead to false negative results, especially when the enzyme to be studied is membrane-bound such as succinate dehydrogenase in mitochrondria. As with cerium (Section 3.1.2) a number of procuredures may improve penetration. Addition of 0.5% Triton X-100 to the incubation media can improve permeability considerably. Another reason for the limited application of tetrazolium salts for electron microscopy is the high sensitivity of dehydrogenases and reductases to fixation.

For light microscopical detection of dehydrogenase activity, unfixed cryostat sections are used to avoid inactivation by fixation. Diffusion of enzyme molecules or intermediate reaction products is prevented either by addition of poly(vinyl alcohol) to the medium, or by interposing a semipermeable membrane in between the unfixed section and the gelled incubation medium (27). These measures enable precise localization. Therefore, we recommend application of these improved media for ultrastructural purposes as well (28). A limited number of tetrazolium salt methods for ultrastructural purposes have been described (see *Protocol 16*) but other dehydrogenases can be demonstrated as well when the appropriate substrates and co-enzymes are added to the incubation medium and provided that these enzymes can withstand fixation (see Section 2.1).

Protocol 16. Tetrazolium salt methods for dehydrogenases (26, 28)

1. For all dehydrogenases, prepare 100 mM phosphate buffer (pH 7.45) containing 18% poly(vinyl alcohol)[a] (hot water soluble; av. mol. wt 80 000; Sigma Chem. Co., St Louis, Missouri, USA). Add 18 g poly(vinyl alcohol) to 100 ml buffer at 80–90 °C under vigorous stirring until the solution is clear. The solution can be easily kept for 2 weeks at 37 °C in an air-tight vial.

2. Dissolve all other compounds with the exception of BPST in distilled water as 100 times concentrated stock solutions.
3. First, add BPST to the medium. Dissolve 4 mg BPST in 20 μl dimethylformamide by gentle heating and add immediately to the medium. Mix BPST thoroughly with the poly(vinyl alcohol)-containing medium with a spatula.
4. Add 10 μl of each of the stock solutions to 1 ml poly(vinyl alcohol)-containing buffer of 37°C.
5. Mix the compounds thoroughly with the viscous poly(vinyl alcohol) by using a spatula. Mixing is much easier at 37°C than at RT because of the lower viscosity of the medium. Final concentrations of compounds in the media for the demonstration of the following enzymes are (the stock solutions contain 100 times as much):

   For succinate dehydrogenase: 50 mM sodium succinate, 0.2 mM phenazine methosulphate, 5 mM sodium azide and 5 mM EDTA.

   For lactate dehydrogenase: 150 mM sodium L-lactate, 3 mM NAD, 0.32 mM 1-methoxyphenazine methosulphate and 5 mM sodium azide.

   For glucose-6-phosphate dehydrogenase: 10 mM glucose-6-phosphate, 0.8 mM NADP, 0.32 mM 1-methoxyphenazine methosulphate and 5 mM sodium azide.

   For NAD(P)H dehydrogenase: 0.5 mM NADH or NADPH and 1 mM menadione.
6. Incubate for 15–60 min at 37°C.
7. Remove the viscous incubation medium with warm 100 mM phosphate buffer (pH 5.3; approx. 50°C).
8. Post-fix in an aqueous solution of 2% osmium tetroxide for 90 min at 4°C.[b]
9. Control incubations:

   For succinate dehydrogenase:

   (a) Omit succinate from the medium.

   (b) Add 10 mM sodium malonate to the medium.

   For lactate dehydrogenase:

   (a) Omit lactate and NAD from the medium.

   (b) Add 20 mM pyruvate to the medium.

   For glucose-6-phosphate dehydrogenase:

   (a) Omit glucose-6-phosphate and NADP from the medium.

   (b) Add 10 mM glucosamine-6-phosphate to the medium.

Protocol 16. *Continued*

For NAD(P)H dehydrogenase:

(a) Omit NADH or NADPH from the medium.

(b) Add 1 mM dicumarol to the medium.

[a] This type of poly(vinyl alcohol) is essential for proper localization of enzyme activity (27).

[b] Perform post-fixation and dehydration very carefully, because: (a) Osmium tetroxide post-fixation may reoxidize formazan into tetrazolium salt leading to false-negative results; (b) Avoid use of propylene oxide during dehydration because it can dissolve formazan leading to false-negative results

---

3.4 Ferricyanide methods

Ferricyanide methods are used for the demonstration of oxidase, dehydrogenase, (cholin)esterase, and arylsulphatase activity. Ferricyanide is an artificial electron acceptor and it can be reduced to ferrocyanide which in the presence of copper ions forms an insoluble, amorphous, electron-dense precipitate of cupric ferrocyanide, also called Hatchett's Brown. This primary reaction product can be amplified by diaminobenzidine and osmium tetroxide treatment but this step is optional.

Methods for the ultrastructural demonstration of oxidases (see *Protocol 17*), dehydrogenases (see *Protocol 18*), and acetylcholinesterase (see *Protocol 19*) are given here. Although the cerium method is preferred at the moment for the demonstration of oxidase activity, the ferricyanide method is still very useful. For electron microscopical demonstration of dehydrogenase activity, the ferricyanide method is preferred to tetrazolium salt methods because the final reaction product can be recognized more easily and non-specific precipitation of final reaction product occurs less frequently. For cholinesterases, it is the method of choice.

3.4.1 Oxidase

The method is given with the example of monoamine oxidase (29), but this method can be adapted for other oxidases by replacing tryptamine·HCl as substrate by the particular substrates given in *Protocol 11*.

Protocol 17. Ferricyanide method for monoamine oxidase (29)

1. Prepare incubation medium containing 40 mM acetate buffer (pH 6.6–6.8), 2.5 mM tryptamine·HCl, 6 mM sodium citrate, 2 mM copper sulphate, 15% (v/v) dimethylsulphoxide (DMSO), and 0.3 mM potassium ferricyanide. Prepare the medium by stirring the components strictly in the following order: tryptamine, acetate, citrate, copper sulphate, DMSO; then adjust pH, and add finally potassium ferricyanide.
2. Incubate sections for 60 min at 37°C.

3. Rinse briefly in 0.22 M sucrose in 100 mM phosphate buffer (pH 7.2).
4. Post-fix with 4% formaldehyde in 100 mM phosphate buffer (pH 7.2) for at least 60 min.
5. Amplify the final reaction product by treating the tissue with 1.5 mM diaminobenzidine in 50 mM acetate buffer (pH 5.6) for 20 min at RT.
6. Rinse tissue thrice (10 min each) in distilled water.
7. Osmicate with an aqueous solution of 2% osmium tetroxide for 90 min at 4°C.

3.4.2 Dehydrogenases

The activity of a series of dehydrogenases has been demonstrated at the ultrastructural level with the ferricyanide method (2). As with the tetrazolium salt methods, the fixation-sensitivity of most dehydrogenases hampered developments at the ultrastructural level. Therefore, fixation should be carried out carefully (see Section 2.1). Diffusion of enzyme molecules and intermediate reaction products from the tissue specimens is avoided by addition of poly(vinyl alcohol) to the medium (see *Protocol 16*). The ferricyanide method for dehydrogenases is described with the example of glucose-6-phosphate dehydrogenase but activity of other dehydrogenases can be demontrated as well when the proper substrates (and co-enzymes) are added to the incubation medium, provided that the proper controls are performed.

Protocol 18. Ferricyanide method for glucose-6-phosphate dehydrogenase (30)

1. Prepare a medium containing 100 mM phosphate buffer (pH 7.2), 18% poly(vinyl alcohol) (see *Protocol 16*), 10 mM glucose-6-phosphate, 0.8 mM NADP, 0.32 mM 1-methoxyphenazine methosulphate, 5 mM sodium azide, 4 mM $MgCl_2$, 10 mM sodium citrate, 1.5 mM copper sulphate, and 0.5 mM potassium ferricyanide. Prepare the (viscous) medium by thoroughly mixing the components with a spatula strictly in the following order: phosphate, poly(vinyl alcohol), glucose-6-phosphate, NADP, $MgCl_2$, azide, citrate, copper sulphate; adjust pH, and finally add potassium ferricyanide and 1-methoxyphenazine methosulphate.
2. Incubate for 60 min at 37°C.
3. Rinse tissue specimens briefly in 0.22 M sucrose in 100 mM phosphate buffer, pH 5.3.
4. Post-fix with 4% formaldehyde in 100 mM phosphate buffer, pH 5.3.
5. Amplify the final reaction product by treating the tissue with 1.5 mM diaminobenzidine in 50 mM acetate buffer (pH 5.6) for 20 min at RT.
6. Rinse tissue thrice (10 min each) in distilled water.

Protocol 18. *Continued*

7. Osmicate with an aqueous solution of 2% osmium tetroxide for 90 min at 4°C.
8. Control incubations:
   - in the absence of either glucose-6-phosphate or NADP
   - include 10 mM glucosamine-6-phosphate in the medium

3.4.3 Cholinesterases

Cholinesterases hydrolyse choline esters such as acetylcholine. A series of inhibitors enable differentiation between the different cholinesterases but also between cholinesterases and nonspecific esterases.

Protocol 19. Ferricyanide method for acetylcholinesterase (31)

1. Prepare incubation medium by dissolving 12.5 mg acetylthiocholine in 2.5 ml distilled water and add 15.8 ml of a 0.82% sodium acetate solution, 0.5 ml of 0.6% acetic acid, 1.5 ml of 2.94% sodium citrate solution, 2.5 ml of 0.75% copper sulphate solution, and 2.5 ml of a 0.165% potassium ferricyanide solution, in strict order during vigorous stirring. The solution is light green and the pH 5.5.
2. Incubate for 30 min at 37°C.
3. Rinse briefly in 0.22 M sucrose in 100 mM phosphate buffer, pH 7.2.
4. Post-fix with 4% formaldehyde in 100 mM phosphate buffer (pH 7.2) for 1 h.
5. Amplify the final reaction product by treating the tissue with 1.5 mM diaminobenzidine in 50 mM acetate buffer (pH 5.6) for 20 min at RT.
6. Rinse tissue specimens thrice (10 min each) in distilled water.
7. Osmicate in an aqueous solution of 2% osmium tetroxide for 90 min at 4°C.
8. Control incubations:
   - in the absence of acetylthiocholine
   - In the presence of 0.01 mM eserine or 0.01 mM diisopropyl fluorophosphate (dissolved in dimethylformamide)

3.5 Diazonium salt methods

Diazonium salt methods are used for the demonstration of phosphatases, esterases, and proteases. Diazonium salts are preferred to metal salt methods for the demonstration of phosphatases at the light microscopical level but this is not the case for electron microscopical investigations. The clearly visible electron-dense precipitates obtained with metal salt methods have made other methods for electron microscopy obsolete. Therefore, diazonium salt methods are used only for the ultrastructural localization of esterase and protease activity. For electron microscopy, hexazotized pararosanilin, or hexazotized New Fuchsin are the best options as coupling agents (see *Protocol 20*). Other diazonium salts yield azo dyes that may suffer from displacement and subsequently give false ultrastructural localization.

Protocol 20. Preparation of hexazotized pararosanilin or New Fuchsin (34)

1. Dissolve 800 mg pararosanilin or New Fuchsin in 2 ml HCl (36% w/v) and 8 ml distilled water. Mix thoroughly until dissolved (2–3 h) and filter off. This solution can be kept in the dark at 4°C for 2–4 weeks.
2. As shortly before enzyme incubation as possible, mix equal volumes of the pararosanilin or New Fuchsin solution with a solution of 8% sodium nitrite in distilled water to hexazotize pararosanilin or New Fuchsin.
3. Add 60 μl of this solution to 1 ml incubation medium.

3.5.1 Esterases

A large number of techniques are available for the demonstration of esterase activity at the ultrastructural level (32). However, no technique can be regarded as ideal, nor do the different procedures give entirely comparable results. The most frequently used methods for electron microscopy are the azo-indoxyl method (see *Protocol 21*) and the azo-coupling method. Esterases cleave the acetate group of indoxyl acetate and this unsubstituted indoxyl forms a dye with hexazotized pararosanilin or New Fuchsin which is strongly osmiophilic.

Protocol 21. Azo-indoxyl method for esterases (33)

1. Dissolve 3 mg indoxyl acetate in 300 μl dimethylformamide and add to 10 ml citrate buffer (30 mM; pH 6.0).
2. Add 600 μl hexazotized pararosanilin or New Fuchsin (see *Protocol 20*).
3. Incubate for 15–60 min at 37°C.
4. Rinse thrice (10 min each) in distilled water.
5. Post-fix specimens in a 2% aqueous solution of osmium tetroxide for 90 min at 4°C.
6. Control incubations:
   (a) In the absence of indoxyl acetate.

Protocol 21. *Continued*

(b) In the presence of either 40 mM sodium fluoride, 0.5 mM diisopropylfluorophosphate, 10 mM diethyl-*para*-nitrophenyl phosphate (E600) or 0.1 mM *para*-chloromercuribenzoate. The different types of non-specific esterases and cholinesterases can be discriminated by incubations in the presence of either one of these inhibitors (34).

The azo-coupling method (35) is very similar to the azo-indoxyl method (see *Protocol 21*), but here α-naphthyl acetate is used as a substrate. The liberated α-naphthol forms an azo dye with hexazotized pararosanilin or New Fuchsin which is osmiophilic. α-Naphthol acetate is hydrolysed more rapidly by esterases than indoxyl acetate, but the indoxyl-based dye is more osmiophilic than the naphthol azo dye. Indoxyl acetate is very useful for the localization of lysosomal esterases.

3.5.2 Proteases

Protease activity is demonstrated with synthetic substrates containing one amino acid or a sequence of two or more amino acids determining the specificity of the reaction. The amino acid sequence is attached to a leaving group, 4-methoxy-2-naphthylamine (MNA), and is cleaved off by the specific protease that recognizes the amino acid sequence. MNA can then bind with a diazonium salt. A wide range of specific proteases can be demonstrated depending on the sequence of amino acids and the incubation conditions as reviewed by Sannes (36).

3.5.3 General considerations

Proteases are often present in tissues and cells in an inactive form. Many proteases are synthesized in a proform and activation may be regulated by another protease by cleaving off a number of amino acids. Activation might also occur during tissue processing. Furthermore, proteases may be inactive because they are bound to their physiological inhibitors. Many protease inhibitors are present in cells and tissues. Binding between protease and inhibitor also can be altered by tissue processing.

Proteases can be divided into two large subclasses: endopeptidases (proteinases) and exopeptidases (peptidases). Endopeptidases recognize a sequence of amino acids within a protein and cleave the protein at that site under suitable conditions. This is simulated with the synthetic substrate by attaching a *N*-amino-terminal blocking moiety such as benzyloxycarbonyl (CBZ) to the amino acid sequence. Exopeptidases cleave terminal amino acid sequences. So, CBZ-Arg-Arg-MNA is recognized by the endopeptidase cathepsin B, whereas Arg-Arg-MNA is cleaved by the exopeptidase dipeptidylpeptidase III.

Four major classes of proteinases are recognized depending on the requirements for their catalytic activity. Serine proteinases are inhibited by diisopropyl fluorophosphate, cysteine proteinases by sulphydryl-blocking agents, aspartic proteinases by pepstatin and metallo proteinases by EDTA.

To discriminate between various protease activities, incubations with selective and specific inhibitors are required. Moreover, incubation conditions are also essential for the specificity of the reaction.

As most proteases (and their physiological inhibitors) are very sensitive to fixation, tissue pre-treatment has to be carried out carefully. Therefore, the same precautions count for proteases as for dehydrogenases. Moreover, some proteases require extra attention. For example, cysteine proteinases such as the lysosomal cathepsins are very sensitive to pH values >7.0. Therefore, pre-treatment of tissue should be performed at acid pH. The method for a specific protease, dipeptidyl peptidase II, is given in *Protocol 22*.

Protocol 22. Diazonium salt method for dipeptidyl peptidase II (37)

1. Prepare 100 mM cacodylate buffer (pH 5.5) and add 60 μl of hexazotized pararosanilin or New Fuchsin (see *Protocol 20*) per millilitre of buffer.
2. Dissolve 1 mg Lys-Pro-MNA (Enzyme Systems Products, USA; Bachem, Switzerland) in 10 μl dimethylformamide and add this to the incubation medium.
3. Incubate for 60 min at 37°C.
4. Addition of 18% poly(vinyl alcohol) to the medium enhances the precision of localization (see *Protocol 16*).
5. Wash sections thrice (10 min each) in distilled water.
6. Osmicate in aqueous 2% osmium tetroxide adjusted to pH 7.8–8.0 with diluted sodium hydroxide for 90 min at 4°C.
7. Wash sections thrice (10 min each) in distilled water, dehydrate in ethanol series and embed in low viscosity resin. Avoid use of propylene oxide because it dissolves the reaction product leading to false-negative results.
8. Control incubations:
   (a) In the absence of substrate.
   (b) A specific inhibitor of dipeptidyl peptidase II (DPP II) is not known, but to establish specificity of the reaction, incubate with Gly-Pro-MNA at pH 7.4. This should demonstrate dipeptidyl peptidase IV which is a closely related protease. DPP II is lysosomal, whereas DPP IV is localized on the plasma membrane. DPP II preferentially cleaves off Lys-Pro at acid pH, and DPP IV Gly-Pro at slightly alkaline pH.

---

4. Post-treatment

Certain methods such as tetrazolium, diazonium, and diaminobenzidine methods need osmication to visualize the final reaction product and this post-fixation step is then mentioned in the protocols. However, osmium tetroxide is also employed as a post-fixative. Post-fixation is carried out after the cytochemical incubations to provide contrast to tissue. Nevertheless, the membrane contrast tends to be low after fixation with osmium tetroxide which is probably due to a decrease in the number of lipid double-bonds produced during the enzyme reactions. A higher contrast can be obtained by post-fixation with osmium tetroxide in combination with potassium ferrocyanide (see *Protocol 23*).

Protocol 23. Post-fixation with osmium tetroxide and potassium ferrocyanide (38)

1. Add 1 vol. of a 2% osmium tetroxide solution in distilled water to 1 vol. of a 3% potassium ferrocyanide solution in 200 mM cacodylate buffer (pH 7.4) at 4°C. The latter solution should not be used before 10 days after preparation, otherwise the membrane contrast may vary.
2. Post-fix tissue samples for 2 h at 4°C.
3. Rinse in 6.8% sucrose in 100 mM cacodylate buffer (pH 7.4) until the solution remains clear.
4. Dehydrate in alcohol series and embed in Epon 812.

References

1. Stoward, P. J. and Pearse, A. G. E. (ed.) (1991). *Histochemistry. Theoretical and Applied*, Vol. 3, 4th edn. Churchill Livingstone, Edinburgh.
2. Borgers, M. and Verheyen, A. (1985). *International Reviews in Cytology*, **95,** 163.
3. Van Duijn, P. (1974). In *Electron microscopy and cytochemistry* (ed. E. Wisse, W. T. Daems, I. Molenaar, and P. Van Duijn), pp. 3–23. North-Holland, Amsterdam.
4. Stoward, P. J. (1980). In *Trends in enzyme histochemistry and cytochemistry* (ed. D. Evered and M. O'Connor), pp. 11–27. Excerpta Medica, Amsterdam.
5. Fahimi, H. D. (1980). In *Trends in enzyme histochemistry and cytochemistry* (ed. D. Evered and M. O'Connor), pp. 33–47. Excerpta Medica, Amsterdam.
6. Dijk, F., Oosterbaan, J. A., and Hulstaert, C. E. (1985). *Histochemistry,* **83,** 573.
7. Beertsen, W., Everts, V., Niehof, A., and Bruins, H. (1982). *Journal of Periodontal Research*, **17,** 640.
8. Hulstaert, C. E., Kalicharan, D., and Hardonk, M. J. (1989). In *Techniques in diagnostic pathology*, Vol. 1 (ed. G. R. Bullock), pp. 133–50. Academic Press, London.

9. Robinson, J. M. and Karnovsky, M. J. (1983). *Journal of Histochemistry and Cytochemistry*, **31,** 1197.
10. Deimann, W., Freeman, R., and Fahimi, H. D. (1981). *Journal of Histochemistry and Cytochemistry,* **29,** 678.
11. Al-Ali, S. A. and Robinson, N. (1979). *Histochemistry,* **61,** 307.
12. Bara, G. and Anderson, W. A. (1973). *Histochemical Journal,* **5,** 437.
13. Gomori, G. (1941). *Archives of Pathology,* **32,** 189.
14. De Jong, A. S. H. (1982). *Histochemical Journal,* **14,** 1.
15. De Valck, V., Geerts, A., Schellinck, P., and Wisse, E. (1988). *Histochemistry,* **89,** 357.
16. Papadimitriou, J. M. and Van Duijn, P. (1970). *Journal of Cell Biology,* **47,** 84.
17. Mizutani, A. (1968). *Journal of Histochemistry and Cytochemistry,* **16,** 172.
18. Briggs, R. T., Draht, D. B., Karnovsky, M. L., and Karnovsky, M. J. (1975). *Journal of Cell Biology,* **67,** 566.
19. Veenhuis, M., Van Dijken, J. P., and Harder, W. (1980). *FEMS Microbiology Letters,* **9,** 285.
20. Hulstaert, C. E., Kalicharan, D., Hardonk, M. J., and Halbhuber, K.-J. (1990). *Transactions of the Royal Microscopical Society,* **1,** 573.
21. Halbhuber, K.-J., Gossrau, R., Moeller, U., Hulstaert, C. E., Zimmermann, N., and Feuerstein, H. (1988). *Histochemistry,* **90,** 289.
22. Kalicharan, D., Hulstaert, C. E., and Hardonk, M. J. (1985). *Histochemistry,* **82,** 287.
23. Angermueller, S. (1989). *Progress in Histochemistry and Cytochemistry* **20**(1), 1.
24. Hopsu-Havu, V. K., Arstila, A. U., Helminen, H. J., and Kalima, H. O. (1967). *Histochemie,* **8,** 54.
25. Angermueller, S. and Fahimi, H. D. (1981). *Histochemistry,* **71,** 33.
26. Fahimi, H. D. and Karnovsky, M. J. (1966). *Journal of Cell Biology,* **29,** 113.
27. Van Noorden, C. J. F. and Vogels, I. M. C. (1989). *Histochemical Journal,* **21,** 373.
28. Van Noorden, C. J. F. (1984). *Progress in Histochemistry and Cytochemistry,* **15**(4), 1.
29. Hanker, J. S., Kusyk, C. J., Bloom, F. E., and Pearse, A. G. E. (1973). *Histochemie,* **33,** 205.
30. Berchtold, J.-P. (1979). *Histochemistry,* **63,** 173.
31. Karnovsky, M. J. and Roots, L. (1964). *Journal of Histochemistry and Cytochemistry,* **12,** 219.
32. Von Deimling, O. and Boecking, A. (1976). *Histochemical Journal,* **8,** 215.
33. Holt, S. J. and Hicks, R. M. (1966). *Journal of Cell Biology,* **29,** 361.
34. Lojda, Z., Gossrau, R., and Schiebler, T. H. (1979). *Enzyme histochemical methods*. Springer-Verlag, New York.
35. Lehrer, G. M. and Ornstein, L. (1959). *Journal of Biophysical and Biochemical Cytology,* **6,** 399.
36. Sannes, P. (1988). *Progress in Histochemistry and Cytochemistry,* **18**(1), 1.
37. Smith, R. E. and Van Frank, R. M. (1975). In *Lysosomes in biology and pathology* (ed. J. T. Dingle and R. J. Dean), pp. 193–249. North-Holland, Amsterdam.
38. Hulstaert, C. E., Kalicharan, D., and Hardonk, M. J. (1983). *Histochemistry,* **78,** 71.

7

High resolution shadowing

HENRY S. SLAYTER

1. Introduction

1.1 Specimen methodology for electron microscopy of individual macromolecules

Modern electron microscopes are optically capable of approaching atomic resolution, but the lack of adequate preparative methods for individual biopolymer specimens has limited useful resolution. Principal difficulties are inadequate specimen contrast, beam damage, and dehydration effects. While methods for extracting excellent structural information from ordered arrays of macromolecules have gradually evolved using both crystallographic and electron microscopic approaches (1, 2, 3), individual macromolecules are still largely analysed using electron microscopy together with metal coating or negative staining. Coating of macromolecules with thin metal films produces very stable specimens with excellent contrast (4, 5, 6) and the technique has contributed extensively to knowledge of biological structures. There is every reason to believe that further technical refinements will aid in macromolecular studies (7).

Parameters useful in describing a thin metal film include grain size, separation, and population density, as well as mass thickness, ρt (where ρ = density of the coating metal and thickness, t). Several factors influence size and distribution of metal particle in a film of calculated thickness 0.2–2 nm: (a) physical characteristics of the metal to be evaporated, such as melting point, crystallographic lattice parameters, and oxidizing ability; (b) temperature of the specimen during evaporation; (c) purity of the evaporated material; (d) similarity of lattice parameters of the substrate with those of the evaporated material; (e) rate of evaporation and amount of metal deposited.

In evaporated films, the number of grains per unit area on a replica first increases versus thickness, levels off, and then finally decreases. The number of grains/nm^2 measured is usually low due to inability to visualize metal particles below the brightfield contrast/resolution limit of about 0.4 nm using conventional transmission electron microscopy (CTEM) brightfield imaging. As films grow thicker, crystallite nuclei ultimately fuse, also resulting in a reduction in grains/nm^2.

It has been reported that platinum nucleation density, as visualized by brightfield CTEM, increases as a function of specimen cooling to 0.1 particles/nm^2 (4). Twice as many crystallites are detected per unit area on a cold (−196°C) substrate as a warm one. Concomitantly, the grains formed are smaller and more homogeneous in size. If appropriately small amounts of metal are used, tungsten produces finer grain than platinum, even when deposited on substrates at room temperature. Tungsten population densities as great as 0.3 particles/nm^2 are observed, but it is probable that additional particles go undetected because of brightfield CTEM contrast limitations. Thus, contrast limitations deny definition of exact crystallite population density, which needs to be determined to optimize contrasting and replication methodologies. STEM (scanning transmission electron microscopy) in the zero energy loss mode offers more contrast than brightfield CTEM, but as yet, very few such instruments are operational in the range of 0.5 nm point-to-point resolution or better.

Fine-grain metal coating has recently been developed to the point that the main limitation is contrast. With the electron optics (such as the JEOL 100-CX; 0.35 nm point-to-point resolution) available to us, we have been able to define grain approaching the resolution limit using both brightfield and high-resolution annular darkfield CTEM imaging. Thus, darkfield CTEM currently provides the best optical mode for visualizing fine grain films, because in brightfield, substrate noise limits detection of the smallest nucleation sites. The shortcomings of darkfield electron microscopy, in terms of beam damage and preparation difficulties, have been discussed elsewhere (7, 8), but applications have generally required use of analytical methods for image interpretation in CTEM or STEM (9, 10). In recent years, we have combined CTEM darkfield imaging with very thin metal film replication, which overcomes several of these difficulties. Ease of macromolecular specimen preparation, specimen stability, ratio of elastic to inelastic scattering, and image contrast are all improved, beam damage is minimized, and resolution is excellent.

Specimen rotation in a plane during metal evaporation generally promotes better information retrieval than is possible with a stationary specimen. By coating all sides of a structural feature, blind spots occurring in shadows are minimized and the addition of a precessing motion to the rotation may produce even more effective coating of large irregular stretches of topography.

Metal-coated molecules have been shown to be relatively insensitive to beam damage (4). Although contrast becomes somewhat diminished as a function of time, due to contamination, the exact features of the metal film are preserved. While it is possible that critical damage may occur during the few seconds prior to first viewing and exposure, no further substantial change occurs, even after 10 min of exposure at a dose rate = 3×10^3 e/nm^2-sec.

The use of darkfield CTEM in conjunction with rotary shadowing using tungsten replica films is the optimal method of choice.

1.2 Optimizing mass thickness of shadowing metal relative to that of substrate material

Optimum thickness is that at which contrast is provided by a maximum number of nucleating crystallites of minimum size per unit area. Thus, control of film thickness is of the utmost importance as a practical matter. One difficulty is that the optimum coating thickness for shadow-casting of an inhomogeneously constructed object may have to be chosen with respect to one domain or another, even with rotary coating.

Film thickness may be usefully estimated from the weight evaporated and the geometry, according to the equation:

$$t_{\text{calc}} = k_{\alpha}\frac{W \sin \theta}{4 \pi R^2 \rho}$$

where W is the weight evaporated in grams, θ is the angle between the plane of the specimen and the vapour beam, R is the distance between the specimen and the source, ρ is the bulk density of the evaporant, and k_{α} is a factor often assumed to be one (4). However, k_{α} is in fact dependent upon the uneven angular distribution of metal vapour in the beam, the sticking fraction on the specimen, which in turn is dependent on specimen height, composition, and the effect of collisions. Thus, k_{α} may approach 3 (*Table 1*), as in the case of a focused electron-gun heater evaporating tungsten, but is close to 1.5 for a straight filament source for platinum evaporation (4).

Despite ultimate limitations on resolution by metal grain size and frequency, combination of use of metal coating of mass thickness in the range 10^{-6} to 10^{-7} g/cm^2, with high resolution brightfield, or 10^{-7} to 10^{-8} g/cm^2 with annular darkfield illumination, permits idealization of visualization of detail in images of macromolecules. These methods should be invaluable in future studies of macromolecular structure.

1.3 Designing continuous metal coatings for biopolymer replication

We utilize apparatus combining several essential features: specimen rotation, low temperature operation, and the ability to select a critical shadowing angle for a given application (*Figure 1*). Thus, thin mica specimen substrates cemented to copper planchets can effectively be cooled in a well-trapped vacuum environment (at 10^{-8} Torr). A cold shroud at −196°C is placed around the specimen area to avoid contamination on replicas made at low temperatures under these conditions. Measurements of film thickness are made during deposition of metal with a water-cooled piezoelectric monitor, carefully calibrated to allow controlled deposition of metal at thickness in the range of 10^{-8} to 10^{-6} g/cm^2 (see section 1.4). Specimens are usually backed with 1–2 nm of carbon, as well as 5–10 nm of nitrocellulose, both for

Table 1.

A. Data relating to platinum evaporation[a,b] from a straight tungsten filament at 6 cm from specimens, for use with brightfield imaging

sin θ	**Angle of incidence**	**Mass evaporated (mg)**	**Calculated[a] mass thickness (ρt g/cm^2) (× 10^{-6}**	**Monitor mass thickness (ρt g/cm^2) (× 10^{-6})**
1.00	90°	0.97	2.2	3.3
0.45	26°45′	1.20	1.2	1.8
0.20	11°35′	3.10	1.4	2.1
0.10	5°45′	4.20	0.9	1.4

B. Data relating to tungsten evaporation from a focused electron-gun source, for use with brightfield imaging[c,d]

sin θ	**Angle of incidence**	**Experimentally useful mass evaporated (mg)**	**Calculated[a] mass thickness (ρt g/cm^2) (× 10^{-6}**	**Monitor mass thickness (ρt g/cm^2) (× 10^{-6})**
0.10	5°45′	50	0.24	0.76
0.10	5°45′	100	0.48	1.50

C. Data relating to tungsten evaporation from a focused electron-gun, for use with darkfield imaging[c,d]

sin θ	**Angle of incidence**		**Measured mass thickness[e] (ρt g/cm^2) (× 10^{-6}**	**Monitor thickness [t,nm]**
0.10	5°45′	Heavy coating 25°C	0.54	0.260
0.10	5°45′	Light coating 25°C	0.09	0.040
0.10	5°45′	Very light coating 25°C	0.05	0.017

[a] Non-spherical deposition resulting in a calculated thickness 1.5 times less than thickness measured by a thickness monitor, using differential weight to obtain mass evaporated which is about 70% of mass loaded on a tungsten filament (ρ is density; t is thickness). We use monitor rather than measured because the piezoelectric monitor is not considered to be an absolute measurement at the specimen plane.

[b] Thickness is calculated for the specimen plane. Clearly, as much as a tenfold increase in thickness could apply to topographical features presenting at normal incidence to the beam. In rotary shadow casting these regions would collect relatively less metal because of distribution of metal around the circumference. If this reduction factor is taken as 4, then features presenting at normal incidence could collect up to 2.5 times the mass thickness of the background.

[c] At 39 cm distance from source to specimen and $\sin \theta = 0.10$ ρt given for plane of specimen. $\rho t = 0.76 \times 10^{-6}$ g/cm^2 produces a low contrast, fine-grain tungsten replica.

[d] Calculated and measured on the normal to the source; $k_\alpha = 2$–3 for tungsten.

[e] Calculated from monitored thickness versus mass evaporated.

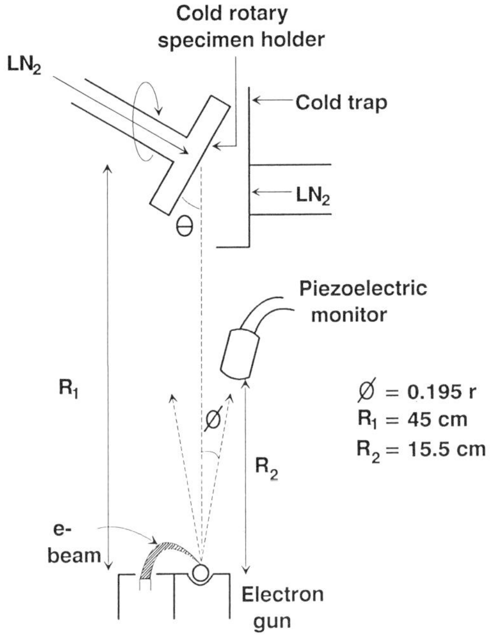

Figure 1. Diagram of apparatus for evaporating tungsten or other refractories at a tilted, cold, rotating specimen, while cold-trapping the space around the specimen. ϕ, the angular spread of the effective evaporant stream, is about 0.195 r. R_1 is about 45 cm to minimize radiant heating of the specimen and R_2 defines the position of the piezoelectric monitor. θ is the angle of incidence of the vapour stream upon the specimen plane.

mechanical stability and in order to minimize migration of atoms in the electron beam during microscopy. These carbon film thicknesses should be minimized in order to reduce background noise.

Figure 2 provides a convenient comparison of mass thickness versus calculated thickness for carbon and tungsten coatings (the plot for platinum is almost identical to that for tungsten). While thickness is a reasonable parameter by which to describe relatively heavy coatings, for thinner coatings (10^{-6} g/cm^2 or thinner), the discontinuous nature of the film suggests mass thickness as a better descriptor. *Figure 3* shows a plot of I_K/I_0, the ratio of forward scattering to illuminating intensity of the beam, versus film thickness for carbon, platinum, and tungsten. This indicates the relative forward scattering power of metal-coat and carbon substrate, and allows estimation of their respective contributions to the image as a function of their relative thicknesses.

It is found experimentally that 2.3-fold less metal thickness is deposited on exposed surfaces of a macromolecule in rotary-coating than is the case in uni-directional shadowing. Since at an angle of shadowing incidence of 5°, or sin 5° = 0.1, in principle, tenfold more metal should coat a molecular aspect

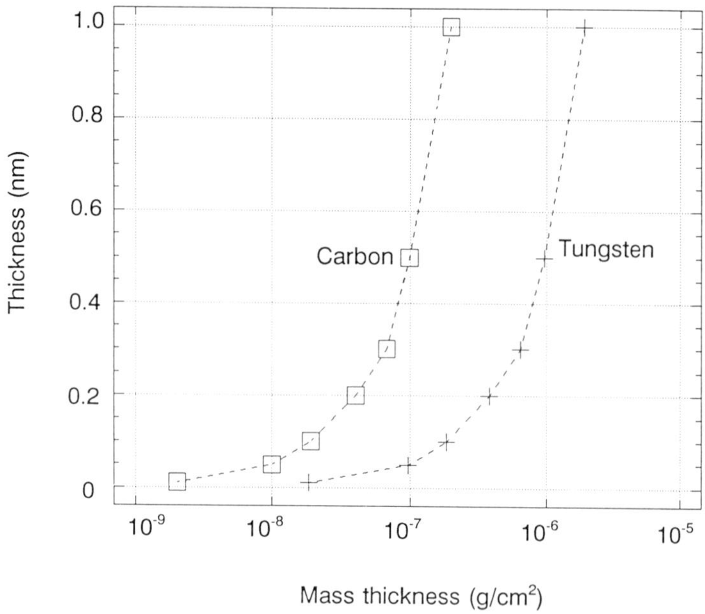

Figure 2. Plot of mass thickness v. film thickness for carbon and platinum. This is a useful comparison of two parameters commonly used in relating film thickness to scattering power.

exposed at normal incidence compared with the substrate specimen plane. There should be 10/2.3 = 4.4-fold more metal coating on a rotary-shadowed molecule than on the specimen plane itself, ignoring the discontinuous crystallite nature of the film. Thus, in *Figure 3*, the figure I_K/I_0 contributed by a 10^{-6} cm carbon film is about 0.08, and that contributed by 10^{-8} cm of tungsten is 0.005. But, since 10^{-8} cm on the substrate projects to 4.4×10^{-8} cm on the molecule (more if the molecule is larger because of the resulting depth of the metal-coated side), the actual I_K/I_0 for that metal contribution would be of the order of 0.02. For a 2.0 nm high side (4.4×10^{-8} cm wide $\times 2 \times 10^{-7}$ cm high), I_K/I_0 would be 0.10, and for a 5×10^{-7} cm high side (4.4×10^{-8} cm $\times 5 \times 10^{-7}$ cm high), 0.16. Comparison of this signal with the background noise signal from substrate carbon is useful, not only in designing coating and substrate thickness, but also in the interpretation of darkfield images at the limit, as is shown by plotting the ratio $(I_K)_W \div [(I_K)_W + (I_K)_C]$ versus t_W at a fixed $t_C = 10$ nm, as in *Figure 4*. The micrograph insert in *Figure 3* shows the S1 (head) region of myosin. The light background structure primarily reflects the 0.4 nm structure and relative contrast close to focus in brightfield, of the 10^{-6} cm thick carbon substate. The two dark myosin heads, about 7 nm high, reflect the contrast induced on them by rotary shadowing with 8×10^{-8} g/cm^2 tungsten.

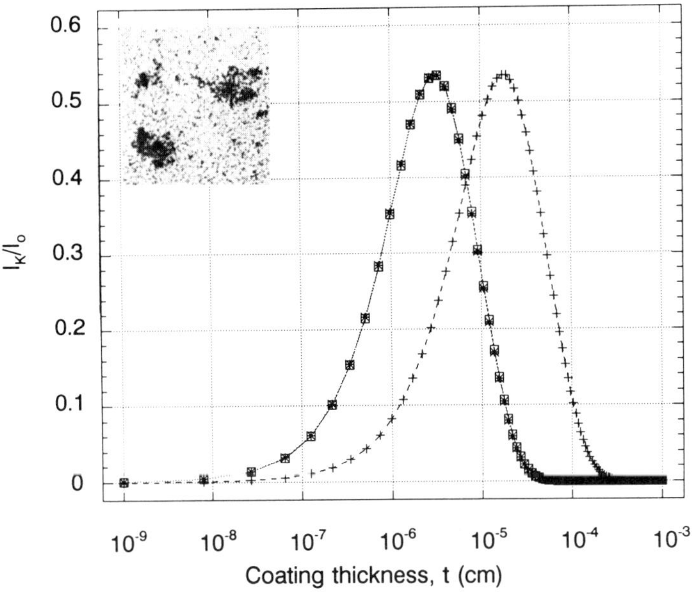

Figure 3. Plot of I_K/I_O. versus film thickness for carbon, platinum, and tungsten. The equation $I_K = I_0 e^{-S_T w}(e^{kS_T w} - 1)$ permits calculation of the extent of forward scattering I_K relative to incident intensity, I_0, using $w = \rho t$, and values of density (ρ), film thickness (t), and specific scattering cross-section (S_T) (15). The dotted curve represents both platinum (□) and tungsten (*), and the dashed curve, carbon (+). Relative contributions to the image of a molecular feature of metal and substrate may be estimated using these curves.

Inset: Control myosin molecule S1 region shadowcast with tungsten, using 8×10^{-8} g-cm^2, photographed in high-resolution brightfield CTEM (×1 220 000). The point-to-point resolution in the background is approximately 0.3–0.4 nm. Divisions within the globular region of the heads are resolved; i.e. a cleft where the lower lobe seems to be split into two parts.

CTEM in both bright and darkfield is the principal method of data collection for analysis of coated macromolecules. The great variety of macromolecular species to which these methods have been applied constitutes a large body of evidence supporting the effectiveness of these methods (11).

1.4 Measurement of film thickness

Measurement of film thickness after evaporation can be accomplished by a variety of methods (microwave resistance, interferometry, film weight, isotope counting, and electron densitometry) (12, 13). To follow the increase in mass thickness during deposition, however, a piezoelectric oscillator is required, especially when mass thicknesses of 10^{-6} g/cm^2 or less are used.

Piezoelectric crystal oscillators quite reliably measure mass changes in relation to integrated area under the waveform of oscillator frequency. For

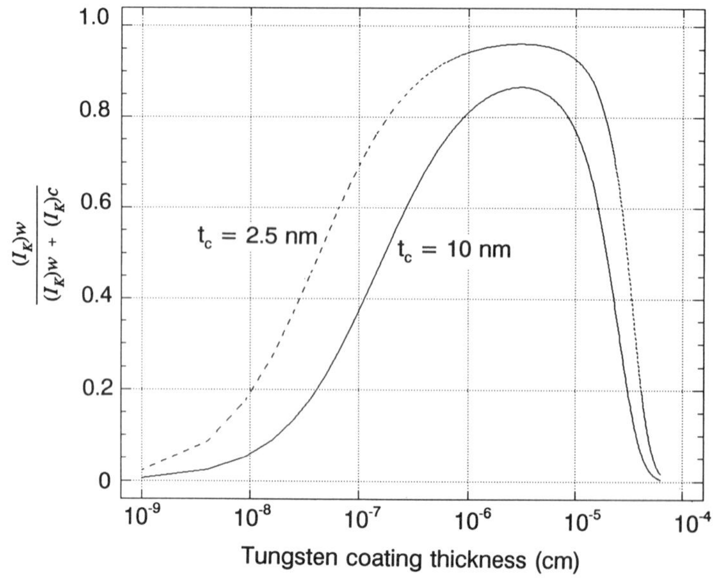

Figure 4. A plot of $(I_K)_W \div [(I_K)_W + (I_K)_C]$ vs. t_W at two fixed values of t_C, the substrate thickness. A plot of $(I_K)_W \div [(I_K)_W + (I_K)_C]$ vs. thickness of tungsten coating indicates the effect of reducing substrate thickness upon signal-to-noise ratio in darkfield electron microscopy. $(I_K)_W$ is forward scattering due to tungsten, and $(I_K)_C$ by carbon. This ratio function is a useful indicator and shows, for example, that the fraction of total forward scattering due to tungsten, for a 2.5 nm vs. a 10 nm carbon substrate, increases by ~2.5-fold at a thickness of 5×10^{-8} cm. $I_K = I_0 e^{-S_T w} (e^{kS_T w} - 1)$ where I_0 is incident intensity, S_T is the scattering constant, w is mass thickness, and $k = 0.8$ (15).

precise measurements, the system must be arranged to provide a geometric advantage. Thus, for example, if the monitor is placed half-way between the metal source and the specimen and very close to the perpendicular vapour path from the source to the specimen, and if the specimen plane is tilted so that $\sin\theta = 0.10$ (where θ is the angle between the incoming vapour stream and the specimen plane), monitor sensitivity can be extended 40-fold, provided that the monitor is placed close to the effective evaporation line of the specimen. Average thicknesses substantially less than 0.1 nm can then be monitored directly (see *Figure 5*).

Figure 5. Plot of thickness monitor reading vs. mass of metal evaporated. Thickness derived from monitor reading (t_{meas}) and the geometry is shown on the ordinate in nanometers: **A** for platinum and **B** for tungsten. The calculated thickness (t_{calc}), based on mass evaporated and the assumption of spherical distribution, is usually less than the measured thickness. It is noteworthy that the thickness parameter (t) is useful but not an exact measure in the case of a dicontinuous film. Thus, t_{meas} is generally less than the average crystallite thickness, so that an empirical correction for metal coating must be

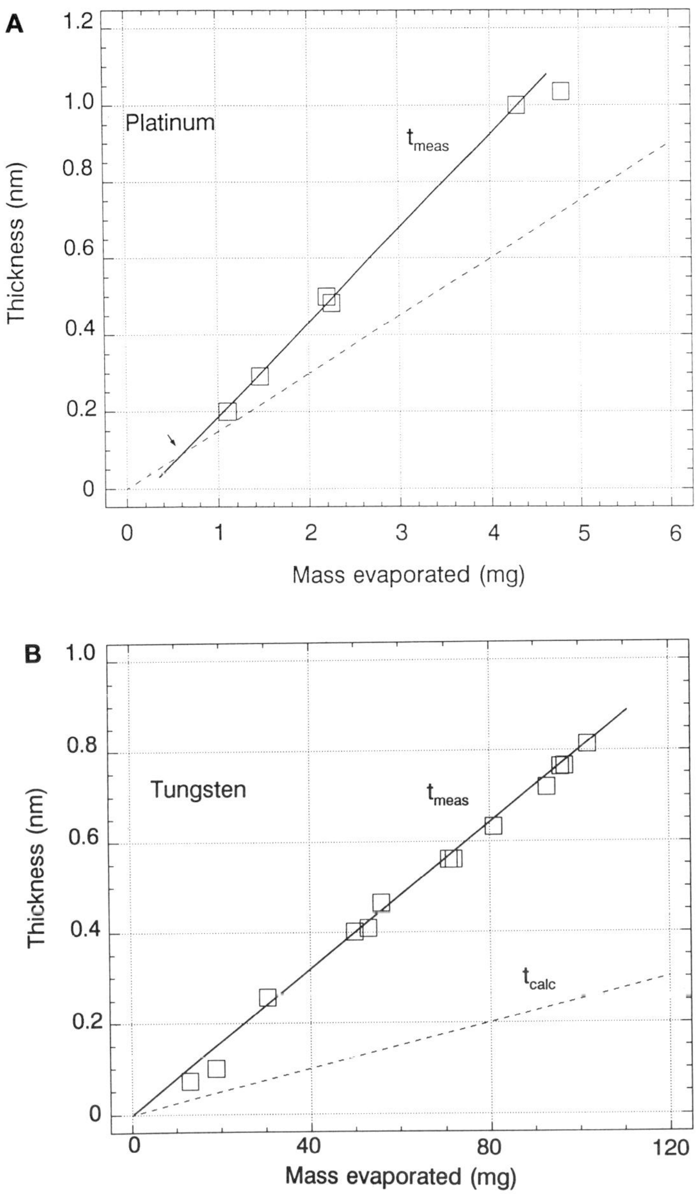

subtracted from measured molecular sizes, for example. On the other hand, t_{calc} in the case of tungsten is threefold less than t_{meas} due to non-spherical distribution from the source, and 1.4-fold less in the case of platinum. Platinum is evaporated from a straight tungsten filament, as shown in *Figure 7*, whereas tungsten is evaporated from a focused electron-beam heater, as shown in *Figure 1*. The experimental data plot in **B** intersects above zero because at smaller evaporated amounts amalgamation of platinum with the tungsten filament reduces the actual amount evaporated.

Although piezoelectric monitor-governed reproducibility of coating thickness is excellent, readings are generally 1.5 to 3 times the value calculated from the geometry and mass evaporated and must be corrected. The discrepancy is largely due to a variable efficiency of deposition away from the perpendicular to the molten source which deviates from assumptions of spherical symmetry of deposition. This finding is consistent with figures obtained from deposition patterns produced by various asymmetric sources (Sloan Technology Corp.; DTM 200 manual).

Because of such non-spherical distribution patterns which are characteristic of most sources, and the dissimilarity of temperature and sticking characteristics between monitor and specimen surfaces, the monitor method should not be considered absolute. However, the method is ideal for performing reproducible evaporations, and thicknesses thus determined do cross-check with direct weight microbalance measurements and interferometry quite well (4).

Coating parameters must be adjusted for surface properties of individual specimens in order to obtain optimal results. Coating thickness must also be designed with the requirements of the electron optical system in mind. That is, a 5×10^{-6} g/cm^2 coating of platinum might be considered thin for high-resolution secondary SEM imaging of a given specimen, a significantly great thickness might be useful to produce a low-loss backscattered SEM image, 8×10^{-7} g/cm^2 might be adequate for CTEM brightfield imaging, and 8×10^{-8} g/cm^2 might be more suitable for CTEM darkfield observation.

2. General methods

2.1 Preparation for electron microscopy

Metal coating of macromolecules is best carried out by the mica-replica method, generally as previously described (4, 6,14), and as amended for thin tungsten coating (6). In darkfield examination, amounts of coating metal may be diminished to the point at which fine macromolecular features disappear from brightfield micrographs due to contrast limitations. In darkfield micrographs, fine features, nearly invisible in brightfield, reappear on specimens, and structural dimensions measured from micrographs are not significantly distorted by the metal coating. Thickness measurements with a piezoelectric monitor become more relevant for thinner films as mass thickness (density × thickness) rather than just thickness at this level due to the discontinuous nature of the metal coating.

Table 2 presents relevant properties of a number of different elements or compounds commonly used, or potentially useful, in the production of substrate films and metal coatings. Clearly, density, boiling point and atomic number are most important since scattering power, ease of use and minimum grain size are prime considerations. We limit our discussion here to platinum, tungsten, and carbon, because these are the most generally used. Toxic materials are usually avoided.

Table 2. Coating materials arranged in order of decreasing refractiveness

Metal	Melting point (°C)	Density (ρ)	Atomic number (Z)	Observed grain and population density
C	3652	2.20	6	Substrate only
W	3410	19.35	74	Excellent
Re	3167	20.50	75	Oxidizes
Os	3000	22.50	76	Toxic!
Ta	2996	16.60	73	Excellent
Be	2970	1.85	4	Toxic!
Mo	2620	10.20	42	Good
Ir	2410	22.40	77	Good but grainy
Al_2O_3	2015	3.90	–	–
Rh	1966	12.40	45	–
Pt	1769	21.45	78	Good but grainy
Pd	1549	11.97	46	Good but grainy
U	1132	18.95	92	Grainy
Au	1063	19.30	79	Poor/grainy
Ge	947	5.40	32	–
Al	659	2.70	13	Substrate

Protocol 1. Preparation for electron microscopy

1. High-grade mica is first cut to rectangular blocks about 7 × 15 × 0.5 nm thick, and then split with sharp forceps to give two clear faces suitable for collecting micron-sized aerosol droplets.
2. Exhaustively dialyse solutions of macromolecules against a volatile buffer system composed of ammonium acetate or carbonate buffered with ammonium hydroxide or acetic acid in double-distilled water.
3. Dilute to 50–100 μg/ml of protein, add 25–50% glycerol, and spray the mixture on freshly cleaved mica, as shown in *Figure 6*. (Redistill glycerol under vacuum to eliminate non-volatile components.)
4. Arrange spraying so that one end of the mica is wet more heavily than the other so that several specimens representing different population densities may be collected.
5. Outgas for 20 h at 10^{-7} Torr to remove as much volatile salt as possible, then coat specimens with a very thin layer of platinum or tungsten by electron-beam evaporation, as shown in *Figure 6*. During this process, rotate specimens in a plane oriented at an angle (such as 5 degrees) to the vapour stream (see Section 3 for details).
6. Select mass thickness of the background film of metal (as determined during evaporation by piezoelectric monitor; see Section 1.4) for a particular application to be 10^{-6} to 10^{-8} g/cm^2 or less (see Section 1.3). An

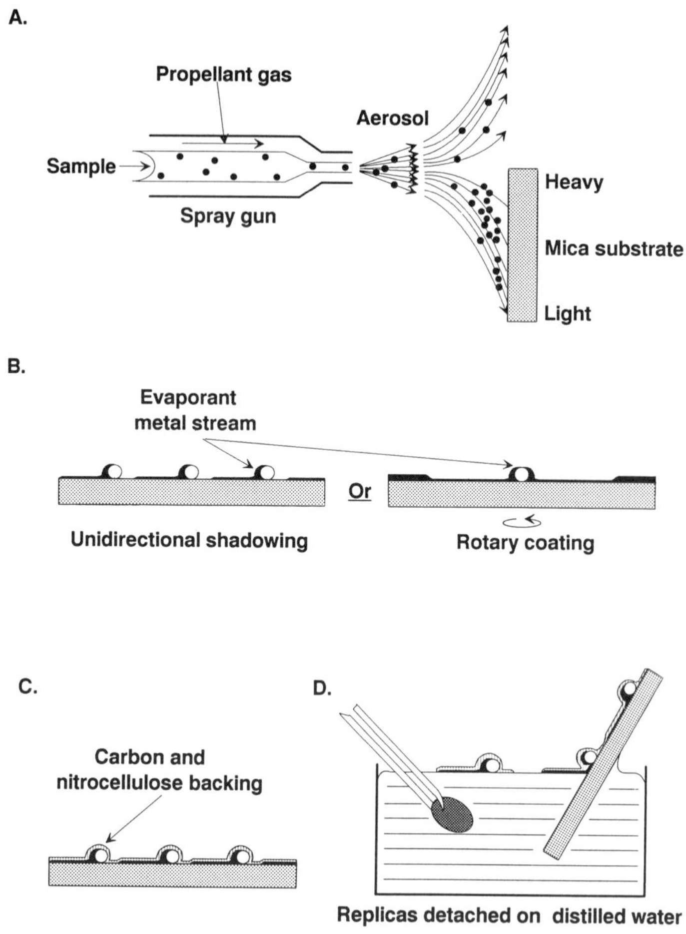

Figure 6. Diagram of steps involved in the mica replica coating method.

Protocol 1. *Continued*

estimate of thickness to be used for replicating metal is 0.5 nm or less; absolute measurement of actual replica thickness at these dimensions is not possible (4).

7. Evaporate carbon at normal incidence to the specimen by visual inspection to produce the desired thickness, using a piezoelectric monitor if possible.
8. Remove specimens from the vacuum and coat with a solution of 0.5% nitrocellulose in *n*-butyl acetate, draining quickly and blotting the excess.

9. After drying 5 min, score the mica with a pin into 3-mm squares, and trim edges.
10. Insert scored mica into a meniscus of double-distilled water to float the replicas off into the water surface; mount on clean grids.

Micrographs are recorded very close to focus on a modern high-resolution CTEM, such as the JEOL 100-CX top-entry stage electron microscope, at 100 kV, generally at a magnification of 53 000× to 300 000×, either in darkfield or brightfield. Micrographs recorded at 390 000× should clearly demonstrate point-to-point instrumental resolution between 0.3 and 0.4 nm in such specimens.

3. Apparatus

3.1 Platinum film deposition

In a typical arrangement, the specimen is placed in the vacuum evaporator on a rotatable platform at a distance of 6 cm (centre-to-centre) from a 0.076 cm diameter tungsten filament 5 cm long (*Figure 7*). A greater source–specimen distance may be necessary if the specimen is not to be coated at grazing incidence in order to minimize specimen heating. An amount of platinum wire chosen as suitable to the geometry and the mass thickness required is wound around the filament. Typically, this utilizes a 2.4 cm length of 0.005 inch (US std) platinum wire. *Table 1* indicates amounts of metal used for different geometries. The mass thickness often emloyed is several times that required for minimum visibility (3–5% contrast). Densitometer traces through CTEM or STEM micrographs of such specimens show that contrast is well in excess of that required for minimum visibility, indicating that, from a signal-to-noise standpoint, thinner coatings with smaller grain size are feasible. (*Figure 8* is traced from a micrograph of a platinum film of 10^{-6} g/cm^2.)

It is important to obtain a reasonably high vacuum in the evaporation in order to produce sharp shadowing and to minimize contamination. For example, to maintain a vacuum of 10^{-6} to 10^{-7} Torr during a typical evaporation, a high-capacity pumping system, of the order of 10 000 litres/sec, is employed. This may, for example, involve using a low vapour pressure diffusion pump fluid, such as polyphenyl ether (Convalex 10, Santovac 5), separated from the evaporation chamber by successive water-cooled and refrigerated traps, together with a high-capacity titanium sublimation pump and a cryotrap in the chamber. With the use of a diffusion pump alone, a vacuum of 3×10^{-6} Torr is usually possible. When cooling the specimen, it is important to maintain the best vacuum possible, as well as to cold-trap the specimen area to reduce contamination to an acceptable level.

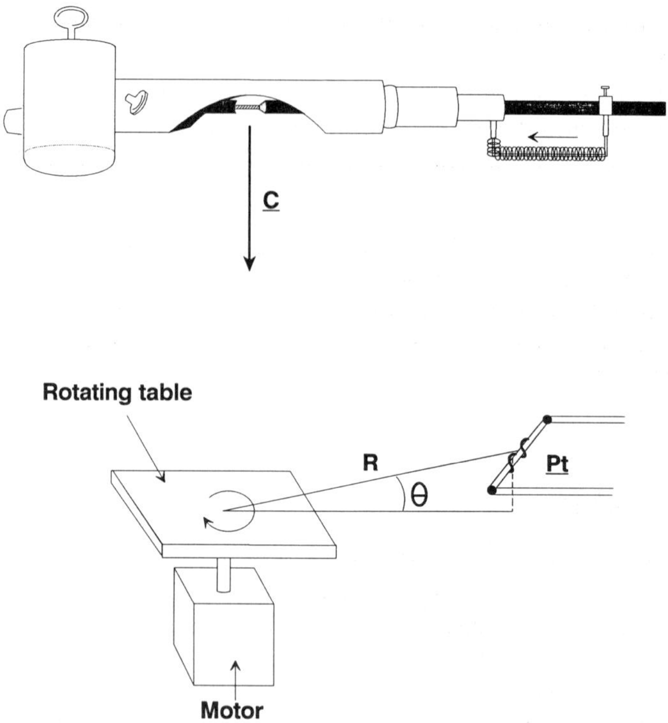

Figure 7. Simple platinum evaporation equipment. A typical carbon arc arrangement is shown at the top. Tip size is about 1.2 mm for evaporation at 50 amp for 3 sec. A low-speed selsyn motor is fitted with a platform mounted on its drive shaft. Typically, it operates at 50–100 r.p.m. at 110 V AC, and must be degreased, lubricated with graphite, and pumped in a vacuum system to equilibrium. *Table 1A* gives relevant parameters for using this arrangement.

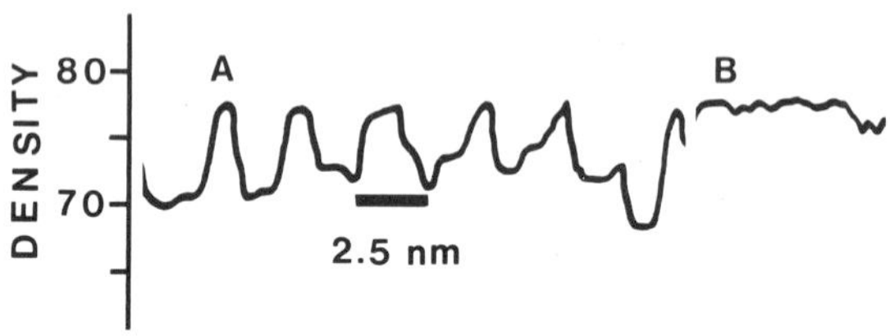

Figure 8. Densitometer trace through a TEM micrograph of a metal-coated film. Specimen consists of 10^{-6} g/cm^2 platinum on a 10^{-6} g/cm^2 carbon film, showing excess contrast arising in the metal coating compared with the substrate. The variation in contrast due to the metal is ~15%, whereas the variation in the carbon is ~3%. Density is ×100.

Protocol 2. Platinum replication

1. Load a 0.076 cm diameter $\times$5 cm tungsten rod, held between two electrodes, by winding a length (usually 1–2 cm; see *Table 1*) of 0.005-inch platinum wire in a spiral at the centre of the rod, as is shown in *Figure 7*.
2. Set up a carbon arc, as shown in *Figure 7*, with a cylinder of about 2 mm diameter by 2 mm long to be evaporated. Only part of this is actually used.
3. Load the freshly sprayed specimen on mica at the appropriate geometry, as shown in *Figure 7*.
4. Close the system and pump to about 10^{-7} Torr overnight.
5. Cool cryo-trap in the chamber with liquid nitrogen to stabilize the vacuum and trap contaminating vapours.
6. Heat the tungsten rod loaded with platinum with a 50 amp current for ~3 sec using a steady (~4 sec) period to reach 50 amp. The rate of evaporation is thus about 1.3 mg/sec.
7. Remaining steps are as in *Protocol 1*, steps 7–10.

3.2 Tungsten film deposition

Evaporation is produced by focused electron bombardment. Shadowing geometry is adjusted to minimize heating of the specimen by the molten tungsten source. Source–specimen distances of 40 cm are useful (*Figure 1*). Since evaporation rate is not constant, a piezoelectric monitor should be used to measure film thickness (see Section 1.4).

Empirically, the measured mass thickness found most generally useful in brightfield for standard CTEM tungsten replicas of most small macromolecules is 0.8×10^{-6} g/cm^2. This value, which is similar to the optimum for platinum replication, produces contrast considerably greater than the 5% limit in both CTEM and STEM, and substantially improved resolution of fine features of macromolecules, over platinum films.

Since tungsten crystallites occur with greater frequency per unit area than those of platinum for a given mass thickness, the scattering power per crystallite is less, resulting in a lower contrast, but a potentially higher-resolution image. As the mass thickness of tungsten is increased, however, granularity rapidly increases to a critical level. Thus, control of mass thickness is of great importance (see Section 1.4) (4).

Extension of present visibility limits by the use of tungsten coating is demonstrated in *Figure 9*, in which the centre frame from *Figure 9A* is a brightfield micrograph of a type VI collagen molecule rotary coated with 10^{-7} g/cm^2 of tungsten. The first frame shows the same molecule in focus in darkfield, demonstrating the higher darkfield contrast. The third frame shows the same molecule about 50 nm underfocus, and demonstrates the need to record darkfield image in focus. *Figure 9B* shows a thrombospondin molecule in brightfield, close to focus, and in darkfield in the second frame, also close to focus. The smallest metal particles in the background are at the theoretical limit of resolution in darkfield of about 0.5–0.7 nm. *Figure 9C* shows a thrombospondin molecule rotary coated with about 10^{-7} g/cm^2 platinum, in brightfield (first frame) and darkfield (second frame). The platinum grain is coarser and in darkfield the image is unusable.

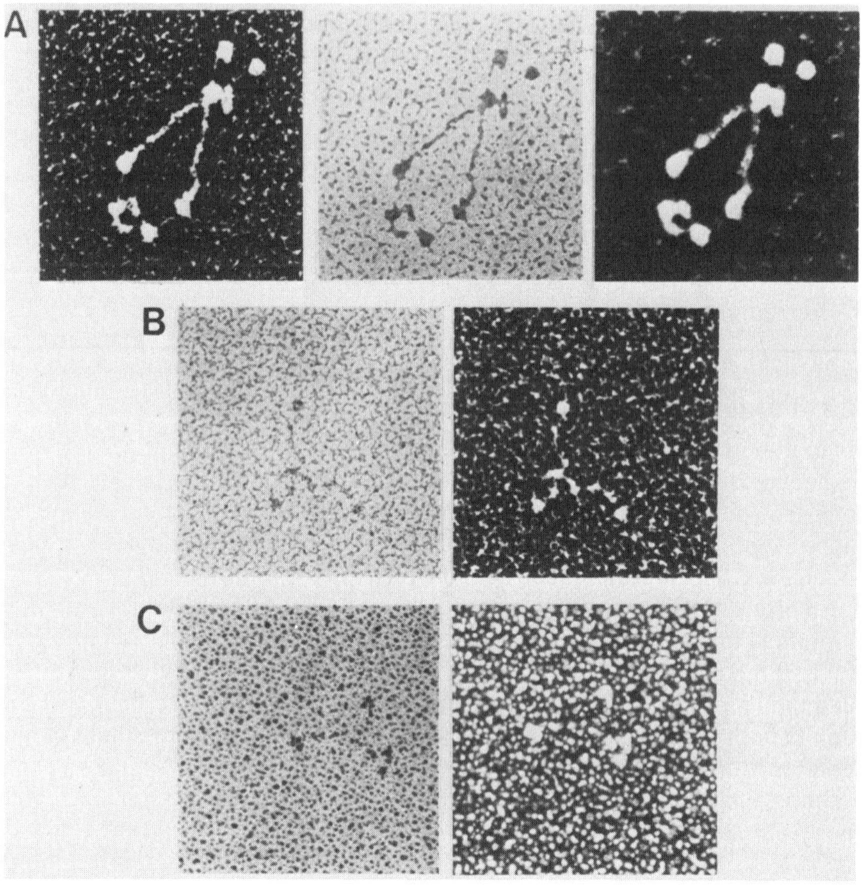

Figure 9. Extension of present visibility limits by the use of tungsten coating and darkfield electron microscopy.

A: Type VI collagen molecule prepared with 10^{-7} g/cm^2 tungsten, close to focus in brightfield (frame 2) vs. darkfield (frame 1). Frame 3 is the same area slightly under focus showing the critical dependency upon exact focus of the image in darkfield. Magnification 196 650×.

B: Thrombospondin molecule in brightfield (left) vs. darkfield (right) prepared by tungsten coating with 10^{-7} g/cm^2. Magnification 242 000×.

C: Thrombospondin molecule in brightfield (left) vs. darkfield (right) prepared by platinum coating with about 10^{-7} g/cm^2. Magnification 242 250×.

An examination of *Figure 3* shows that selection of metal coating thickness at the specimen plane of 10^{-8} cm produces forward scattering of 0.5% of I_0. The side of a 5 nm globular molecule coated at $\sin\theta = 0.1$, therefore collects $10 \times 10^{-8} = 10^{-7}$ cm of metal coat, but the depth of the metal pile on a molecular side is 5 nm. This then should produce forward scattering of the order of 15%. The background carbon substrate should by a similar analysis produce a contribution of 8% I_0 for a 10 nm substrate. Thus, in the vicinity of the molecular replica, roughly 2/3 of the signal will be contributed by the metal replica, due to a mass thickness (ρt) on the specimen plane of only 2×10^{-8} g/cm^2 (see *Figure 4*). Clearly, one may expect to reduce mass thickness from this point while retaining a useful darkfield signal, the limit depending upon the fine features being examined and metal grain separation.

Attempting to combine essential features of specimen rotation, low-temperature operation and the ability to select the angle at which shadowing occurs, we have recently constructed a device which accomplishes all of these functions, as described below.

3.3 Description of cooling rotating specimen holder

As discussed earlier (Section 1.3), the importance of specimen cooling during evaporation is significant. Basically, this equipment consists of a laminated stainless steel tube with a blanked end inside the vacuum system and an open end outside (see *Figure 10*). This device is sealed into a 'Ferrofluidic' rotary vacuum seal, with provision for tilting by a series of wedge-shaped flanges. A tube carrying liquid nitrogen can be inserted to within millimetres of a thin steel wall, on the other side of which the specimen is affixed. Thus, the specimen is cooled in times short enough (20–30 min) that the rotary seal is well-protected from freezing, by laminated construction which provides insulating layers. Electron-beam-induced evaporation may thus be executed at large distances (~50 cm) from the specimen, which may be simultaneously rotated, tilted and cooled with liquid nitrogen in a well-trapped system.

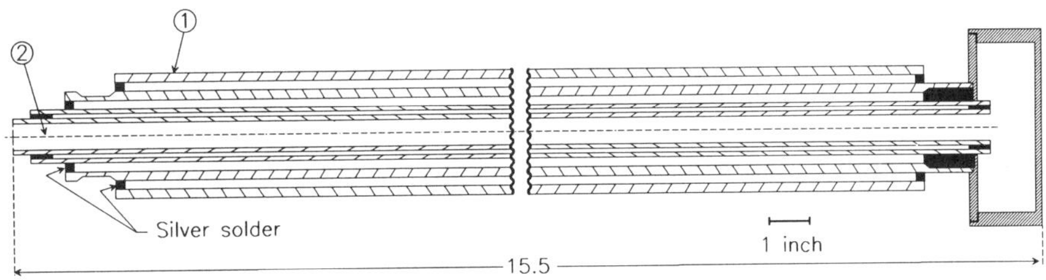

Figure 10. Cross-section of rotating-tilting cold specimen hardware which is inserted through the wall of a vacuum system using a 'Ferrofluidic' seal which can be rotated while cold and tilted.

Protocol 3. Tungsten replication

1. Place ingot of pre-melted tungsten in the electron beam crucible.
2. Load the specimen as shown in *Figure 1*, using either double-sided Scotch tape to affix the mica to the rotary specimen holder, or else attaching mica pre-mounted on a copper planchet.
3. Pump the system overnight to 10^{-7} Torr, cooling the cryo-trap in the chamber with liquid nitrogen, and cooling the specimen to the required temperature.
4. Evaporate at 5 kV and 150–200 milliamps, using an electron-beam system (e.g. Denton Mini-Gun, Denton Vacuum); monitor evaporation by piezoelectric oscillator (e.g. DTM 200, Sloan Technology Corp.).
5. Remaining steps are as in *Protocol 2*, steps 7–10.

3.4 Correction curves

Having plotted experimentally determined correction curves obtained from plots of measured thickness versus approximate true dimension, we may then plot correction versus mass thickness for typical particles whose dimensions are in the range of the measurements made. While the corrections suggested in *Figure 11* illustrate the magnitude of correction, the individual investigator should develop a curve for a given experimental situation.

When minimal coating thickness is used with darkfield, the result is that corrections are small enough to be ignored in most situations. *Figure 11* shows correction curves derived as indicated, where the diference between the line of identity and the experimental curve is used as the correction for measured dimensions using either platinum or tungsten rotary coating at a given mass thickness. The inset in *Figure 11* compares the correction applied to a 5-nm dimension, as a function of mass thickness applied, measured from a micrograph of a metal-coated specimen prepared by several common methods. This shows the extreme variation of over an order of magnitude between methods. In *Figure 12*, for rotary tungsten coating, the dotted line (representing 10^{-8} g/cm^2) falls practically on the line of identity—meaning that the actual correction is nil using that quantity by calculation. Similarly, the solid line is very close to the line of identity, and this represents 10^{-7} g/cm^2. Experimental points, plotted as open squares, scatter generally around both curves, suggesting that an attempt to correct dimensions for these very thin coatings is probably subject to greater error than the measurements themselves. The nearly horizontal lines at the bottom of this chart show that the actual correction for 10^{-8} g/cm^2 (dotted line) is close to 0. The amount of correction (solid line) for 10^{-7} g/cm^2 is less than 0.05 nm. Since this is a fraction of an atomic width, the necessity to correct for metal at this point is essentially not present. In fact, at this point, corrections are largely related to

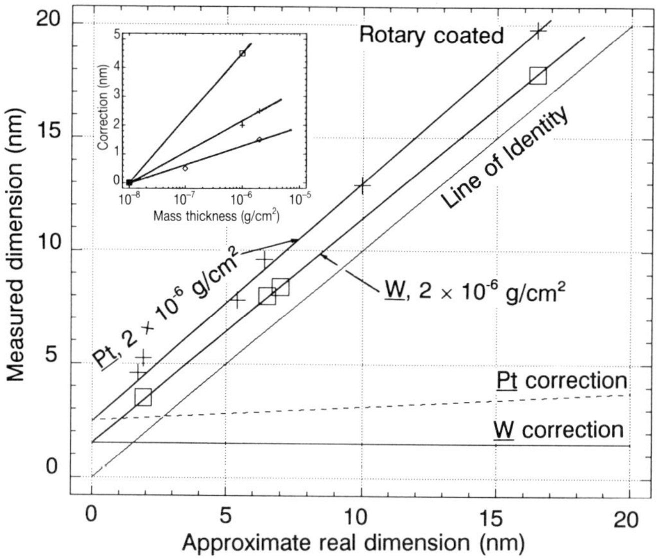

Figure 11. Plots of measured width vs. approximate true dimensions and derived corrections for rotary shadowing of 2×10^{-6} g/cm^2 at 5 degrees (platinum and tungsten). Measured dimension is plotted vs. an approximate real dimension, which is obtained from other structural information (e.g. X-ray diffraction). Correction curves are obtained by subtracting the appropriate plotted line from the line of identity. These corrections are about one-half of those found appropriate for unidirectional shadowing with the same mass thickness, and of course apply only in a given set of experimental circumstances. (Data from ref. 4.)

Inset: Chart comparing typical correction applied to a 5 nm dimension, measured from a micrograph of a metal-coated specimen as a function of mass thickness of metal coating at 5 degree incidence. (□) indicates unidirectional platinum coating; (+) indicates rotary platinum coating; (◇) indicates rotary tungsten coating.

residual salt exaggerating small dimensions and over exposure of the negative in darkfield operation, leading to expanded dimensions and fine features, due to haloing effects.

However, a significant reality is that in darkfield microscopy the photographic process is of great importance. Because of haloing effects, overexposure happens easily in darkfield microscopy since the background is black, and this can result in variations in apparent size of the structures photographed, which far exceed any correction for metal.

Optical diffraction analysis of micrographs and/or measurement of the Scherrer breadth (15) of reflections in electron diffraction patterns of coated specimens are useful means of estimating approximate average size of larger crystallites. Images can be filtered effectively by optical diffractometry analysis (16) with respect either to removal of background noise from metal grain

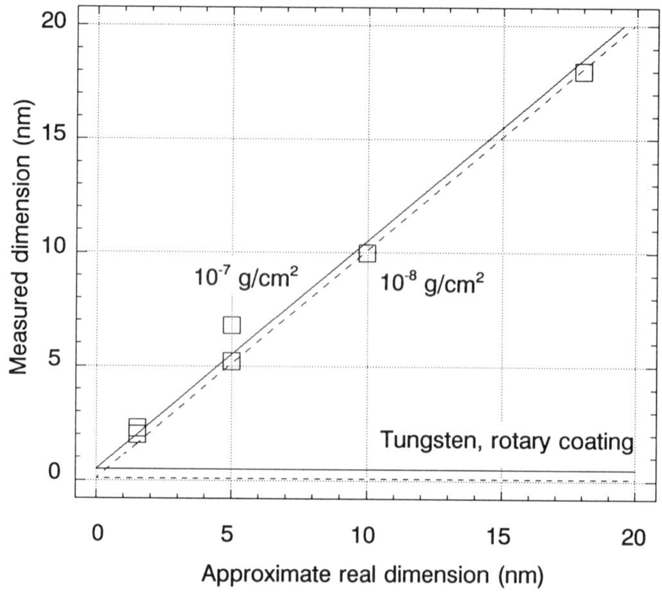

Figure 12. Plots of measured width vs. approximate true dimension, with derived corrections for tungsten at 10^{-7} and 10^{-8} g/cm^2, rotary coated at 5 degrees. There is virtually no correction at 10^{-7}, and no correction at 10^8 g/cm^2, which falls on the line of identity. Data points are found to scatter around both curves, indicating the difficulty of making meaningful corrections with very thin coatings.

or to enhancement of symmetry of macromolecular projections. Appropriate masking of the Fourier plane of the diffractometer is used to select for one or the other type of optical filtration.

4. Electron microscopy

Brightfield and darkfield electron microscopy is best carried out in a modern transmission electron microscope (CTEM) such as a JEOL 100-CX, using a top entry stage. Annular darkfield illumination must be used in place of the standard tilted-beam mode in order to achieve optimal resolution. Detailed discussion of the theory and specifics of these methodologies will be found elsewhere (7, 11).

4.1 Specifics of darkfield imaging

Since, at the current level of 0.3–0.4 nm point-to-point instrumental resolution in brightfield, contrast becomes the limiting factor, re-evaluation of prospects for high resolution darkfield methods is in order (see *Figure 3*, inset: a high magnification micrograph of a single myosin head region; *Figure 9* shows high magnification brightfield and darkfield micrographs of the same type VI collagen macromolecule). Relevant theory as well as practical limitations are discussed elsewhere (7, 8, 11, 15, 17). However, in darkfield, intensity normally transmitted through the objective lens without intersection with the specimen in brightfield, is removed so that only intensity scattered forward though the lens aperture is utilized in the image. With background intensity thus removed, contrast is greatly increased.

Using minimal-thickness refractory metal coatings, Bragg reflections do not confuse the image, permitting optimal application of currently-available annular darkfield illumination. The development of techniques for routine deposition of almost continuous metal films in which crystallite size is limited to 0.4–1.0 nm, has made darkfield imaging of metal-coated macromolecules feasible. Charge bleed-off and stability are assisted both by the refractory metal film and the overlying carbon film, which, with adequate specimen contrast, it is now feasible to apply. Finally, these specimens are almost as easy to prepare and handle as conventional specimens for high-resolution brightfield microscopy. The use of annular darkfield illumination, produced, for example, by a 1600μm annular aperture placed at the level of condenser lens II in conjunction with a 40 μm objective aperture, eliminates instrumental factors which have previously interfered with darkfield operation. Whereas only one sector of a tilted beam reaches the specimen in tilted illumination, adequate intensity is readily obtained with the use of the full annulus of illumination. At the same time, asymmetric contamination and/or charging of the objective aperture, with consequent deterioration of resolving power, are avoided.

Application of these methods to a number of macromolecular species shows, in general, that features only occasionally or marginally detected in brightfield may be visualized consistently with darkfield illumination. Distortion of dimensions or shapes of molecular features by accumulation of coating metal, while not avoided altogether, are also minimized.

In summary, we have described rationale and methods for metal coating of biological specimens, particularly macromolecules currently in use in our laboratory. This is intended to be a practical guide, and not a review of the literature.

Acknowledgements

The author is indebted to Dr Simon C. Watkins for helpful comments on the manuscript, and to Ms Deirdre M. Ryan and Ms Marielle Langlois for comments and assistance. This work was supported by Public Health Service Grant GM 14237 from the National Institute of General Medical Sciences, United States Public Health Service.

References

1. Ohtsuki, M. and Crewe, A. V. (1980). *Proceedings of the National Academy of Sciences, USA,* **77,** 4051.
2. Rachel, R., Jakubowsky, U., and Baumeister, W. (1985). *Journal of Microscopy,* **141,** 179.
3. Vogel, R. H., Provencher, S. W., von Bonsdorff, C.-H., Adrian, M., and Dubochet, J. (1986). *Nature,* **320,** 533.
4. Slayter, H. S. (1976). *Ultramicroscopy,* **1,** 341.
5. Slayter, H. S. (1978). In *Principles and techniques of electron microscopy: Biological applications*, Vol. 9 (ed. M. A. Hayat), pp. 175–245. Van Nostrand Reinhold, New York.
6. Slayter, H. S. (1983). *Annals of the New York Academy of Sciences,* **408,** 131.
7. Slayter, H. S. (1989). In *Methods in enzymology*, Vol. 169 (ed. J. Hawiger), pp. 326–35. Academic Press, New York and London.
8. Dubochet, J. (1973). In *Principles and techniques of electron microscopy* (ed. M. A. Hayat), pp. 115–51. Van Nostrand Reinhold, New York.
9. Fiskin, A. M., Cohn, D. V., and Peterson, G. S. (1987. *Journal of Biological Chemistry,* **252,** 8261.
10. Ottensmeyer, F. P., Bazett-Jones, J. H., and Price, G. B. (1978). *Ultramicroscopy,* **3,** 303.
11. Slayter, H. S. (1988). In *Advances in cell biology* (ed. K. R. Miller), Vol. 2, pp. 235–61. JAI Press, Greenwich, Connecticut.
12. Slayter, H. S. (1980). In *SEM/1988/I*, pp. 171–182. SEM Inc., AMF O'Hare, Illinois.
13. Flood, P. R. (1980). In *SEM/1988/I*, pp. 183–200. SEM Inc., AMF O'Hare, Illinois.
14. Margossian, S. S., Bhan, A., and Slayter, H. S. (1983). *Journal of Biological Chemistry,* **258,** 13359.
15. Hall, C. E. (1953). In *Introduction to electron microscopy*, p. 284. McGraw-Hill, New York.
16. Lake, J. A. and Slayter, H. S. (1970). *Nature,* **227,** 1032.
17. Klug, A. (1979). *Chemica Scripta,* **14,** 291.

8

Freeze-fracture techniques

SEIZO FUJIKAWA

1. Introduction

Freeze-fracture techniques were developed originally to prepare cellular ultrastructures in a frozen-hydrated and living state for electron microscopy (1). The replicas made from freeze-fractured biological specimens have provided a new three-dimensional ultrastructure of cells (2). Since the elegant proof by Branton and his co-workers (3) that the freeze-fracture technique reveals the macromolecular organization of the hydrophobic interior of bilayer membranes, the technique has been extensively used in the field of biological science and has made a major contribution to our understanding of biological ultrastructure. (For nomenclature of freeze-fractured membranes, see ref. 4.)

The basic procedure for freeze-fracture technique includes:

(a) cryofixation of materials;
(b) fracturing (and etching) of specimens;
(c) replication of the freeze-fracture plane by evaporation of heavy metals under high vacuum;
(d) removal of the specimen from replica; and
(e) observation of replicas with the transmission electron microscope (TEM).

Recent studies have demonstrated many new methods for freeze-fracture electron microscopy which open further possibilities for the application of freeze-fracture techniques in biological sciences. This chapter details the procedure for the standard freeze-fracture technique, and also presents new methods which are thought to be of value for the study of cellular ultrastructure.

2. Standard technique

For the freeze-fracture technique, materials are first cryofixed. The ideal method for cryofixation is to freeze fresh living materials in such a way that they retain a well-preserved ultrastructure, similar to that in the living state. It is easy to freeze some biological materials which have a low water content, without damage due to the formation of ice-crystals (5). However, most biological materials that have a high water content require some type of pre-treatment to prevent artefacts caused by formation of ice-crystals during freezing. In the standard freeze-fracture technique, in which observation of the three-dimensional structure of cells, especially observation of membrane internal ultrastructure, is the main purpose, materials are first chemically fixed, treated with a cryoprotectant, frozen, and then the freeze-fracture replica is produced. It is possible to prepare freeze-replicas from almost all biological materials including single cells, cell components, and bulk tissues with a high water content. The standard technique for freeze-fracture, which is the basis for other modified techniques described in this chapter, is outlined in *Protocol 1*.

Protocol 1. Standard freeze-fracture technique

1. In the case of bulk tissues, excise them to proper size for chemical fixation. In the case of cellular suspensions, sediment them by centrifugation.
2. Fix materials by chemical fixatives.
   (a) Chemical fixation is necessary to prevent artefacts arising from subsequent cryoprotectant treatment. The lower permeability of the cryoprotectant into cells or its toxicity may cause serious artefacts without pre-fixation. The minimum fixation which will prevent formation of artefacts arising from subsequent cryoprotectant treatment is best.
   (b) Glutaraldehyde is the most widely used fixative for freeze-fracture. The specifics of glutaraldehyde fixation, i.e. the concentration of glutaraldehyde, the kind, concentration, and pH of buffer, and duration and temperature of the fixation period, should be changed according to the type of material. In general, the same conditions are used for glutaraldehyde fixation as are used for the preparation of ultra-thin sections (ref 6; see also Chapter 1).
   (c) For many biological materials, fixation with 0.5–2% (v/v) glutaraldehyde (25% aqueous stock solution) in isotonic buffer (pH 7.2) for 1–6h at 4°C is widely employed.
   (d) Combinations of paraformaldehyde [0.5–4% (w/v), always prepared from its *para*-polymer] with glutaraldehyde (0.5–1% v/v) or paraformaldehyde alone (0.5–4% w/v) are also used as chemical fixatives (6).
   (e) Fixation with osmium tetraoxide should be avoided for freeze-fracture samples, because it prevents fracture along the interior of membranes.

3. Adjust the size of the specimens, in the case of bulk tissues, to the final size (1–2 mm^3) to be mounted on the specimen-holder for freeze-fracture.
In case of isolated cells, sediment them by centrifugation.
4. Immerse the specimens immediately in 20–30% glycerol (v/v) in the buffer solution kept at 4°C.[a]
   (a) The time for immersion of specimens in glycerol solution will depend on the glycerol permeability of specimens, generally ranged between 1 h to overnight at 4°C. The specimens, which float into the glycerol after transfer, sediment naturally following penetration of the cryoprotectant. The minimum concentration of glycerol and minimum immersion time found to be necessary to prevent ice-crystal formation is best.
   (b) Dimethyl sulphoxide (DMSO), ethylene glycol (EG), and ethanol (penetrating cryoprotectants), and polyvinyl pyrrolidone (PVP), dextran, and hydroxyethyl starch (HES) (non-penetrating cryoprotectants), are also useful for ultrastructural studies for special purposes (7).
5. Mount cryoprotected specimens on specimen-holders before freezing.
   (a) Use a specimen-holder appropriate for the particular purpose. The design of specimen-holders differs according to:
      - the type of freeze-fracture machine
      - materials, such as single-cell suspensions, monolayer cell cultures, and bulk tissues
      - the type of replica, such as complementary replicas and ordinary replicas
   (b) The adhesion between specimen and holder must be firm enough to resist fracturing stress, and to ensure good thermal conductivity.
6. Freeze specimens by plunge-freezing, i.e. freezing by abrupt immersion in liquid coolant. The coolants usually used for plunge-freezing are the following (8):
   - propane (cooling rate 19 100°C/sec)
   - Freon 22 (9000°C/sec)
   - Freon 12 (6580°C/sec)
   - nitrogen slush (1700°C/sec)
   (a) From the point of view of cooling rate the use of propane as coolant, which has the fastest cooling rate, is recommended under proper conditions. (For specimen-freezing procedure, see *Protocol 2*.)
   (b) Other coolants can also be used for plunge-freezing of cryoprotected specimens and may provide good freezing, if a proper concentration

Protocol 1. *Continued*

of cryoprotectant and proper size of specimen is used. The preparation using Freon 12 and 22 can be carried out by a procedure similar to that of propane, but more safely and easily than with propane, which carries a risk of explosion.

(c) Nitrogen slush can easily be produced as a coolant in most laboratories, safely and without the problem of pollution. (For the preparation of nitrogen slush, see *Protocol 3*.)

7. Store frozen specimens in large Dewar vessels filled with liquid nitrogen (LN_2) until use. When they are used, transfer them into a small storage container with LN_2.

8. Transfer and mount frozen specimens on the cold stage of a freeze-fracture machine.[b]

(a) Because each freeze-fracture machine has a different construction, follow the procedure described in the manufacturer's instruction manual. The actual manner of transferring and mounting frozen specimens on the cold stage of a freeze-fracture machine will differ according to the construction of each machine.

(b) For machines with a pre-evacuation chamber, mounting the specimens on the cold stage is possible without breaking the chamber vacuum. Before mounting the sample on the cold stage in this type of machine, the cold stage, cold knife, and cold shroud surrounding the cold stage should be pre-cooled to the lowest temperature.[c]

(c) In other machines, which have no pre-evacuation chamber, only the cold stage should be pre-cooled to the lowest temperature under high vacuum before loading. The knife (and cold shroud) must be cooled after specimen loading to prevent contamination from breaking the vacuum.

(d) The important principle for transferring and mounting specimens on the cold stage is to keep specimen temperature during these processes as low as possible and prevent specimen contamination due to frost. Obviously, a quick transfer is best.

9. Equilibrate specimen temperature under high vacuum within the chamber of a freeze-fracture machine to the temperature for fracturing. Decide the fracture temperature according to purpose, by referring to the explanations below.

(a) Vapour pressure of ice differs in relation to temperature (7,9). When the temperature of the specimens has a higher vapour pressure than the vapour pressure of the chamber (i.e. vacuum in the chamber), etching (by sublimation of ice) takes place. On the other hand, when the temperature of the specimens has a lower vapour pressure than vacuum, no etching takes place.

(b) Although it increases the possibility of contamination on fracture planes, fracturing at lower temperatures may minimize plastic deformation of macromolecular structure in the specimens (10).

(c) In almost all freeze-fracture machines, the displayed temperatures relate to the cold stage, and not to the specimen itself. Hence, even after the temperature as displayed reaches the set value, it is necessary to wait at least another 10 min to obtain exact equlibration before freeze-fracturing.

10. Fracture the specimens by one-cut using a cold knife.

(a) In some machines, the cold knife is constructed as a microtome assembly, in which repeated micro-cuts are possible to make a comparatively flat fracture plane.

(b) In other machines such micro-cuts are impossible, and essentially only one large-cut is possible.

(c) Replicas made by micro-cut are more stable during the thawing and cleaning process, even with a smaller amount of metal evaporation, than replicas made by one-cut. Additionally, the comparatively flat fracture plane made by micro-cuts assists the floating of replicas during digestion of cellular materials. However, there is a general belief that one large-cut can provide an extensive fracture plane without any knife mark, while many areas in the fracture plane made by micro-cut may have knife marks. The choice between micro-cut and one-cut will depend upon the results required and the acceptability of knife marks.

11. After fracturing, etch the fracture plane (if necessary) by keeping specimens for 30 sec to 1 min at 1 or 2°C higher temperature than the critical temperature[d] for sublimation of ice (9) while under covering with the cold knife.

12. To replicate the fracture plane of the frozen specimens, evaporate an electron-opaque metal, usually by simultaneous evaporation of platinum–carbon (Pt–C), on to the fracture plane to provide contrast.[e] The metal is typically applied at an oblique angle (usually around 45 degrees) to the average orientation of the fracture plane.

(a) Pt–C can be evaporated either by resistance-heating or by electron-beam evaporation.

(b) For routine freeze-fracture replication, a Pt–C film of 1.5–2.0 nm thickness should be made. The approximate film thickness can be determined:

- using quartz crytal film-thickness monitors
- according to the amount of platinum wire wrapped around the carbon electrodes (in case of resistance-heating)

Protocol 1. *Continued*

- empirically, by varying time, current, and voltages used for evaporation.

(c) Although many machines have a rotary-shadowing apparatus (see Section 3.1), it is recommended that replicas are made first by using unidirectional shadowing, especially when an unfamiliar specimen is to be examined. The structural organization in the fracture plane can be more easily comprehended following unidirectional-shadowing than is the case with rotary-shadowing.

(d) Tantalum–tungsten (Ta–W) alloy is also used for high-resolution shadowing (see Chapter 7). For this purpose, a specially designed electron-beam gun is necessary (7).

13. Immediately after Pt–C shadowing, reinforce the fracture plane replica by evaporation of a low electron-scattering supporting layer of carbon, evaporated perpendicular to the fracture plane.

(a) Carbon is also evaporated either by resistance-heating or electron-beam evaporation.

(b) The optimum thickness of the carbon backing layer is strongly dependent on the type of specimen and required stability during thawing and cleaning, but is generally in the range of 10–35 nm. For the highest resolution, carbon film should be as thin as possible, but to obtain large replicas a thicker film of carbon is necessary, so as to prevent fragmentation of the replicas during cleaning.

14. After replication, remove the specimens from the machine by reversing the procedure for loading.

15. Thaw specimens on their holders by gradual immersion in distilled water at an angle of 45 degrees, in depressions on the porcelain spotting plates,[f] using forceps to grip the holders.

(a) In this way the specimens are thawed before the fracture plane with replicas is immersed in the water.

(b) When the fracture plane is immersed, replicas from the isolated cells will separate from the specimen and float to the surface of the water. However, the separation of replicas from specimens does not result from this simple procedure alone with other types of material, in which replicas will sink in distilled water together with the materials.

16. Transfer the specimens to the cleaning solution.

(a) Replicas which continue to float during thawing are transferred from thawing to cleaning fluids using a 2.0–4.0 mm diameter platinum loop made of 0.2 mm diameter wire.

(b) Replicas which sink together with the material can be carefully transferred by means of a pipette.

17. Digest biological materials with cleaning solution.

(a) A variety of replica-cleaning schedules have been proposed and it is often necessary to employ an empirical approach to determine the best method for a particular biological material.

(b) With most biological materials, clean replicas are obtained by treatment with 6% sodium hypochloride (or commercial bleach), or a 30% aqueous solution of chromic acid for 1–2h at room temp.

(c) With more difficult cleaning materials, such as hard tissues from higher plants and bacterial spores, see *Protocol 4*.

18. Wash replicas finally in distilled water.

19. Mount replicas on clean grids of 100–400 mesh, depending upon their size.

(a) Use tweezers of a self closing type to hold the grid.

(b) Rinse replicas on the grids in acetone for a final cleaning and to reduce the surface tension stress during drying of replicas.

(c) Remove acetone from the grids by touching the edge of a small filter paper to the tip of the tweezers holding the grid.

20. Store grids temporarily in a Petri dish (preferably glass to prevent formation of static), and more permanently in a grid-holder for several months in a clean, dark environment.

21. Observe and photograph replicas at an accelerating voltage from 60–120kV by TEM.

(a) The freeze-fractured cells show extensive membrane interior structure, in which intramembrane protein particles are distributed on the smooth lipid fracture face.

(b) The outer and inner surface of membranes cannot be revealed by the standard procedure. Furthermore, non-membraneous structures within cells are also difficult to reveal clearly by the standard procedure.

[a] Unlike preparation of specimens for thin sectioning, washing of specimens with buffer after fixation can be omitted.

[b] The procedures from here to step 14 are carried out inside the freeze-fracture machine.

[c] Some machines with a pre-evacuation chamber have a special transfer holder to prevent contamination of specimens during transfer from LN_2 to the pre-evacuation chamber.

[d] At vacuum of 1×10^{-6} Torr, for example, the proper etching temperature is -109°C (the vapour pressure is 1.4×10^{-6}).

[e] Before fracturing, the evaporation sources must be degassed by heating them inside the chamber.

[f] Spotting plates have the advantage that the white background makes it easier to see the replica fragments.

Protocol 2. Procedure for freezing specimens by propane (M. V. Parthasarathy, personal communication)

1. Inside a fume hood,[a] fill a 1-litre Dewar with LN_2.
2. Slowly insert the propane well into the LN_2 by grasping the propane well firmly with crucible tongs. Avoid getting any LN_2 into the propane well.
3. Attach a Pasteur pipette to the hose from the commercial propane bottle.
4. Insert the end of the Pasteur pipette into the bottom of the propane well and slowly open the propane bottle valve.
5. When the propane well is filled, remove the pipette, close the valve, and quickly and carefully cover the well with a clean glass slide.
6. Allow the propane to cool for half an hour. The Dewar must be kept filled with LN_2.
7. Line the bottom of the specimen basket with filter paper cut to the appropriate shape. The filter paper serves to absorb any liquid propane on specimens before it freezes.
8. Fill a second Dewar with LN_2 and insert the basket that will hold the frozen specimens.
9. Using forceps, quickly plunge the specimens into the liquid propane for at least 5 sec. Then quickly transfer the holder to the basket in the LN_2 and release it from the forceps.

[a] When using propane, all work should be done inside a fume hood and no flames should be present until the propane has completely evaporated and drawn up the hood.

Protocol 3. Preparation of nitrogen slush (7)

1. Put 100–200 ml LN_2 into a well-insulated container (such as an expanded polystyrene cup).
2. Evacuate the LN_2 within a chamber pumped by a rotary pump until the nitrogen solidifies at a temperature of about −210°C.
3. Leak chamber and remove the cup. As soon as air leaks into the vacuum chamber, the nitrogen starts to warm up, producing a mixture of liquid and solid (nitrogen slush).
4. Freeze specimens by rapid immersion in the nitrogen slush as soon as possible, because such a mixture is not in a state of stable equilibrium.

Protocol 4. Cleaning procedure for specimens which are difficult to digest

1. Immerse specimens in concentrated H_2SO_4 by transfer through a gentle graded dilution series at RT.
2. Keep specimens in concentrated H_2SO_4 at least overnight, until all specimens become soft and change to a dark brown colour.
3. Transfer softened specimens to distilled water by means of a platinum loop to dilute and remove H_2SO_4 so that specimens are floated.
4. Transfer floating specimens by platinum loop into 6% sodium hypochloride at RT for 1 h or more, until the softened specimens adhering to the replicas become first white in colour and then completely digested.
5. Wash replicas with distilled water for 10 min.

3. Modified techniques

3.1 Rapid-freezing, deep-etching, and rotary-shadowing technique

Apart from providing images of membrane interior structure, freeze-fracture of cells also provides cross-fractured images of cytoplasmic structures. The removal of water by sublimation (deep-etching) from rapidly frozen and fractured specimens reveals the three-dimensional nature of cellular structure, especially non-membranous structures within the cytoplasm. Rotary-shadowing is particularly suitable for demonstrating the fine structure of filamentous structures. All types of materials, including single cells, cell components, and bulk tissues, can be used as starting materials. However, different rapid-freezing techniques should be used, depending upon the material.

Protocol 5. Procedure for rapid-freezing, deep-etching, and rotary-shadowing

1. Use fresh materials (without pre-treatment by chemical fixation and cryoprotectants), principally, because in this technique fine molecular structures are observed.[a]
2. Freeze materials rapidly by using one of the rapid-freezing techniques described below,[b] to reduce the size of ice-crystals to less than 10 nm.

   (a) *Propane jet freezing* (11): The smallest volume of specimen (preferably a cell suspension) is placed in the plane of symmetry between two thin sandwich foils to produce a layer of minimum thickness. Jets of liquid propane are directed from the single or double nozzles on to

Protocol 5. *Continued*

the specimen. Freeze-fracture of the sandwich specimen is easily accomplished by simply taking off the sandwich supports.

(b) *Spray freezing* (12): Spray cellular suspensions with low viscosity through thin capillaries or orifices. On ejection the suspensions form fine droplets with diameters below 100 μm. These immediately freeze on contact with the liquid propane used as cryogen. After evaporation of the propane at −85°C, the frozen droplets are mixed with butyl benzene having a melting point around −90°C to −100°C. The mixture is then transferred to the specimen-holder and re-cooled by immersion in LN_2. The specimen can be fractured with a cold knife.

(c) *Impact freezing* (13): The tissue block or cell suspension is frozen by abrupt contact with a solid surface of high thermal conductivity cooled with LN_2 or liquid helium temperature. High cooling velocity can be achieved within the first 1–2 nm depth from the surface of specimen upon contact. For fracturing, a cold knife, as used for a micro-cut, is convenient.

(d) *High-pressure freezing* (14): A cell suspension or tissue block is placed between two sandwich holders. A high pressure (2100 bar) is applied to the specimen within the holder for 15 ms and then the material is frozen by pressurized LN_2. Due to supercooling of specimens to less than −70°C under high pressure, it was shown that in some materials layers of 600 μm thickness could be frozen with ice-crystals less than 10 nm in diameter. One sandwich holder is removed under LN_2 for fracturing with a cold knife.

3. Transfer and mount frozen specimens on the cold stage of a freeze-fracture machine, following the standard procedure.

4. (a) Fracture and etch specimens. Etching should be done at between −97°C and −105°C for 8 min to 1 h under cover of the LN_2-cooled cold knife. Depending upon the purpose and type of material, the required degree of etching can be achieved. The desired etching is always the minimum etching necessary to expose the structure of interest.

(b) Re-cool specimens before shadowing to below −115°C in order to stop sublimation.

5. Shadow specimens by evaporation of Pt–C while rotating them at 60–150 r.p.m. with the evaporation source at an angle of 6–45 degrees to the specimen surface.

6. Soon after Pt–C shadowing, evaporate carbon to reinforce replicas following standard procedure.

7. Remove specimens from the freeze-fracture machine.

8. Thaw specimens following the procedure described below (7).[c]
   (a) Place replicated specimens, still frozen, in a plastic scintillation vial containing cooled methanol (−80°C).
   (b) Allow the methanol to thaw so that the specimens are simultaneously freeze-substituted.
   (c) Rinse thawed specimens once in water.

9. Digest materials with cleaning solution following standard procedure.

10. Mount replicas on grids with a supporting film of formvar or collodion (15).[d]

11. Observe replicas by TEM.
   (a) The three-dimensional structure is enhanced by photographing replicas with TEM at a tilt of ±10 degrees and by observing the micrographs as a stereo-pair.
   (b) Perception of the image in rotary-shadowed specimens is often enhanced by photographic reversal.

[a] In some cases, however, pre-treatments such as mild cellular rupture, chemical fixation, or cryoprotectant treatment with ethanol (a cryoprotectant which will sublime), are also employed. (See Section 5.1.2.)
[b] All this rapid freezing equipment is commercially available.
[c] To inhibit disintegration of replicas by distortion during thawing.
[d] The use of a grid with supporting film is desirable, due to fragmentation of replicas into small pieces during cleaning. When such a grid is used, an acetone rinse of the grid should be avoided.

3.2 Membrane surface observation

The true external surface of the plasma membrane and isolated membranes of organelles can be revealed by deep-etching (Section 3.1). With the standard deep-etching procedure, however, the presence of eutectic substances around the membranes reduces the surface regions which can be observed. For observing extensive surfaces of membranes by freeze-fracture at high resolution, the 'fracture-flip replication' technique is suggested (16). In this (17, 18) and other related methods (19, 20), the extensive surface image is obtained from freeze-fractured membrane halves which are retained on an evaporated carbon film (*Figure 1A-C*). The membrane surfaces which can be observed by this method vary, depending upon differences in pre-treatment of materials.

Protocol 6. Procedure for observing membrane surface

1. (a) For observing the outer surface of plasma membrane in single cells (protoplasts), vesicle suspensions, and isolated membranes (*Figure 1A*), fix materials by chemical fixatives, treat with cryoprotectant (glycerol), and freeze by plunging, following the standard procedure.

Protocol 6. *Continued*

(b) For observing outer (PS) or inner surfaces (ES) of endomembranes within single cells (*Figure 1B*), freeze single cells as rapidly as possible. Do not fix materials by chemical fixatives, because chemical fixation may cross-link cytoplasmic substances to the endomembrane surfaces, after which removal of such substances from the membrane by subsequent washing becomes difficult. In unfixed, isolated cells, the inner protoplasmic surface (PS) of the plasma membranes can also be

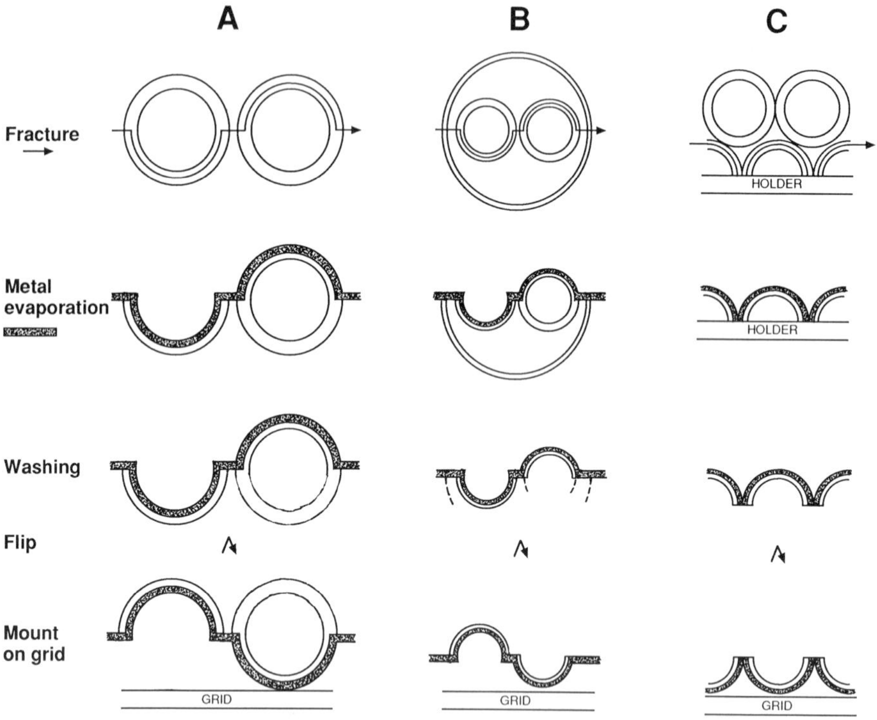

Figure 1. Schematic drawing for retaining freeze-fractured membrane halves on evaporated metal film. Materials are freeze-fractured (in some cases, membrane label is done before freezing), evaporated by metal (either Pt–C and carbon or by carbon alone), washed (not digested) by distilled water to remove bulk materials leaving only membrane halves, flipped and mounted on grid. The membrane halves retained on evaporated metal film are, in some cases, further processed either by shadowing with Pt–C or by labelling. **A**: Chemically fixed isolated cells. Due to chemical fixation, removal of cytoplasmic material is difficult by washing (right-hand cell). Consequently, only outer convex surface of membrane can be observed (left-hand cell). **B**: Unfixed, rapidly frozen isolated cell. The part of cells, except for membrane halves which are stabilized by metal film, can be removed by washing. Consequently, both inner and outer surfaces of endomembranes can be observed. **C**: Unfixed, rapidly frozen tissues adhering on holder. By fracturing through outermost cellular layer in tissue adhering on holder, the inner surface of plasma membrane can be observed.

revealed in a similar manner to endomembranes (for example, think of the right-hand endomembrane in *Figure 1B* as a plasma membrane of a single cell). However, for observing the concave surface (PS of plasma membranes and ES of endomembranes), a special washing schedule, such as changing pH or ionic strength of washing solution, will be necessary to remove cytoplasmic materials associating with membrane surface.

(c) For observing inner surface (PS) of plasma membranes in bulk tissues (*Figure 3C*), pre-treat material (without chemical fixation) in order to fracture and remove the bulk of the specimens, leaving the half-membranes attached to the holder, following the methods described below:

i. Mount a piece of fresh tissue with one face, on which the cells have previously been gently ruptured, pressed against a flat specimen holder (19).

ii. Mount fresh tissue between a pair of specimen-holders with a one-sided application of polyvinyl alcohol. If the down-facing side of the material is mounted with polyvinyl alcohol, the fracture frequently follows the superficial membranes on its top surface (21).

2. Transfer and mount frozen specimens on cold stage of a freeze-fracture machine.

3. Fracture specimens at −120°C to −130°C by micro-cut (if available).[a] Because etching tends to bring about ultrastructural change in the freeze-fractured half-membranes, fracturing only should be done.

4. Evaporate carbon on the fracture plane, following the standard procedure.

5. Remove specimens from the freeze-fracture machine.

6. Thaw specimens following the standard procedure. The replicas must be separated from the bulk specimens by floating on to a water surface.

7. Wash replicas (not digestion) with distilled water using at least six successive changes (30 min per wash). Decide the optimal condition for washing by an empirical approach for each particular specimens of biological material. It is most important to completely remove specimens from the replicas, leaving the half-membranes stabilized by evaporated carbon.

8. Mount the floating replicas on grids.

(a) Bring the grid down on top of the floating replica. The fluid surface should not be broken.

(b) Lift off grid horizontally.

(c) Turn the grid over so that the plasma membrane surfaces stabilized by carbon are on top (*Figure 1A–C*).

Protocol 6. *Continued*

(d) Dry replica by touching the edge of small piece of filter paper to the grid (if freeze-drying is necessary for a fragile membrane, see Section 3.3).

9. Immediately after drying, return the grid to the freeze-fracture machine or other vacuum coating unit for Pt–C-shadowing of the membrane surface, which is retained on the replica.
   (a) One way of mounting the grid on the (cold) stage is to apply two-sided adhesive tape, attached on one side to the stage, to a small part of grid's edge.
   (b) It is not essentially necessary to keep replicas at low temperature during shadowing. However, high resolution is obtained by shadowing on to a cooler surface (22).
10. Shadow (unidirectionary or rotary) replicas by evaporation of Pt–C, following the standard procedure.
11. Remove replicas from the machine.
12. Observe replicas by TEM, following the standard procedure. For one example, see *Figure 2*.

[a] In order to make replica floats during washing.

3.3 Composite replica

This method is an application of the method in Section 3.2, in which the freeze-fractured membrane halves retained on the replica are used. The method, known as the 'composite replica', allows observation of membrane ultrastructure from both fracture face and true surface on the same replica (23, 24). In the application of this technique, the intramembrane particles on the fractured membrane face and glycocalyx on the membrane outer surface can be seen superimposed within the replica. The membranes which can be observed by this method vary, depending upon differences in pre-treatment, but are essentially the same as those described in Section 3.2, as shown in *Figure 1A–C*.

Protocol 7. Procedure for preparation of composite replicas

1. Freeze materials by appropriate methods, depending upon pre-treatment, following Section 3.2.
2. Transfer and mount frozen specimens on the cold stage of a freeze-fracture machine.
3. Fracture specimens (without etching) at −120°C to −130°C.

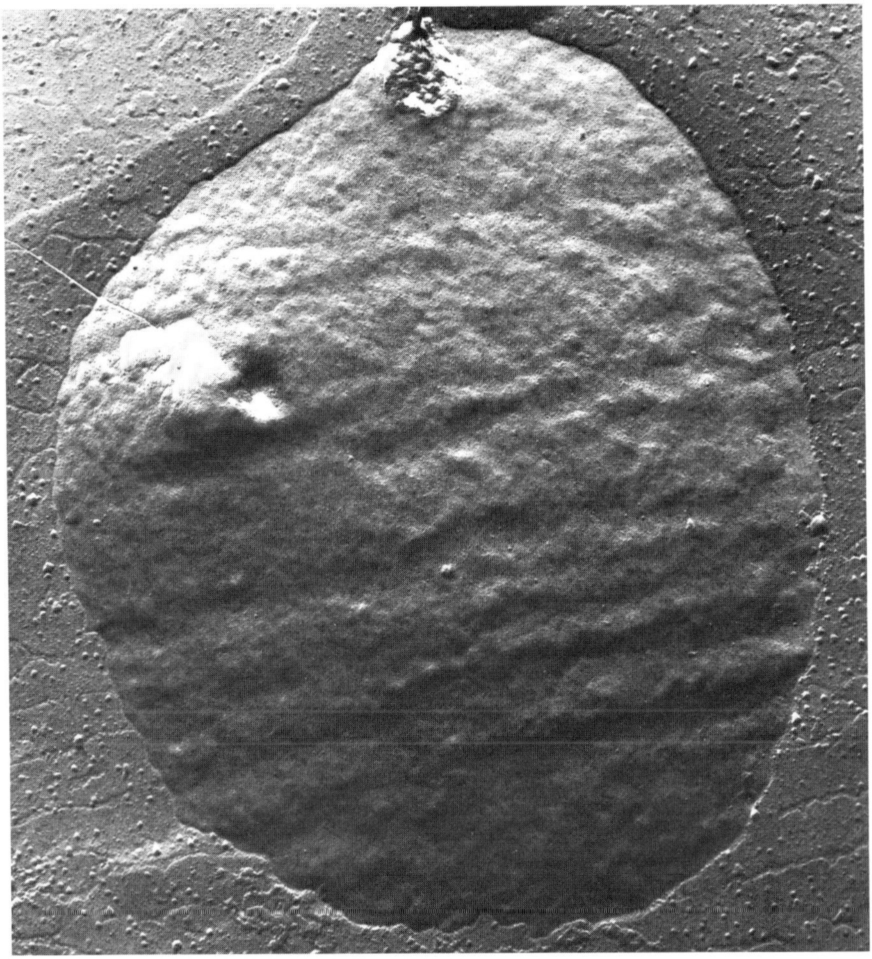

Figure 2. Protoplasmic surface (PS) of chloroplast envelope within a protoplast isolated from rye leaves, prepared by *Protocol 6*. × 21 500. (From S. Fujikawa and P. L. Steponkus, unpublished data.)

4. Uni-directionally shadow the fracture plane with Pt–C. Reduce the amount of Pt–C shadowing to about one-half the usual thickness, because in the composite replica method another replica will be superimposed over the original one.
5. Soon after shadowing, evaporate carbon on to the fracture plane for reinforcement of replicas, following the standard procedure.
6. Remove specimens from the freeze-fracture machine.
7. Thaw and wash (not digestion) replicas with distilled water, leaving only

Protocol 7. *Continued*

the half-membrane leaflet stabilized by the replica, following the procedure of Section 3.2.

8. (a) Mount replica on the grid, following the procedure of Section 3–2.
   (b) Alternatively, sandwich the replica between two tabbed 200-mesh nickel grids to prevent deformation of replica in the subsequent freezing process.
9. Soon after washing, freeze the replicas on grids by plunging into nitrogen slush.
10. Transfer and mount frozen replicas on to the cold stage in a freeze-fracture machine kept at −150°C.
11. Freeze-dry replicas in the freeze-fracture machine.
    (a) Raise the stage temperature gradually to −95°C.
    (b) Keep specimens at this temperature for 45 min to accomplish the freeze-drying.
12. Rotary-shadow replicas by evaporation of Pt–C (for distinction of the fracture face, on which unidirectional shadowing is to be performed).
13. Warm replicas to room temp. inside the freeze-fracture machine, to prevent structural changes through reabsorption of water.
14. Remove replicas from the freeze-fracture machine.
15. Observe replicas by TEM, following the standard procedure. Observation by TEM should be carried out as soon as possible, because of the presence of biological materials between the composite replicas.

4. Combined freeze-fracture and cryo-scanning electron microscope (SEM) techniques

The structures observed by freeze-fracture techniques are generally confined to very small areas due to fragmentation of replicas into small pieces during cleaning. This characteristic of the freeze-fracture technique makes it difficult to know the original position of the cell exactly, observed in the freeze-replica, within the actual specimen. While several methods have been proposed for obtaining freeze-replicas without disintegration by using a backing layer on the replicas before digestion (25), a new technique has been proposed that uses a special freeze-fracture machine to which a SEM with a cold stage is directly connected (26–28). By using this machine, continuous observation from total fracture plane in frozen specimens (by cryo-SEM) to the part of interest with largest resolution (by freeze-replica) is possible. All types of materials, including single cells, cell components, and bulk tissues, can be observed as starting materials.

Protocol 8. Procedure for combined freeze-fracture and cryo-SEM technique

1. Pre-treat and freeze materials.
   (a) With materials that are to be processed by chemical fixation, cryoprotectant treatment, and plunge-freezing, follow the standard procedure.
   (b) With fresh materials that need rapid-freezing, follow the procedure of Section 3.1.
2. Transfer and mount specimens on the cold stage of a freeze-fracture machine.
3. Fracture specimens at appropriate temperature, following the standard procedure. Before shadowing, however, fracture planes can be pre-observed by SEM.
   (a) Soon after fracturing, transfer specimens, using a transfer rod, to a SEM cold stage kept at −165 °C.
   (b) Cover specimens with cold shroud kept at −180 °C.
   (c) Observe specimens by secondary emission electron mode of the SEM at low voltage (2–3 kV).
   (d) Unless the fracture plane of interest is exposed, repeat fracturing by transferring specimens again to the cold stage of the freeze-fracture machine until the fracture plane of interest is exposed.[a]
4. Rotary-shadow specimens by evaporation of Pt–C on to the cold stage of the freeze-fracture machine. Rotary-shadowing is better for the production of a good cryo-SEM image.
5. Evaporate carbon on to the fracture plane for reinforcement.
6. Transfer evaporated specimens to the SEM cold stage, kept at −165 °C.
7. Observe and photograph total fracture plane or area of interest by secondary emission in SEM (29).
8. If special areas of interest are found, from which collection of the replica is planned, the regions surrounding the areas of interest can be cut lightly using a thin needle attached to a micromanipulator, while observing the specimens with the SEM (27).
9. Remove specimens from the freeze-fracture machine.
10. Thaw and digest materials,
   (a) For collection of replicas from specific areas of interest, which were previously cut by the micromanipulator needle, place the specimen on a Petri dish and gradually add the cleaning solution to the same level as the specimen height.
   (b) After partial digestion of the specimens, collect the replicas, includ-

Protocol 8. *Continued*

ing the specific areas together with the partially digested specimens, by pushing a platinum loop (with a diameter slightly larger than the cut area) horizontally on replicas to include desired part inside the loop, under observation by binocular microscope.

(c) After complete cleaning of replicas, mount them on single hole grids with supporting films of formval or collodion.

11. After printing of electron micrographs from the SEM, look for the corresponding areas on replicas under observation with TEM and photograph (see *Figure 3*).

[a] Observation of the fracture plane by SEM before shadowing is possible only in specimens without glycerol treatment. In glycerinated specimens, direct irradiation of electron beam on the fracture plane causes slight melting of glycerol, which obscures the ultrastructure of the fracture plane. After shadowing, however, no detectable structural changes occur in glycerinated specimens following observation with SEM for long periods (more than 3 h).

5. Cytochemistry

One of the cytochemical procedures initially proposed for freeze-fracturing is a technique to label single-cell surfaces with ferritin-conjugated ligands (30, 31). When a suspension of cells is deeply etched after fracturing, a narrow margin of the true surface of the membrane can be exposed adjacent to the fracture faces. Because label is attached to the membrane surface, this area is revealed. However, because of the presence of extracellular eutectic substances, the existence of ligands is often obscured. Other cytochemical techniques applicable for freeze-fracture have been proposed recently. The main approach to labelling materials for freeze-fracture is divided primarily between two procedures utilizing labelling before fracture (Section 5.1) and labelling after fracture (Section 5.2). Furthermore, depending upon the technique, the structures that can be labelled vary. In one case, labelling only occurs on the outer surface of membranes of isolated cells, but in the other case labelling occurs for almost all cellular structures exposed by the fracture.

5.1 Labelling before fracture

5.1.1 Label technique for membrane

The method of labelling an extensive membrane surface before freeze-fracture, to show its relation with structure on the membrane fracture face, was first proposed by Pinto da Silva and Kan (32) and is now widely used as an effective cytochemical technique for freeze-fracture (33–35). The principle of this technique is similar to the method set out in Sections 3.2 and 3.3, in which fractured half-membrane leaflets that would normally be digested away during the replica-cleaning process are retained on the replica.[a] In this technique, the label with high electron density attached to the membrane surface, stabil-

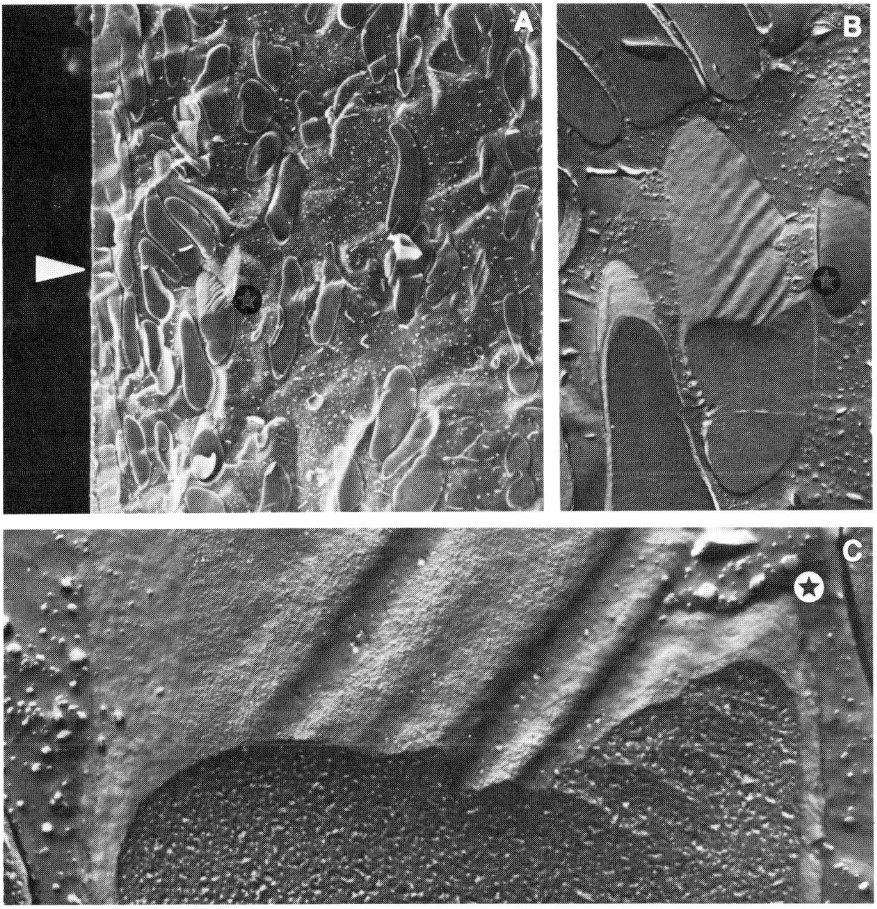

Figure 3. Human red blood cell suspension frozen by impact-freezing using liquid helium, prepared by *Protocol 8*. **A**: Cryo-SEM photograph showing extensive fracture plane perpendicular to metal contact surface (a white arrowhead). × 1100. **B** and **C**: Freeze replica photograph showing a part of cryo-SEM photograph (**A**). **B**: × 4500. **C**: × 30 000. A star in each photograph is placed on the same area.

ized by the replica, can be seen superimposed upon the structure of fracture face.[b] In the method described below, only the outer surface of the plasma membrane of isolated cells and outer surface of isolated membrane vesicles can be labelled (*Figure 1A*).

[a] The label technique using freeze-fractured membrane halves which are retained on replica was historically reported earlier than the techniques in Sections 3.2 and 3.3.

[b] With slight modification of this technique, i.e. by coating the fracture plane with carbon alone followed by washing the replica and then shadowing the replica with Pt–C, (i.e. the same procedure as with the fracture-flip technique), the label that is applied before freezing can be seen superimposed upon the structure of the outer surface (36).

Protocol 9. Procedure for preparing replica showing labelled membrane surface before fracturing

1. Incubate glutaraldehyde fixed cells with labelling solution. (See *Protocol 10*.)
2. Follow completely the same process as described in *Protocol 1*, from cryo-protection to removal of the specimens from the freeze-fracture machine.[a]
3. Follow completely the same process as in *Protocol 6*, from thawing of the specimens to drying the replicas.
4. Observe replicas with TEM as soon as possible, following standard procedure.

[a] Etching after fracturing should be omitted.

Protocol 10. Procedure for labelling wheat germ agglutinin (WGA) or Con A binding sites in outer surface of the plasma membrane of human leukocytes by colloidal gold (CG) (32)

1. Incubate 0.1 ml packed cells [fixed in 1.5% glutaraldehyde in phosphate buffered saline (PBS), pH 7.4, for 1 h at 4°C and wash twice in PBS] in 1 ml PBS containing 0.25 mg WGA or 0.25 mg Con A for 1 h at 37°C.
2. For a negative control of WGA binding, pre-incubate 0.1 ml packed cells in a 1 ml PBS solution with 0.4 M *N*-acetyl-D-glucosamine for 15 min at 37°C followed by incubation in a WGA solution as above, but in the presence of 0.4 M *N*-acetyl-D-glucosamine. For negative control of Con A binding, pre-incubate cells in 0.5 M methyl-α-D-mannopyranoside for 15 min followed by incubation in a Con A solution as above in the presence of 0.5 M methyl-α-D-mannopyranoside.
3. Wash cells twice in ice-cold PBS.
4. Incubate cells in ovomucoid-coated CG (37) for 3 h at room temp. or incubate in a suspension of horseradish-peroxidase-coated CG.
5. Wash cells with PBS to remove unbound label.

5.1.2 Label technique for membranes and other intracellular components

The method for labelling cellular interior structure, especially filamentous structures, followed by rapid-freezing, deep-etching and rotary-shadowing, was originally outlined by Heuser *et al.* (13) and is now used widely for biological studies (38, 39). In this labelling technique, the gold particles can be retained partially embedded within the replica after digestion of specimen.

Particles with high density stand out clearly against the replicated structures (40). All types of materials, including single cell, cellular components, and bulk tissues, can be processed as starting materials.

Protocol 11. Procedure for preparing replicas containing labelled intracellular structures

1. (a) Mildly disrupt biological material[a] by osmotic shock, detergents, or sonication in order to allow the label access to the cells' interiors, following *Protocol 12* or *13*.

(b) After mild disruption, fix material using chemical fixatives and label following *Protocol 12* or *13*.

2. Follow completely the same process as in *Protocol 5*, from rapid-freezing to observation of specimens by TEM.

[a] These mild disruptive techniques also remove some cellular components.

Protocol 12. Procedure for the disruption, fixation, and antibody (IgG) labelling of neuronal cytoskeleton proteins in axons of mammalian spinal cord neurons (41)

1. Agitate small tissue pieces [(2–3 mm long, 0.5 mm thickness) with 1% Triton X-100 in K^+-buffer [70 mM KCl, 5 mM $MgCl_2$, 3 mM EGTA, 0.25 mM phenyl methylsulphonyl fluoride (PMSF), 30 mM Hepes buffer, pH 7.4] for 3 h at 4°C.

2. Wash twice in K^+-buffer and fix with 1% paraformaldehyde in K^+-buffer for 1 h at 4°C.

3. Trim samples into smaller pieces and incubate overnight at 4°C in K^+-buffer containing bovine serum albumin (BSA) (1 mg/ml) and adsorbed affinity-purified primary antibodies (goat) to the cytoskeletal proteins.

4. Wash in K^+-buffer and incubate overnight with rabbit anti-goat IgG (150 μg/ml), or monovalent Fab fragments (100 μg/ml).

5. Wash in K^+-buffer and fix with 2% paraformaldehyde for 3 h at 4°C.

6. Rinse with distilled water followed by extensive soaking with 15% ethanol.

Protocol 13. Procedure for disruption and labelling (in this case without chemical fixation) of the acetylcholine receptor on the plasma membrane inner surface (PS) in purified *Torpeda* post-synaptic membranes (42).

1. Immobilize membranes by loading membrane suspension on glass discs

Protocol 13. *Continued*

coated with 0.3% polyornithine (dissolved in distilled water) for 1–10 min at RT.

2. To expose the cytoplasmic surfaces of membranes, rinse once with distilled water and place the glass discs in the bottom of a beaker containing distilled water.
3. Sonicate (23 KHz) for 4–5 sec at a power output of 5 W using a probe sonicator with a ⅛-inch tip.
4. Label membranes with monoclonal antibodies (ascites diluted with PBS to the appropriate concentration) to the acetylcholine receptor for 1–2 h at RT.
5. Rinse several times with PBS.
6. Apply protein A–colloidal gold for 30 min at 1:20 dilution at RT.
7. Wash with distilled water.

5.2 Labelling after fracturing

5.2.1 The label technique for membranes

This cytochemical technique was introduced by various laboratories under different names: namely, freeze-fracture replica immunogold labelling (19), replica-staining label fracture (43), and replica-label-whole-mount methods (20). However, all these methods employ essentially the same technique, in that half-membrane leaflets retained on the replica are labelled with immunogold after the replica has been made. In this technique, label bound to the outer surface of fractured membrane halves is viewed superimposed upon the replica showing ultrastructure of fracture faces.[a] Because membrane fragments adhering to the replica are directly exposed to immunogold reagents, a high level of specific labelling can occur by using small amount of incubation solution. As the labelling step is done after fracturing, however, the availability of surface binding sites will depend on how the integral proteins bearing these sites have partitioned. The membranes which can be observed by this method are similar to those of Section 3.2 as shown in *Figure 1A–C*.

[a] With slight modification of this technique (for similar procedure with fracture-flip, see Section 3.2), i.e. by coating the fracture plane with carbon alone followed by washing the replica (not digestion) and then shadowing of replica with Pt–C, the label which is applied after washing can be seen superimposed on the structure of the outer surface.

Protocol 14. Procedure for preparing replicas showing labelled membrane surface after fracture

1. Pre-treat materials in different ways depending upon the membranes which are to be observed, following Section 3.2.

2. Freeze materials, following the standard procedure (using cryoprotected sample) or Section 3.1 (unprotected sample).
3. Follow completely the same processes as in *Protocol 7*, from mounting specimens on the cold stage of the freeze-fracture machine to washing of replicas.
4. Mount replicas on grids as in *Protocol 6*.[a]
5. Incubate replicas (half-membrane leaflets retained on replica) on a small volume of cytochemical reagents. During incubation the replica must float on the solution. For incubation, see *Protocol 15*.
6. Observe replicas by TEM, following the standard procedure.

[a] Mounting replicas on grids is also possible after labelling.

Protocol 15. Procedure for labelling the outer surface of the plasma membrane of human erythrocytes by protein-A-coated colloidal gold (PA-CG) or immunoglobulin-coated colloidal gold (Ig-CG) (43)

1. Incubate replicas with rabbit anti-human erythrocyte membrane antibody, 1:50 or 1:100 in PBS containing 1% BSA for 30 min at 37°C. For reducing non-specific labelling by CG on the replica surface (without retained membrane), the incubation solution should also contain high salt (500 mM NaCl) and a wetting agent (0.1% Tween-20).
2. Wash replicas three times in PBS.
3. Incubate replicas with PA-CG or with goat anti-rabbit Ig-CG for 60 min at room temp.
4. Wash replicas three times for at least 30 min in PBS.
5. Dry replicas on the grid.

5.2.2 Labelling technique for membranes and other intracellular components

This labelling technique, proposed by Pinto da Silva and his co-workers (44–47), uses a quite different approach to the standard freeze-fracture technique. The freeze-fracture plane with attached label is observed either as a thin section or as a replica made from a critical point dried specimen. This method and a few other methods (20) are the only techniques in which the freeze-fracture plane is directly labelled. However, it must be borne in mind that due to drying of the fracture plane, severe structural deformation may be produced by freeze-fracture from the normal membrane organization (21). This technique has been the most extensively applied of all freeze-fracture cytochemical techniques (48, 49), and all types of materials, including single cells, cell components, and bulk tissue, can be used as starting materials.

Protocol 16. Procedure for preparing specimens showing labelled freeze-fractured plane

1. Fix materials with glutaraldehyde, following the standard procedure. For cell suspension, gently spin down in 15–30% BSA, which is then cross-linked with glutaraldehyde to form a gel.
2. Cryoprotect specimens by glycerol solution, following the standard procedure.
3. Freeze specimens by plunging, following the standard procedure.[a]
4. Fracture samples under LN_2.[b] How this fracturing under LN_2 is best carried out depends on the type of material and the route by which it is to be examined (i.e. by thin sectioning or replicas).
   (a) For the route to thin sectioning, crush repeatedly by forcefully prodding specimens with a LN_2 cooled metal rod in a small container including LN_2, or by putting in a very loose fitting Potter–Elvehjem type homogenizer.
   (b) For the replica route, in which a large, relatively flat fracture face obtained from an easily recognized fracture plane of specific orientation is required, fracture specimens individually with a fine scalpel by holding in forceps.
5. Thaw fractured specimens in the presence of glutaraldehyde and glycerol in buffer solution, and then deglycerinate by washing with buffer containing 1 mM glycylglycine (45).
6. Follow with the cytochemical labelling sequence of choice. (See *Protocol 17*.)
7. From this point in the procedure, there are two alternatve routes before specimen examination.
   (a) Post-fixation with OsO_4, dehydration, embedding, sectioning, and staining, for making thin sections (6).
   (b) Post-fixation with OsO_4, critical point drying (50), formation of replicas by Pt–C shadowing, reinforcement by carbon evaporation, and digestion of specimen, for making replicas.
8. Observe specimens by TEM following the standard procedure.

[a] It is not necessary to mount materials in specimen-holder.
[b] It is not necessary to fracture frozen specimens in the vacuum chamber of a freeze-fracture machine.

Protocol 17. Incubation conditions for labelling pancreatic and testicular tissues by lectin-gold (51)

1. Incubate tissue samples for 1 h at room temp. with either *Ricinus communis* I-colloidal gold complex (RCA I-CG) or *Helix pomatia* lectin-colloidal gold complex (HPL-CG) (52), which have specificity for the sugar residues D-galactose and *N*-acetyl-D-galactosamine, respectively.
2. For negative control treatments, incubate samples for 30 min at RT in a solution of 0.2 M D-galactose (for RCA I-CG) or 0.2 M *N*-acetyl-D-galactosamine (for HPL-CG). They are then incubated in RCA I-CG, which was previously mixed with the corresponding sugar for 1 h before use, in the presence of 0.2 M D-galactose or in HPL-CG, which was also previously mixed with corresponding sugar for 1 h before use, in the presence of 0.2 M *N*-acetyl-D-galactosamine.
3. Wash specimens in PBS, and in order to prevent displacement of the label, fix in 1% glutaraldehyde in PBS.

6. Autoradiography

Methods for producing autoradiography from freeze-fractured specimens were proposed by Fisher and Branton (53), Schiller *et al.* (54) and Carpentier *et al.* (55). Among these, the methods proposed by Carpentier *et al.* (55) using fracture membrane-halves retained on replicas (*Figure 1A–C*) has the highest resolution when relating label and the structure of freeze-fractured membranes. In this method, autoradiographic label from the half-membrane leaflet retained on the replica can be seen superimposed on ultrastructure of the freeze-fractured face of the membranes. Cells and cellular membranes, in which a radioactive source can be incorporated before freezing, are used as starting materials.

Protocol 18. Procedure for autoradiography of freeze-fractured membranes

1. Incubate cells in buffer with a radioactive source. (See *Protocols 19* and *20*.)
2. Follow completely the same process, from freezing to washing of replicas, as in *Protocol 7*.
3. Mount replicas on Parlodion-coated grids in such a way that the attached cell fragments remaining are sandwiched between the replicas and the copper grids.

Protocol 18. *Continued*

4. Process grids with replicas for autoradiography by the method of Caro *et al.* (56). (See *Protocol 2* and Chapter 5.)
5. Observe replicas by TEM, following the standard procedure.

Protocol 19. Incubation condition for radio-labelling HL-60 myelomonocytic leukemia cells (55)

1. Wash cells with balanced salt solution binding buffer (pH 7.4) supplemented with 1% BSA.
2. Incubate cells with 30 μg/ml ^{125}I-transferrin for 30 min at 4°C in the same buffer.
3. Wash cells three times with ice-cold buffer for removing unbound transferrin.
4. Fix with 2.5% glutaraldehyde in 0.1 M cacodylate buffer, pH 7.4.
5. Impregnate in 30% glycerol for 2 h at 4°C.

Protocol 20. Incubation condition for radio-labelling 3T3 L1 fibroblasts adhering to plastic coverslip (55)

1. Incubate in 1 ml of Krebs–Ringer bicarbonate buffer with ~20 ng/ml ^{125}I-α_2-macroglobulin (which will preferentially associate with coated pits on the plasma membrane) or with ^{125}I-cholera toxin (which will associate with the non-coated pits area of the plasma membrane) for 2 h at 4°C.
2. Wash cells three times with same buffer at 4°C.
3. Fix with 2.5% glutaraldehyde in 0.1 M cacodylate buffer, pH 7.4.
4. Impregnate in 30% glycerol solution for 2 h at 4°C.

Protocol 21. Autoradiographic development of replicas (56)

1. Mount the grid with replica on a clean glass slide.
2. Cover with a layer of emulsion.
3. Dry overnight at RT and then store at 4°C for 1–2 weeks.
4. Photographically develop the grid, rinse in water, fix, and wash four times in distilled water.
5. Remove grid from the glass slide and dry for 24 h.

References

1. Storinski, C. and Breathnach, A. S. (ed.) (1975). *Freeze-fracture replication of biological tissues: Techniques, interpretation and applications*. Academic Press, London.
2. Moor, H. (1964). *Zeitschrift für Zellforschung*, **63,** 546.
3. Branton, D. (1966). *Proceedings of the National Academy of Sciences, USA*, **55,** 1048.
4. Branton, D., Bullivant, S., Gilula, N. B., Karnovsky, M. J., Moor, H., Muhlethaler, K., Northcote, D. H., Packer, L., Satir, B., Satir, P., Speth, V., Staehelin, L. A., Steere, R. L., and Weinstein, R. S. (1975). *Science,* **190,** 54.
5. Fujikawa, S. (1988). *Electron Microscopy Reviews*, **1,** 113.
6. Glauert, A. M. (ed.) (1975). *Practical methods in electron microscopy: Fixation, dehydration and embedding of biological specimens*, Vol. 3. North-Holland, Amsterdam.
7. Robards, A. W. and Sleytr, U. B. (ed.) (1985). *Practical methods in electron microscopy: Low temperature methods in biological electron microscopy*, Vol. 10. Elsevier, Amsterdam.
8. Elder, H. Y., Gray, C. C., Jardine, A. G., Chapman, J. N., and Biddlecombe, W. H. (1982). *Journal of Microscopy,* **126,** 45.
9. Umrath, W. (1983). *Mikroskopie,* **33,** 11.
10. Ting-Beall, H. P., Burgess, F. M., and Robertson, J. D. (1986). *Journal of Microscopy,* **142,** 311.
11. Muller, M., Meister, N., and Moor, H. (1980). *Mikroskopie,* **36,** 129.
12. Bachmann, L. and Schmitt, W. W. (1971). *Proceedings of the National Academy of Sciences, USA,* **68,** 2149.
13. Heuser, J. E., Reese, T. E., Dennis, M. J., Jan, Y., Jan, L., and Evans, L. (1979). *Journal of Cell Biology,* **81,** 275.
14. Moor, H., Bellin, G., Sandri, C., and Akert, K. (1980). *Cell and Tissue Research,* **209,** 201.
15. Coggins, L. W. (1987). In *Electron microscopy in molecular biology: A practical approach* (ed. J. Sommerville and U. Scheer), pp. 3. IRL Press at Oxford University Press, Oxford.
16. Forsman, C. A. and Pinto da Silva, P. (1988). *Journal of Cell Science,* **90,** 531.
17. Fujimoto, K. and Pinto da Silva, P. (1988). *Journal of Cell Science,* **91,** 161.
18. Pimenta, P. F. P., da Silva, R. P., Sacks, D. L., and Pinto da Silva, P. (1989). *European Journal of Cell Biology,* **48,** 180.
19. Gruijters, W. T. M., Kistler, J., Bullivant, S., and Goodenough, D. A. (1987). *Journal of Cell Biology,* **104,** 565.
20. Rash, J. E., Johnson, T. J. A., Dinchuk, J. E., Duch, D. S., and Revinson, S. R. (1989). In *Freeze-fracture studies of membrane structure* (ed. S. K. Hui). CRC Press, Boca Raton, Florida. (In press.)
21. Severs, N. J. (1989). *Journal of Electron Microscopy Technique,* **13,** 175.
22. Abermann, R., Salpeter, M. M., and Bachmann, L. (1972). In *Principle and techniques of electron microscopy* (ed. M. A. Hayat), Vol. 2, pp. 197–217. Van Nostrand Reinhold, New York.
23. Wade, J. B., and Coleman, R. A. (1989). *Proceedings of the National Academy of Sciences, USA,* **86,** 2723.

24. Coleman, R. A. and Wade, J. B. (1989). *Journal of Electron Microscopy Technique,* **13,** 216.
25. Stolinski, C., Gabriel, G., and Martin, B. (1983). *Journal of Microscopy,* **132,** 149.
26. Fujikawa, S., Suzuki, T., Ishikawa, T., Sakurai, S., and Hasegawa, Y. (1988). *Journal of Electron Microscopy,* **37,** 315.
27. Fujikawa, S., Suzuki, T., and Sakurai, S. (1990). *Scanning,* **12,** 99.
28. Fujikawa, S. (1990). *Journal of Electron Microscopy,* **39,** 80.
29. Gabriel, B. L. (ed.) (1982). *Biological scanning electron microscopy.* Van Nostrand Reinhold, New York.
30. Pinto da Silva, P., Douglas, S. D., and Branton, D. (1971). *Nature,* **232,** 194.
31. Shotton, D., Thompson, K., Wofsy, L., and Branton, D. (1978). *Journal of Cell Biology,* **76,** 5/2.
32. Pinto da Silva, P. and Kan, F. W. K. (1984). *Journal of Cell Biology,* **99,** 1156.
33. Brown, D. and Orci, L. (1986). *Journal of Histochemistry and Cytochemistry,* **34,** 1057.
34. Boonstra, J., Van Belzen, N., Van Maurik, P., Hage, W. J., Blok, F. J., Wiegart, F. A. C., and Verkleij, A. J. (1985). *Journal of Microscopy,* **140,** 119.
35. Chevalier, J. and Pinto da Silva, P. (1986). *Biologie Cellulaire,* **55,** 181.
36. Pavan, A., Mancini, P., Lucania, G., and Frati, L. (1990). *Journal of Cell Science,* **96,** 151.
37. Geoghegan, W. D. and Ackerman, G. A. (1977). *Journal of Histochemistry and Cytochemistry,* **25,** 1187.
38. Harding, C., Heuser, J., and Stahl, P. (1983). *Journal of Cell Biology,* **97,** 329.
39. Nakata, T. and Hirokawa, N. (1987). *Journal of Cell Biology,* **105,** 1771.
40. Mannweiller, K., Hohenberg, H., Bohn, W., and Rutter, G. (1982). *Journal of Microscopy,* **126,** 145.
41. Hirokawa, N., Glicksman, M. A., and Willard, M. B. (1984). *Journal of Cell Biology,* **98,** 1523.
42. Bridgman, P. C., Carr, C., Pedersen, S. E., and Cohen, J. B. (1987). *Journal of Cell Biology,* **105,** 1829.
43. Forsman, C. A. and Pinto da Silva, P. (1988). *Journal of Histochemistry and Cytochemistry,* **36,** 1413.
44. Pinto da Silva, P., Kachar, S., Torrisi, M. R., Brown, C., and Parkison, C. (1981). *Science,* **213,** 230.
45. Pinto da Silva, P., Parkison, C., and Dwyer, N. (1981). *Proceedings of the National Academy of Sciences, USA,* **78,** 343.
46. Pinto da Silva, P., Parkison, P., and Dwyer, N. (1981). *Journal of Histochemistry and Cytochemistry,* **29,** 917.
47. Pinto da Silva, P., Torrisi, M. R., and Kachar, B. (1981). *Journal of Cell Biology,* **91,** 361.
48. Torrisi, M. R. and Pinto da Silva, P. (1984). *Journal of Cell Biology,* **98,** 29.
49. Ruggiero, F., Lethias, C., and Garrone, R. (1987). *European Journal of Cell Biology,* **44,** 90.
50. Hayat, M. A. and Zirkin, B. R. (1973). In *Principle and techniques of electron microscopy: Biological applications*, Vol. 3 (ed. M. A. Hayat), pp. 297–313. Van Nostrand Reinhold, New York and Toronto.
51. Kan, F. W. K. and Nanci, A. (1988). *Journal of Electron Microscopy Techniques,* **8,** 363.

52. Roth, J. (1983). *Journal of Histochemistry and Cytochemistry,* **31,** 987.
53. Fisher, K. A. and Branton, D. (1976). *Journal of Cell Biology,* **70,** 453.
54. Schiller, A., Rix, E., and Taugner, R. (1976). *Histochemistry,* **59,** 9.
55. Carpentier, J. E., Brown, D., Iacopetta, B., and Orci, L. (1985). *Journal of Cell Biology,* **101,** 887.
56. Caro, L. G., Van Tubberghen, R. P., and Roll, J. A. (1962). *Journal of Cell Biology*, **15,** 173.

9

Negative staining

ROBIN HARRIS and ROBERT HORNE

1. Introduction

The fundamental concept of negative staining is rather old and extremely simple, yet, as with many microscopical specimen preparation techniques, there are numerous subtleties, intricacies, and technical hazards which have to be carefully observed by the user. Gross visual examples of 'negative staining' are calcified fossils within dark-brown/black mud, stone and clear-glass particles of varying shape surrounded by dye or ink solution, the shape of which is made visible in a 'negative' sense. In descriptive terms for transmission electron microscopy (TEM), salts containing heavy metal atoms, which have the property of drying from aqueous solution as an amorphous 'glass' rather than a crystalline layer, can surround and permeate small biological particles. Experiments on the relative drying rates between commonly used negative stains (1) indicated that both protein and lipids retained water for longer periods than the surrounding negative stain during the dehydration process. When illuminated by the electron beam inside a TEM, very much less of the transmitted electron beam will pass directly through the stain-containing regions than through the regions containing biological molecules (containing atoms of much lower atomic weight). This selective scattering of the electron beam induced by the distribution of heavy metal atoms in the layer of dried negative stain surrounding biological particles is responsible for image production. Thus, the external outline and aqueous cavities within biological particles can be readily revealed in bright field transmission electron optical images (*Figure 1*).

In this chapter techniques will be presented for the negative staining of whole cells and detergent-extracted cytoskeletons, membrane fractions, organelles and subcomponents, liposomes, viruses, and macromolecules. The application of negative staining to protein molecules and filaments was dealt with by Speiss *et al.* (2) in an earlier volume of the *Practical Approach* series. Direct technical repetition will be avoided where possible, but some relevant extensions within this important area will be included, in particular the formation and negative staining of two-dimensional protein crystals.

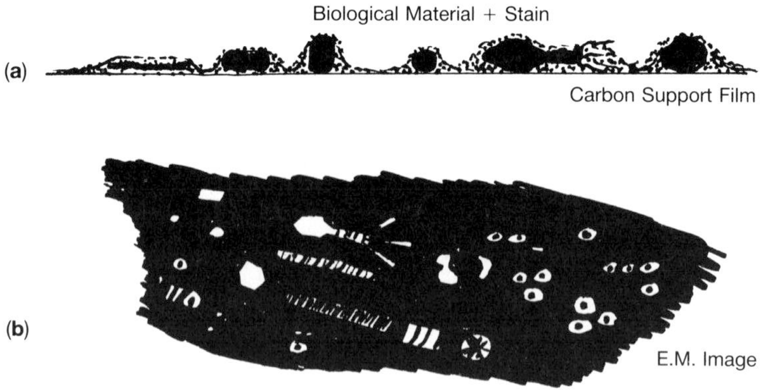

Figure 1. A diagrammatic explanation of negative staining.

1.1 Historical background

It is of interest to mention at this point that the idea of surrounding a biological specimen with stain for light microscopy was noted as early as 1884 and followed by another observation in 1892 (see historical account of negative staining by Horne and Wildy, ref. 3). In the case of electron microscopy it was reported by Hall (4) that certain 'anomolous images' seen in some of his preparations of positively stained virus particles possessed a reversed contrast and revealed the structure to advantage. Hall wrote: 'Although the effect shown in Fig. 8 (bushy stunt virus stained with phosphotungstic acid at pH 4.6 insufficiently washed) is the opposite of what is usually sought by the use of electron stains the visibility of particles of low scattering power can be enhanced as well, if not better, by surrounding them with dense material, rather than impregnating them with dense material.'

The second specific reference to this type of image was made by Huxley (5), using specimens of tobacco mosaic virus (TMV) rods, stained with 40% phosphotungstic acid. In these specimens there was an 'outlining' effect where the stain had not been completely removed by washing. Moreover, in some areas the electron-dense stain had penetrated along the axial central hole of the TMV rods. The development of the technique currently referred to as 'negative staining', initially used to reveal virus particles by high-resolution TEM, was first presented at the Fourth International Congress on Electron Microscopy by Horne and Brenner (6), and in greater detail by Brenner and Horne (7).

This was rapidly followed, in the early 1960s, by the widespread use of negative staining to reveal many different viral particles, cellular organelles, membrane fractions, and liposomes, as well as individual macromolecules and supramolecular structures (8–11). The topic has been thoroughly reviewed by Horne (12), Horne and Wildy (3), Oliver (13), Harris (ref. 14, and EMSA Teaching Video No. 126, 1989) and recently by Kiselev *et al.* (15).

2. Materials

2.1 Negative stains

Details will be provided for only the four principal negative staining salts listed in *Table 1*. Application of the lesser used negative stains (*Table 2*) will, in general, closely parallel those in *Table 1*.

Uranyl acetate is used as an unbuffered 1% to 2% aqueous solution (pH 4.2 to 4.5). Ammonium molybdate (2%) should be adjusted with ammonium or sodium hydroxide from its initial acidic pH to pH 7.0 (generally do not exceed pH 7, as crystallization may occur on drying). Likewise, 2% aqueous solutions of phosphotungstic acid and silicotungstic acid are adjusted to neutrality with sodium hydroxide (or KOH). Because of its inherent radioactivity, bottles of solid uranyl acetate and uranyl acetate solutions should be stored in lead containers. All stains should be kept at room temperature and the bottles should not be inverted or shaken before use, as particulate sediment will interfere with the evenness of staining. Filtration of negative staining solutions may be called for if a problem with particulates is encountered.

Table 1. The principal negative stains

Negative stain	Chemical formula	pH range for use
Uranyl acetate	$UO_2\ [CH_3COO]_2$	4.2–4.5
Sodium (K) phosphotungstate	$Na_3PO_412WO_3$	5–8
Sodium silicotungstate	$Na_3SiO_212WO_3$	5–8
Ammonium molybdate	$[NH_4]_6Mo_7O_{26}$	5–7

It should be mentioned that precautions must be taken when handling pathogenic specimen material in the presence of negative stains, particularly where the pH of the stains has been adjusted to neutrality to prevent or reduce the positive interaction between the heavy metal salts and the biological material. It has been shown that certain viruses and organisms can remain infective when mixed with negative staining solutions and also following drying at room temperature. UV irradiation of specimen grids should therefore be performed, when appropriate.

Ammonium molybdate is particularly valuable as a negative stain for osmotically sensitive organelles, since the molarity of the 2% solution is approximately equivalent to isotonic (16, 17), it does, however, impart a somewhat lower electron density than the other stains. Sodium and potassium phosphotungstate have been widely used for negatively staining viral particles, protein molecules, and liposomes, but it should be borne in mind that this stain does have a marked disruptive effect on many membrane systems. Phosphotungstate is known to interact with lipoproteins and produces 'myelin forms' from initially smooth erythrocyte ghosts (18). Sodium silicotungstate is useful because of its particularly fine grain. Whilst it is highly desirable that negative stains should not interact with biological material in a 'positive staining' manner, it is to be expected that the cationic uranyl ion will bind to protein and sialic acid carboxyl groups and to lipid and nucleic acid phosphate groups. In doing so, uranyl acetate may very often induce aggregation of the material, unless there is firm adsorption onto the carbon support film prior to the addition of the negative stain. The benefit to be gained from the use of uranyl acetate and uranyl formate is that they induce the highest electron density and image contrast, although it has been shown that for high-resolution studies the microcrystalites of dried uranyl salts are mobile when irradiated by electrons in the TEM (19). Thus, the signal-to-noise ratio in electron micrographs from negatively stained specimens is likely to be influenced by the effects of radiation on the redistribution of negative stain. Also, the ability of negative stain to fill localized undulations or surface troughs in the carbon film will add to the noise level.

Of the lesser used negative stains (*Table 2*), methylamine tungstate has been found to be useful for negative staining macromolecules, membranes, and viruses, and the low contrast induced by sodium tetraborate has been used to distinguish the empty shell of apoferritin in the presence of iron-filled ferritin (20). Ammonium uranyl oxalate can be brought to neutral pH by the addition of ammonium hydroxide, but this solution is photolabile and though of excellent electron density it does impart a coarse image grain. High-resolution studies using glucose, glucose-negative stain mixtures, and goldthioglucose can only be performed under reduced electron dose conditions, but these have been valuable approaches since protein structure is preserved very much better than in the conventional negative stains alone (15).

Table 2. Some lesser-used negative stains

Negative stain	**Comments**
Uranyl acetate–EDTA	Imparts grainy image
Uranyl magnesium acetate	
Uranyl nitrate	
Uranyl formate	Imparts fine grain to image
Ammonium uranyl oxalate	Can be neutralized; photolabile; imparts grainy image
Sodium tungstate	
Methylamine tungstate	Useful for viruses and membranes
Sodium tetraborate	Induces low contrast
Cadmium iodide	
Glucose	Low electron dose only
Aurothioglucose	Low electron dose only
Glucose–UAc/PTA	Low electron dose only

2.2 Support films

The preparation of support films has been fully presented elsewhere (2) and will not be repeated here. For high magnification negative staining studies 5–10 nm carbon films are the most suitable, but they are inherently fragile. From the point of view of achieving maximum signal-to-noise ratio from negatively stained specimens it is desirable to use very thin carbon films (*Figure 2*). Extremely thin carbon films can be supported across thicker 'holey' carbon (microgrid) films. The regions of thin carbon spanning the 'holes' are then selected for study; unfortunately, this approach often severely limits the available area of suitably stained specimen. Surface activation of carbon films by glow discharge in the presence of amine vapour, irradiation with UV light or immersion in a solution of cationic substances such as Alcian Blue or polylysine, will considerably assist the adsorption of negatively charged biological material, prior to addition of negative stain. Whilst of obvious advantage, in certain circumstances this adsorption may not always be desirable, since free mobility of protein molecules and supramolecular complexes in the aqueous negative stain before it dries is more likely to give rise to varying orientations of such particles (21). For diagnostic virology or studies on larger cellular particulates and bacteria, the more robust formvar/ butvar plastic films or carbon-coated plastic films have been found to be

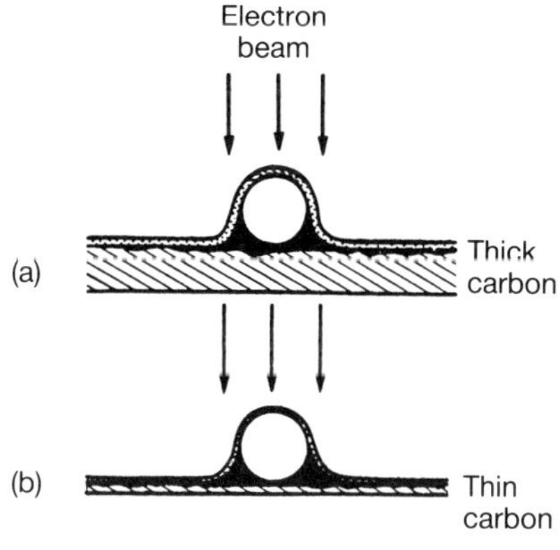

Figure 2. Diagram (**a**) shows a hypothetical biological particle of diameter approximately 10 nm supported on a carbon film of thickness approximately 10 nm. In addition to the negative staining of the particle, some unevenness of negative staining on the support film will occur due to irregularities or contaminants deposited at the time of preparing the carbon film. During irradiation by the electron beam (arrows) a continuous layer of carbonaceous contamination is also built up over the specimen. Under these conditions the signal-to-noise level ratio in the final image is likely to be relatively low. In diagram (**b**) there is a reduced support film thickness, with a more even surface and with a lower electron dose there will be less surface contamination. Thus, an improved signal-to-noise ratio is obtained, resulting in superior resolution in the final recorded image.

perfectly satisfactory, but it should be borne in mind that such films are not perfectly level (13) and that they will be more mobile within the electron beam than carbon alone. The ultimate in 'thinness' of supporting film might reasonably be considered to be a layer of negative stain alone. This can be achieved by spreading a mixture of material + spreading agent such as bacitracin + negative stain on holey carbon films (22) or even across clean naked 400- or 500-mesh copper grids (23). These approaches have not, however, found widespread application.

Negative stain depth on the support film is clearly of great importance since there is the possibility that particulates may be selectively orientated by the surface tension of very thin aqueous layers of stain and that staining from beneath rather than staining by total immersion may also occur (21, 24). To counter this problem, if indeed it is a problem, attempts have been made to produce more even negative staining by creating a layer of material + negative stain between two or more layers of carbon (25). This approach, whilst successful in the technical sense, has an unfortunate loss of resolution because of the double thickness of carbon, and in many instances there is marked compression of biological material when the two layers of carbon come together as the negative staining solution dries (26); see *Figure 3*. Indeed, the

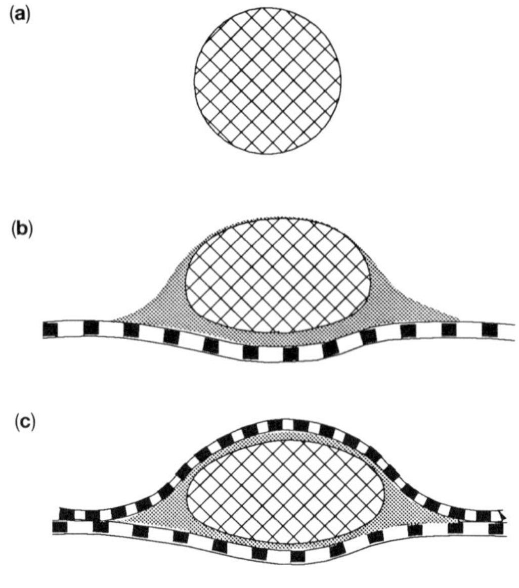

Figure 3. A diagram depicting the hypothetical negative staining of a large macromolecule or organelle. In (**a**) the original shape of the particle is shown. In (**b**) the particle is negatively stained on a single layer of carbon. Some deformation of the particle is present, depending upon the stain depth and flexibility of the particle. Often there is little stain on the upper surface of the particle. In (**c**) the particle is sandwiched between two layers of carbon. Although there is stain all around the particle, there is a greater amount of flattening. (From J. Frank, ref. 26.)

problem of particle or molecular flattening is one that in many instances cannot be overcome by the negative staining technique, unless the particles are freely suspended and undistorted within the thickness of the dried negative stain.

3. Negative staining procedures

For negative staining, some general guide-lines can be given as to the optimal particle/protein concentration within the aqueous suspension. For conventional negative staining, the concentration of viral particles should be in the order of 10^6 to 10^7 virus particles/millilitre, protein solutions should be at a concentration of 0.1 to 0.2 mg/ml, lipid suspensions should be in the range 0.1 to 0.5 mg/ml and cellular and cellular membrane suspensions in the range 0.5 to 1.0 mg/ml. In all cases it is necessary to avoid particle superimposition on the support film, by having an excessive concentration of material. For the 'negative staining–carbon film' procedure (see below) an approximately 10- to 20-fold greater concentration of viral particles and protein molecules is desirable.

In almost all cases it is highly desirable to have the biological material suspended in low concentration (5 to 10 mM) saline or buffer, but procedures are available to reduce isotonic saline or high salt to an acceptable concentration, such as by placing grids with saline-containing specimen material on distilled water-moistened filter paper, or simply by introducing a repeat application of negative stain to reduce the concentration of the interfering salt or buffer. The preparation of specimens from osmotically sensitive biological structures, such as mitochondria, chloroplasts, and other membrane-bound organelles, isolated in buffers or present in fractions taken from sucrose density gradients, present a number of difficulties during negative staining. The buffer salts and sucrose will interfere with the negative staining process, so it is necessary to dialyse against a volatile isotonic buffer, such as ammonium acetate or ammonium carbonate, for a period of 12 to 24 h prior to negative staining.

Procedures utilizing spraying suspensions of biological material and stain on to support films have been described elsewhere (2) and will not be expounded here since they have largely been superseded by the droplet techniques. Nevertheless, in some studies there are advantages to be gained from the use of spray-droplet methods where small stain-droplet patterns can be deposited directly on to carbon films. The distribution of biological particles within such droplets will depend upon the quantity of material mixed with negative stain before spraying, adjustment of which to obtain evenly distributed droplets with an adequate stain density has to be achieved by trial and error. The spray-droplet methods have been found to be particularly useful for the preparation of specimens of viruses, macromolecules, and liposomes, but attention should be drawn to the necessary safety precautions that must be taken when spraying pathogenic material.

3.1 Droplet procedures

3.1.1 'On-grid' droplet procedures

The 'on-grid' droplet procedures (27) are particularly useful if one has to rapidly prepare an individual rather than multiple specimens, and it requires some manual dexterity in the application of specimen material and stain directly to the carbon film from a pipette without damaging the film.

Protocol 1. The 'on-grid' droplet procedure

1. Apply a 5–10 μl droplet of biological material directly from a pipette to the support film on 300- or 400-mesh grids, which is held (by a rubber band) in forceps. Spreading can be assisted by the inclusion of bovine serum albumin (0.1 mg/ml) or the peptide bacitracin (0.3 mg/ml) (28).
2. After a few seconds (depending on the material, particle concentration, and level of adsorption required), touch the edge of the droplet to the edge of a filter paper, leaving a thin aqueous film containing the adsorbed and freely suspended biological material.
3. Apply a 10 μl droplet of negative stain to the grid and after a few seconds remove it, similarly, by touching to the edge of a filter paper.
4. Allow the layer of biological material + negative stain to dry at room temperature, after placing onto a filter paper in a covered Petri dish.
5. Forceps should be carefully cleaned in distilled water or methanol and wiped dry, to prevent carry-over and contamination of subsequent specimens.

3.1.2 The 'single-droplet' parafilm procedure (refs 11 and 29)

This procedure has the advantage that multiple specimens can be prepared somewhat more rapidly, using different materials and different negative stains, than is the case with the 'on-grid' procedure. In addition, the specimen material and negative stain can be held at varying temperatures (4°C to 80°C), the specimen prepared and the stain dried at this temperature in a thermostatically controlled incubator (30).

Protocol 2. The 'single-droplet' procedure

1. Pipette 10–20 μl droplets of aqueous suspension of biological material on to a clean 'Parafilm'/paraffin wax surface, alongside 20 μl droplets of negative stain.
2. Using carbon or carbon-coated formvar 300- or 400-mesh grids, pick up a droplet of the biological suspension on to the support film and after a few

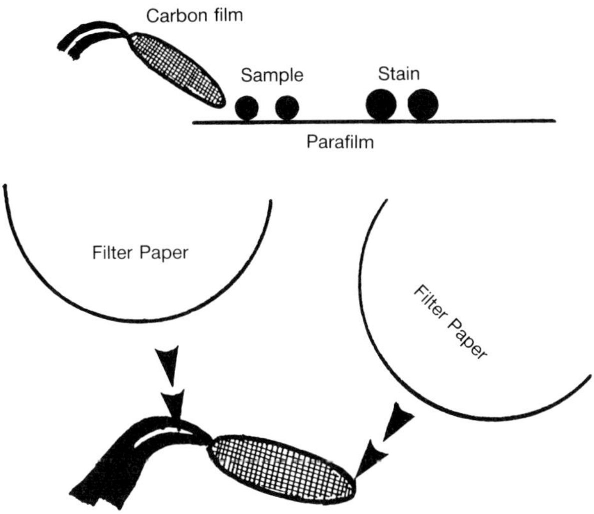

Figure 4. The single-droplet procedure for negative staining.

seconds remove the excess fluid by touching on to the edge of a filter paper, which is also placed between the forcep tips (*Figure 4*).

3. Immediately pick up a droplet of negative stain and after a few seconds remove the excess stain. To remove excess salt, buffer, sucrose, or glycerol, etc. in the biological sample, this step can be repeated.
4. Place the grid on to a filter paper in a Petri dish, cover, and allow the thin layer of biological material + negative stain to dry.
5. Duplicate specimens, a series of specimens obtained from a purification procedure or specimens stained with different negative stains can be rapidly prepared after carefully washing and wiping the forcep tips between each specimen.

3.2 The mica spreading procedures

3.2.1 The 'negative staining–carbon film' procedure (31, 32)

This procedure was developed using highly purified icosahedral and filamentous viral particles, but has been shown to be more widely applicable to purified protein molecules. The principal aim of this negative staining technique is to produce regular or crystalline two-dimensional mono-viral or mono-molecular arrays, which will then produce electron optical images suitable for digital image analysis with the production of two- and three-dimensional image reconstructions. The ability of particles to form two-dimensional crystalline arrays on the mica surface is critically dependent upon the purity and concentration of the material in suspension. The studies of one of us (R.W.H.) has shown that the best results with both viruses and protein molecules is achieved from freshly purified material, which will contain minimal amounts of small-molecular-weight dissociation products that are likely to inhibit the formation of crystalline arrays. Storage at 4°C of a number of icosahedral and filamentous plant viruses over periods ranging from 2 h to 48 h has been found to result in a dramatic reduction in the size of the crystalline areas formed on the mica surface, which is undoubtedly due to progressive viral dissociation. In contradistinction to the above comments, it should also be borne in mind that the gradual dissociation of certain types of viral particles into their subcomponents can be achieved by using a suitable negative stain other than uranyl acetate (see below). The gradual breakdown of adenovirus particles has been achieved by finally immersing the mica into 3% ammonium molybdate (pH 7.5 to 8.0) (33). Samples of the carbon film containing the adsorbed viral particles floating at an ammonium molybdate liquid/air interface were taken at intervals from 30 min to 48 h.

Protocol 3. The 'negative staining–carbon film' procedure

1. Mix 10 μl of a solution of purified protein or viral particle suspension, at a protein concentration of 1 to 2 mg/ml, with 10 μl of ammonium molybdate (pH usually 7.0, but can be varied), with or without polyethylene glycol (PEG).[a]
2. Apply the mixture of ammonium molybdate and biological material to the surface of small pieces of freshly cleaved mica (*Figure 5*) and remove the excess fluid by touching onto the edge of a filter paper.
3. Allow the film of fluid on the mica surface to slowly dry within a covered Petri dish. It is at this stage that the two-dimensional crystals of protein or virus are formed. Variations, such as the PEG concentrations, pH, the drying temperature, and drying time can be introduced (i.e. by placing the Petri dish containing the mica into a temperature-controlled incubator or on to a moist filter paper).
4. Coat the dried layer of ammonium molybdate + protein or virus vertically *in vacuo* with a thin layer (5 to 10 nm) of carbon. This layer of carbon actively adsorbs to the dried layer of protein molecules or viral particles on the mica.
5. Float the layer of carbon + adsorbed protein or virus off the mica on to the surface of 2% aqueous uranyl acetate solution.[b] No protein aggregation occurs at this stage, because of the firm adsorption on to the nascent carbon layer.
5. Pick up small pieces of the floating carbon film from beneath the uranyl acetate surface, on 400-mesh clean grids or on 'holey' carbon support films.
6. Carefully remove excess uranyl acetate and carbon film by wiping the grid,

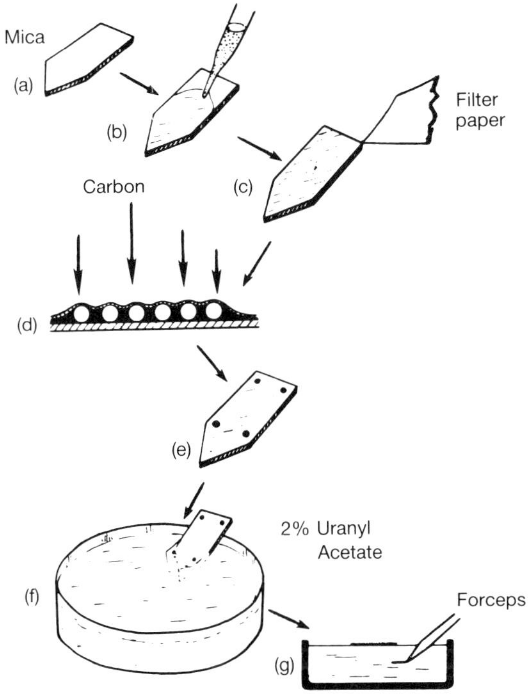

Figure 5. The negative staining–carbon film procedure.

carbon uppermost, on a filter paper and allow the remaining stain to dry at room temp. On average, 5 to 10 specimen grids can be prepared, depending upon the dimensions of the piece of mica.

[a] Polyethylene glycol (PEG) M_r 1000 to 10 000 (0.1% to 0.2% w/v) can be added to the ammonium molybdate prior to mixing with the biological material, as this reagent potentiates the formation of two-dimensional crystals (34, 35). Also, *note* that for solutions containing high concentrations of glycerol, with or without a volatile buffer salt such as ammonium acetate, it is possible to prepare negatively stained specimens by the above procedure. The glycerol-containing suspension is spread directly on pieces of mica, as in Section 2 above, and is dried *in vacuo* on to the mica surface. The film of carbon is then deposited on to the dried protein, organelles, viral particles, etc., and the subsequent steps (4 to 6) of the procedure above followed (36).

[b] Other negative stains can be used.

3.2.2 The immobilized cell 'negative staining–carbon film' procedure (36)

This technique has recently been successfully developed, as shown diagramatically in *Figure 6*, for the negative staining of single layers of red blood cell (RBC) and blood platelet plasma membrane obtained directly from intact cells (as opposed to the double layer plasma membrane of collapsed red blood cell ghosts, obtained by osmotic lysis and washing of cell suspensions, pre-

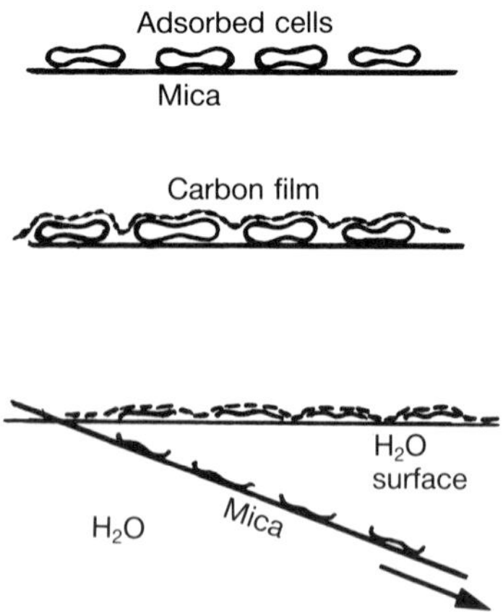

Figure 6. The immobilized cell negative staining–carbon film procedure.

pared by the single-droplet technique, *Protocol 2*, above). Extension of the procedure to cultured cells has indicated the considerable potential of this technique (J.R.H. unpublished observations).

Protocol 4. The immobilized cell 'negative staining–carbon film' procedure

1. Prepare small pieces of freshly cleaved mica, as in *Figure 5*. Impart a positive charge by dipping the mica into 0.01% Alcian Blue (AB), followed by thorough washing in distilled water.
2. Apply 20 μl of a suspension of PBS-washed RBCs, platelets, or other cells to the AB–mica and leave the pool of cells for a few minutes for the cells to become attached. Remove unbound cells by 10 rapid vertical immersions into three 50 ml vol. of PBS and drain on to a filter paper between volumes. At this stage the presence of the bound/immobilized cells can be assessed by light microscopy.
3. Remove the PBS by flooding the mica surface with 30% glycerol–0.155 M ammonium acetate (a volatile salt), drain the mica on to the edge of a filter paper, and wipe dry the under surface. Remove the glycerol *in vacuo* at 10^{-5} Torr for 1 h, and coat the dried layer of cells vertically with a thin layer of carbon.
4. Float off the carbon film + adsorbed upper plasma membrane on to the surface of distilled water. This splits open the cells and releases their contents, as the lower portion of the cells remain attached to the mica, as shown in *Figure 6*.
5. Pick up small pieces of the floating carbon film and adsorbed plasma membrane from beneath the water surface on 400-mesh clean grids, and carefully wipe on a filter paper to remove excess carbon film.

6. Immediately touch the lower surface of the carbon with adsorbed plasma membrane on to a droplet of negative stain (2% uranyl acetate, ammonium molybdate, sodium phosphotungstate) on a paraffin wax surface, as in *Protocol 2* above.
7. Remove the excess negative stain by touching the grid on to the edge of a filter paper and allow the residual stain to dry at room temp. in a covered Petri dish.

It should be noted that plasma membranes prepared from intact cells by the procedure given in *Protocol 4* are, in fact, negatively stained from the internal/cytoplasmic surface (the external surface is adsorbed to the carbon film). This being the case, the procedure is ideal for the assessment of *in situ* membrane skeleton or cytoskeleton attachment to the plasma membrane. In addition, it provides a simple procedure for the assessment of the frequency and distribution of externally-bound ferritin or colloidal gold labels, applied to cells when immobilized on the strips of mica or of cytoskeletally-bound labels applied after the cell lysis stage.

4. Some applications of negative staining

Whilst the emphasis of this chapter is upon the technical aspects of negative staining, it is appropriate to include a few representative examples of biological materials prepared for TEM by each of the four protocols included above.

In reality, there is little or no difference between the TEM images of materials spread and negatively stained by the 'on-grid' droplet procedure of *Protocol 1* and the 'single-droplet' procedure of *Protocol 2*; the authors have a preference for the latter approach, because of its simplicity and the facility for producing numerous specimens relatively quickly.

Figure 7 shows a shape-transformed human blood platelet, negatively stained by *Protocol 1*, using 2% ammonium molybdate. The finger-like projections are brought about by bundles of actin fibres, which are generally not observed by negative staining, but surface features predominate (*Figure 7b*). Comparable TEM images of human erythrocyte ghosts, again negatively stained with ammonium molybdate, but by the single droplet procedure outlined in *Protocol 2*, are shown in *Figure 8*. In *Figure 8a*, smooth-surfaced collapsed, i.e. double membrane thickness, haemoglobin-free erythrocyte ghosts are shown, with surface folds clearly apparent. A very different image is obtained when there is residual haemoglobin, in which case numerous tube-like projections extend from the main body of the erythrocyte ghost (*Figure 8b*). It should be emphasized that these projections are not induced by the ammonium molybdate, but sodium phosphotungstate does induce erythrocyte ghost fragmentation and images of this type (18). Liposome suspensions, larger multilammelar lipid bodies, and lipid bilayer sheets can readily be revealed by negative staining, as shown in *Figures 9* and *10*, which

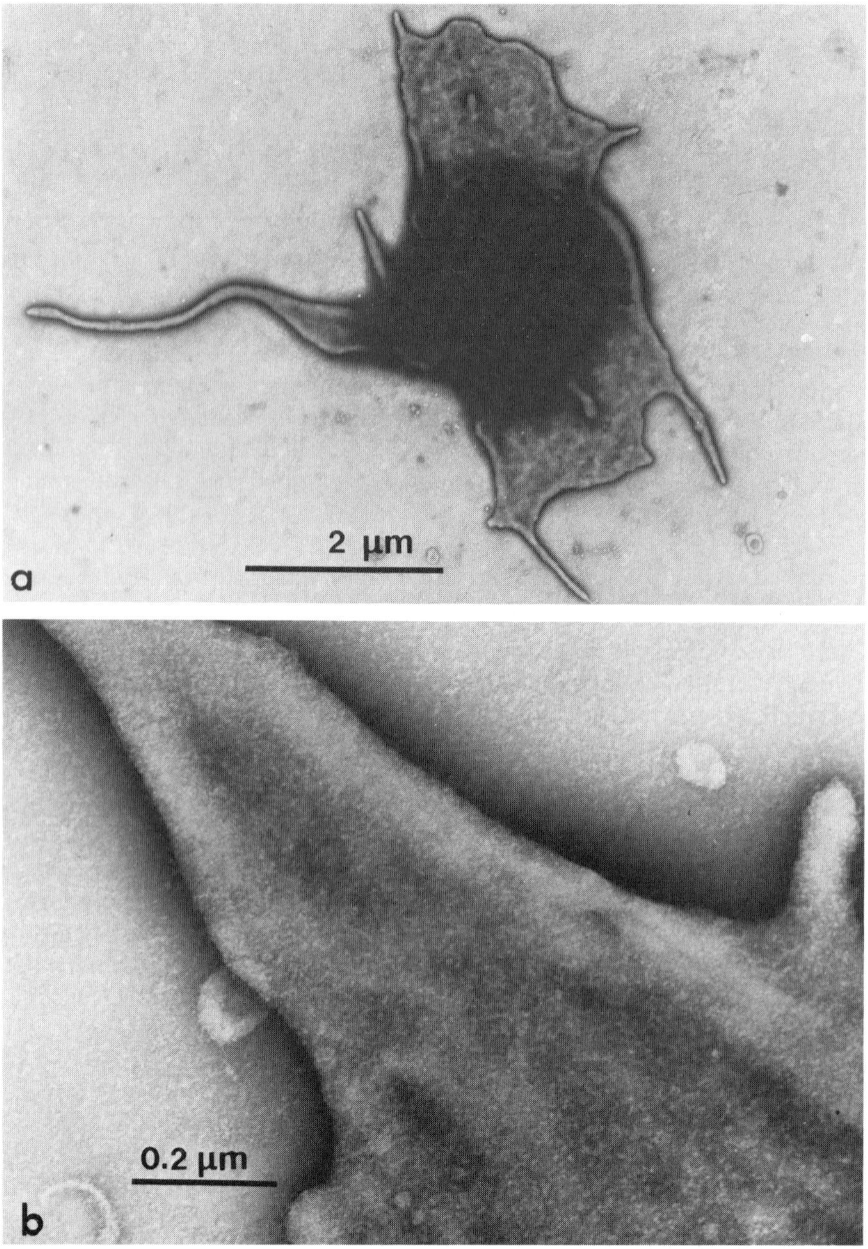

Figure 7. Shape transformed human blood platelets, negatively stained with 2% ammonium molybdate (pH 6.9), by the on-grid droplet technique (*Protocol 1*). The electron-dense central region in (**a**), which contains the internal granular elements of the platelet is too thick for the resolution of ultrastructural detail, but the surrounding thinner regions reveal the particulate detail of the platelet surface glycocalyx (**b**).

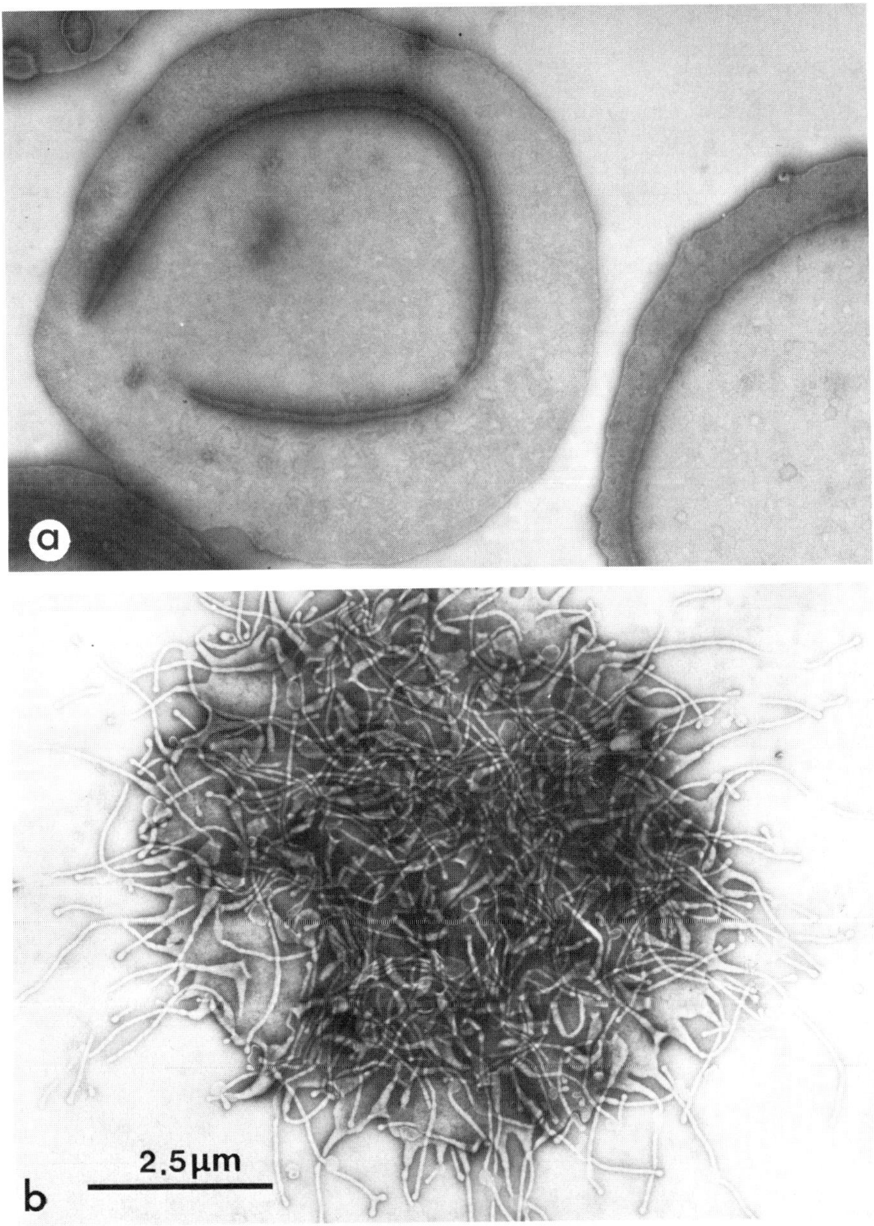

Figure 8. Human erythrocyte ghosts negatively stained with ammonium molybdate by the single-droplet technique (*Protocol 2*). In (**a**) the erythrocyte ghosts have been washed free of all haemoglobin. The images of the collapsed ghosts result predominantly from two superimposed membranes, with thicker folded regions. In (**b**) the erythrocyte ghosts still contains some haemoglobin. The membrane has been transformed into numerous tube-like extensions which flow out from the main body of the erythrocyte ghots.

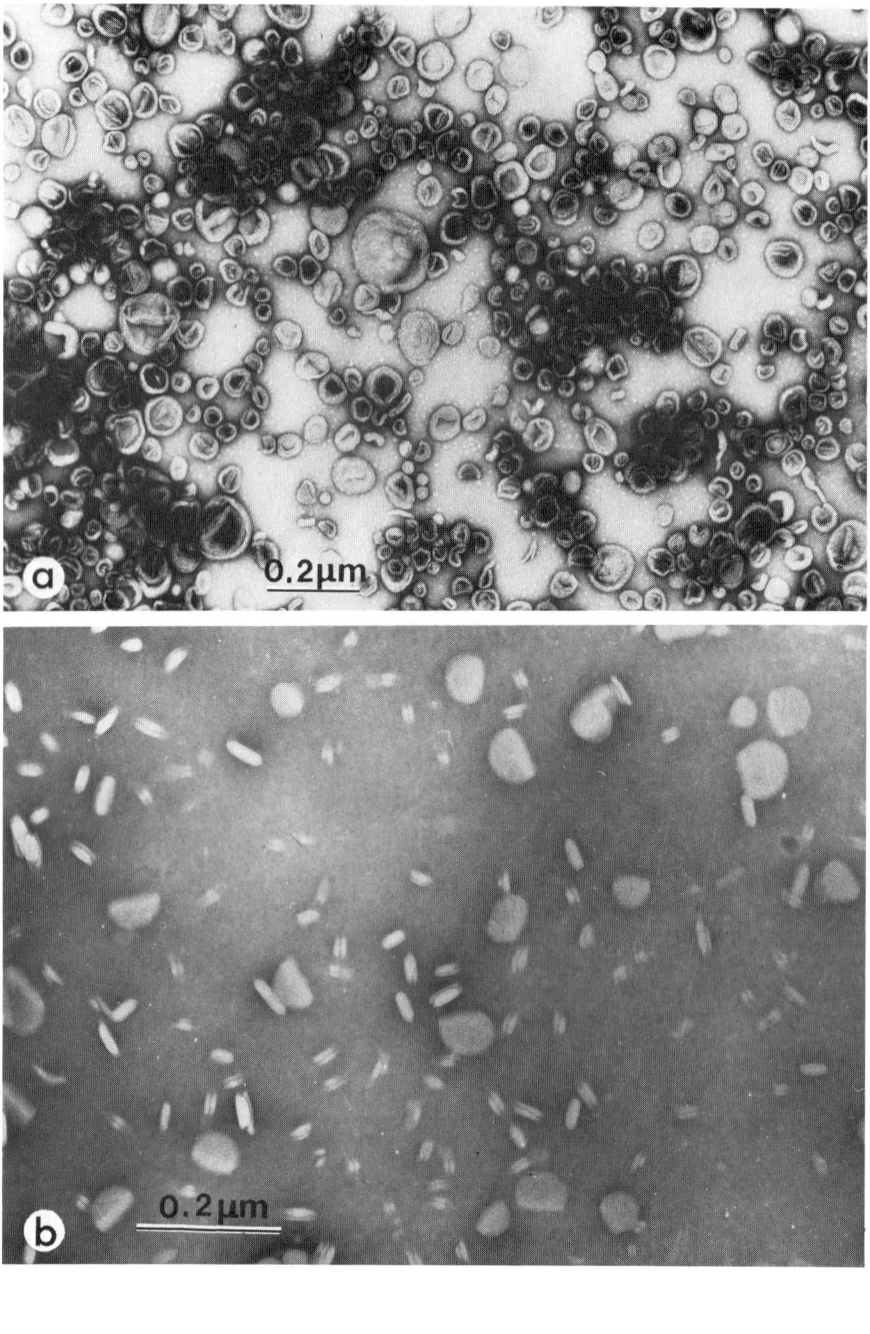

Figure 9. Bovine brain sphyngomyelein liposomes, negatively stained with uranyl acetate (**a**) and sodium phosphotungstate (**b**) by the single-droplet technique (*Protocol 2*).

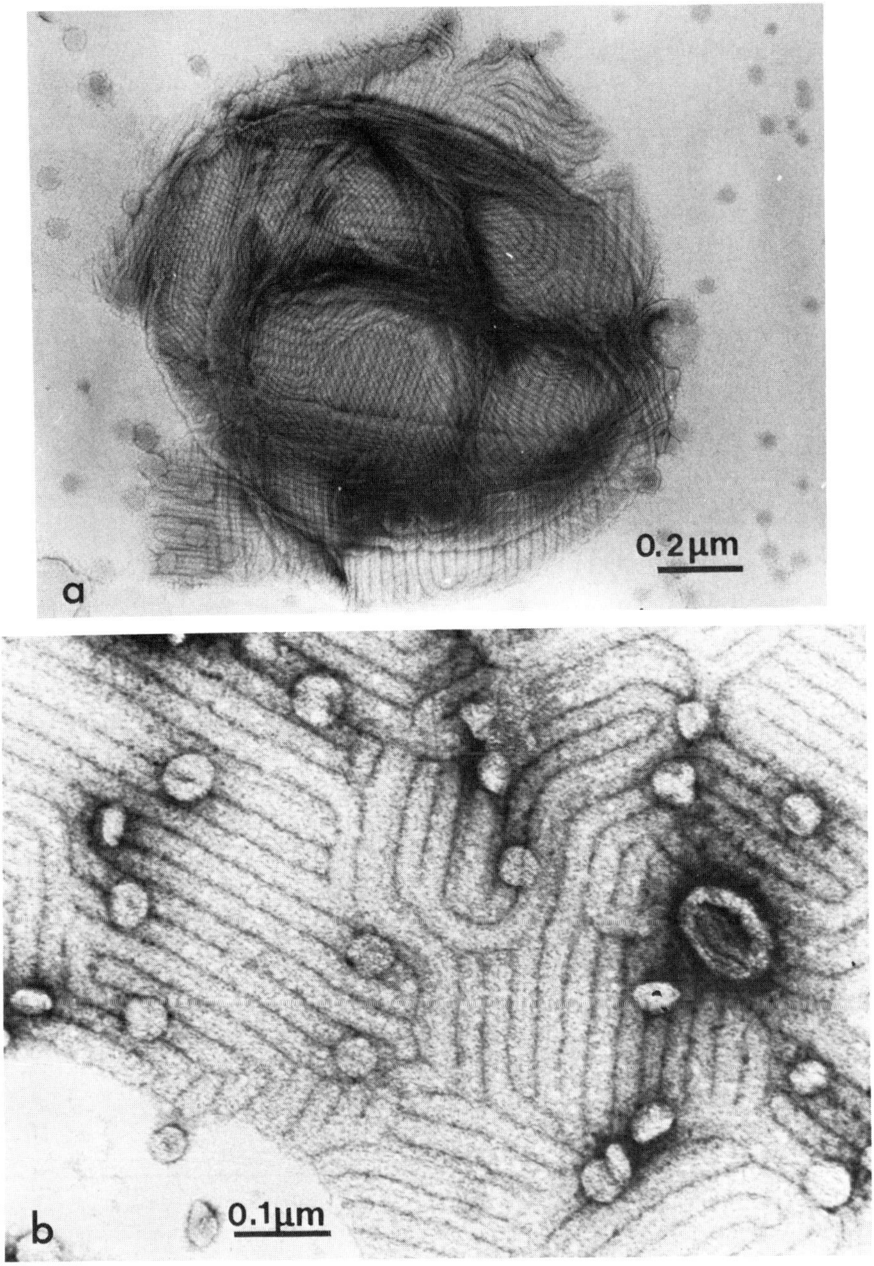

Figure 10. (**a**) A large sphingomyelin multi-lamellar body showing the Pβ′ pre-transition phase of the phospholipid bilayer. (**b**) Part of a single bilayer sheet of sphingomyelin, showing the undulatory Pβ′ pretransition phase. Individual liposomes superimpose the undulatory bilayer. Both (**a**) and (**b**) are negatively stained with uranyl acetate by the single-droplet technique (*Protocol 2*).

present examples of aqueous suspensions of bovine brain sphingomyelin (30), prepared by the single-droplet technique (*Protocol 2*), using sodium phosphotungstate and uranyl acetate. Cellular organelles and subcomponents can likewise be revealed by negative staining, after removal of sucrose and isotonic buffer salts, if present (see Section 3 above). *Figure 11* shows examples of a fragment of rat liver hepatoma nuclear envelope negatively stained with ammonium uranyl oxalate (*Figure 11a*) and disrupted *Xenopus* oocyte nuclear envelope negatively stained with ammonium molybdate (*Figure 11b*), both specimens prepared by the single-droplet technique.

No examples of protein molecules negatively stained by the droplet techniques (*Protocols 1* and *2*) will be given, as these have been thoroughly dealt with by Speiss *et al.* (2), but it must be emphasized that these approaches have been used extensively and very successfully by numerous workers. The application of the negative staining–carbon film procedure (*Protocol 3*) to protein molecules has been limited to relatively few workers, but considerable success has nevertheless been achieved (32, 34, 35). An example of the M_r 800 000 protein cylindrin (identical to the 22S multicatalytic proteinase complex) prepared by *Protocol 3* in the absence of polyethylene glycol is shown in *Figure 12*. The extremely thin carbon support film spanning a hole in a thicker carbon support film is relatively mobile and has produced some molecular elongation and compression, a hazard which must be avoided when making accurate measurements of the molecular dimensions. In the presence of polyethylene glycol (M_r 10 000) the cylindrin molecule does form small two-dimensional crystalline regions, but greater success has been achieved with human erythrocyte catalase (35), as shown in *Figure 13*. Two virological examples of the negative staining carbon film technique are shown in *Figure 14* (tobacco mosaic virus) and *Figure 15* (cowpea chlorotic mottle virus).

Protocol 4, which describes the 'immobilized cell negative staining–carbon film procedure' has not been used extensively to date. An example of a single thickness of erythrocyte membrane (cf. *Figure 8a*) negatively stained by this procedure is shown in *Figure 16*. At higher magnification the unextended membrane skeleton network is seen (*Figure 16b*), within which it is possible to define individual elongated structures, which in all probability represent the unstressed and structurally intact spectrin–actin network (membrane skeleton) on the inner surface of the erythrocyte membrane.

Figure 11. Examples of negatively stained nuclear envelope. In (**a**) a fragment of rat hepatoma nuclear envelope is negatively stained with ammonium uranyl oxalate (*Protocol 2*). The octagonal symmetry of the nuclear pore complexes is revealed with great clarity, together with the ribosome particles attached to the outer nuclear membrane. In (**b**) part of a nuclear envelope dissected from a *Xenopus* oocyte has undergone disintegration during the spreading process of the single-droplet negative staining technique, with ammonium molybdate. Individual nuclear pore complexes have separated from the surrounding nuclear membranes. Traces of the fibrous nuclear lamina are present.

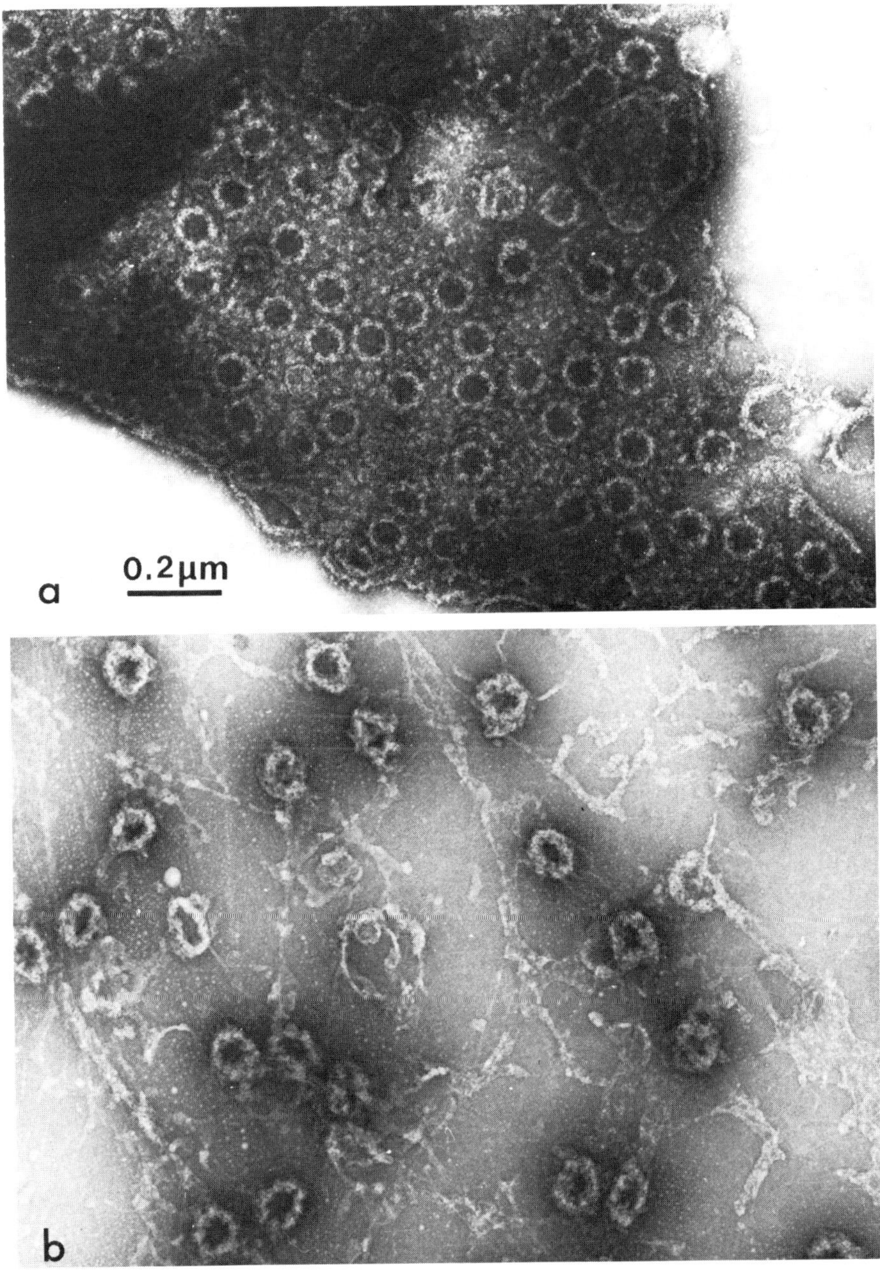
0.2 µm
a
b

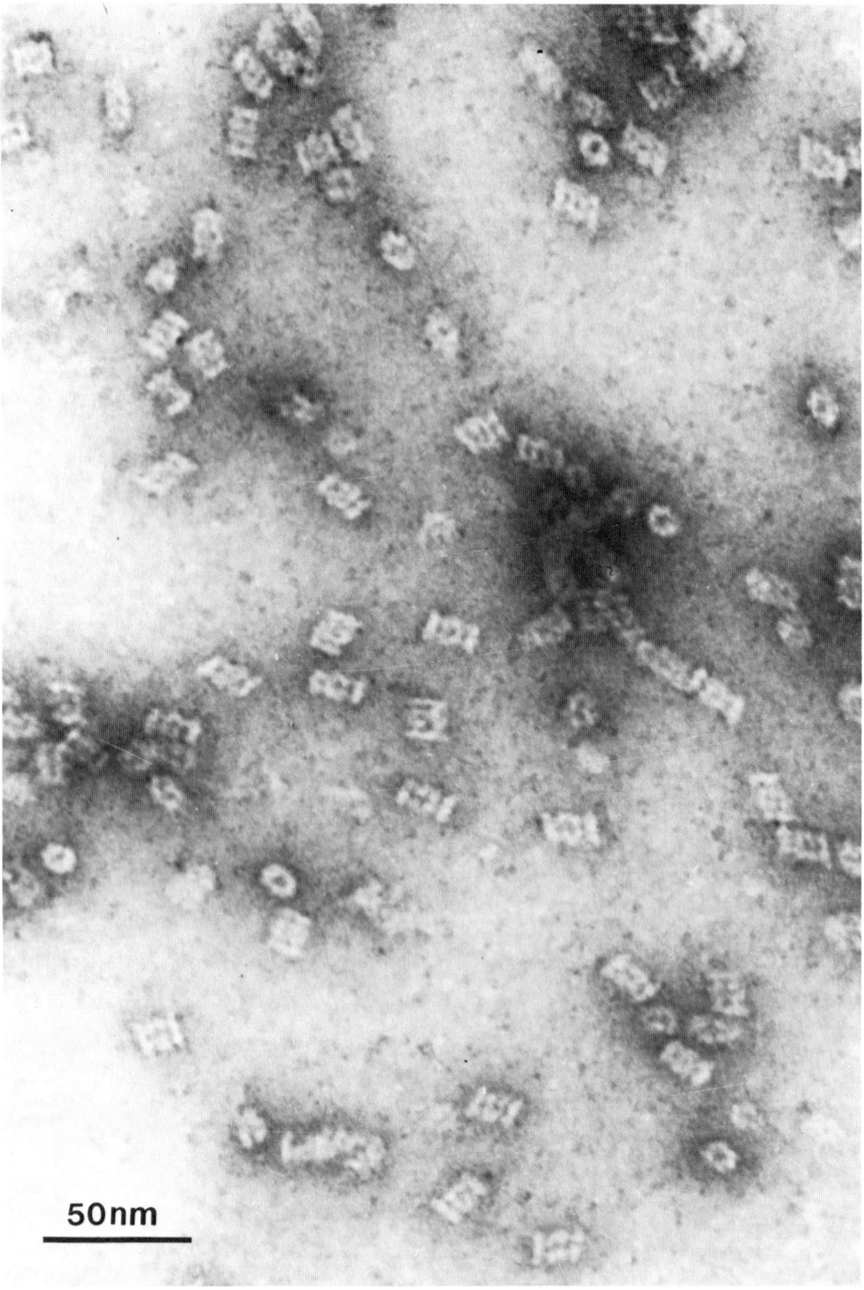

Figure 12. Individual molecules of human erythrocyte cylindrin (multicatalytic proteinase complex) prepared by the negative staining–carbon film procedure (*Protocol 3*), in the absence of polyethylene glycol.

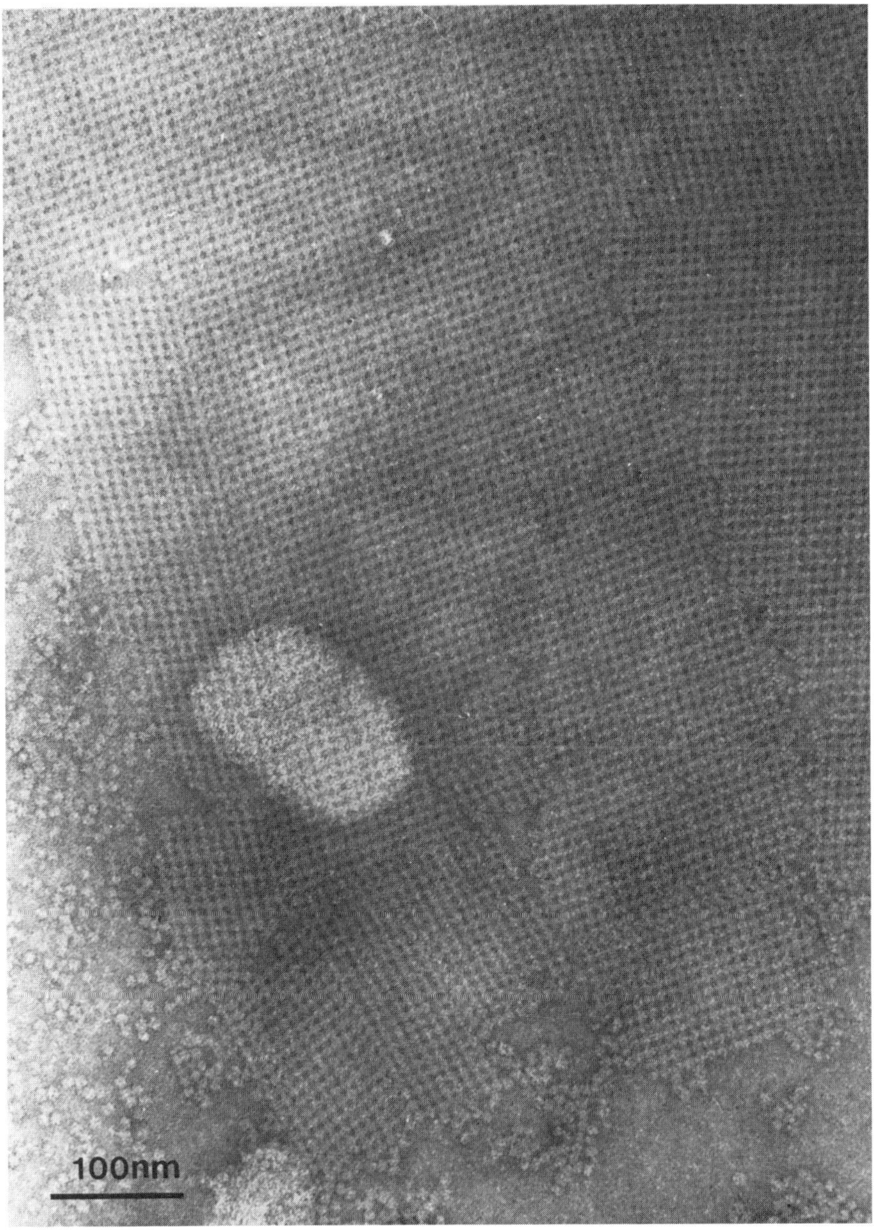

Figure 13. Two-dimensional crystals of human erythrocyte catalase prepared by the negative staining–carbon film procedure (*Protocol 3*) in the presence of 0.05% polyethylene glycol *Mr* 10 000. (From ref. 35.)

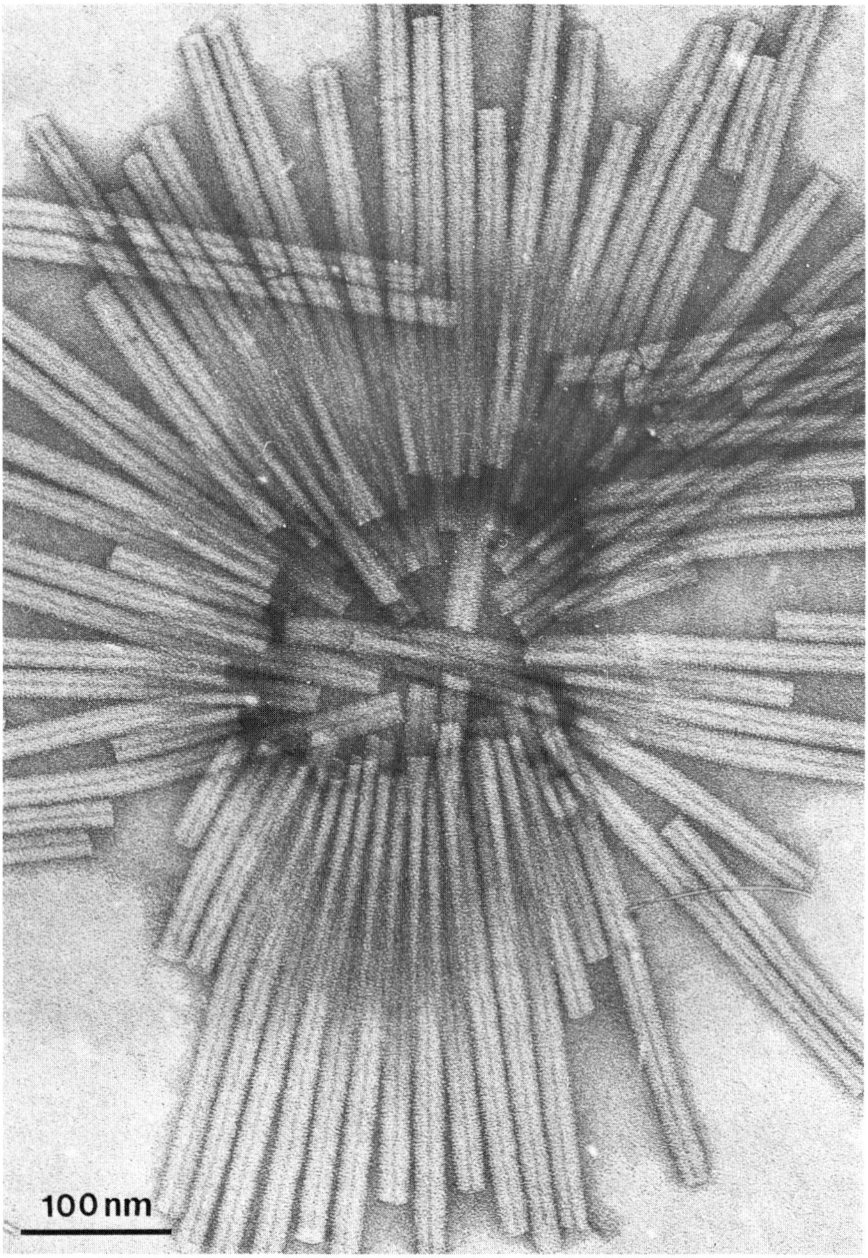

Figure 14. Tobacco mosaic virus prepared by the negative staining–carbon film procedure (*Protocol 3*), in the absence of polyethylene glycol.

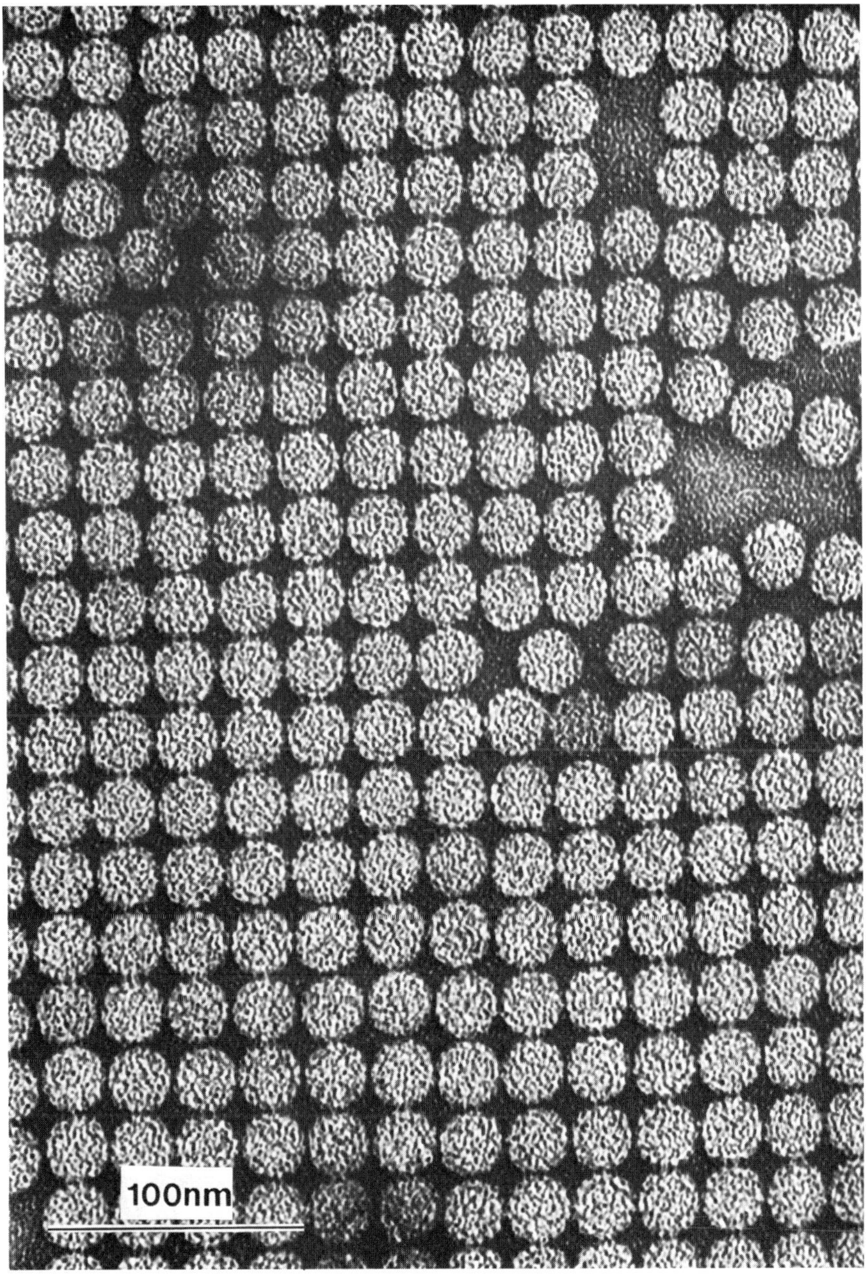

Figure 15. Cowpea chlorotic mottle virus prepared by the negative staining–carbon film procedure (*Protocol 3*), in the absence of polyethylene glycol.

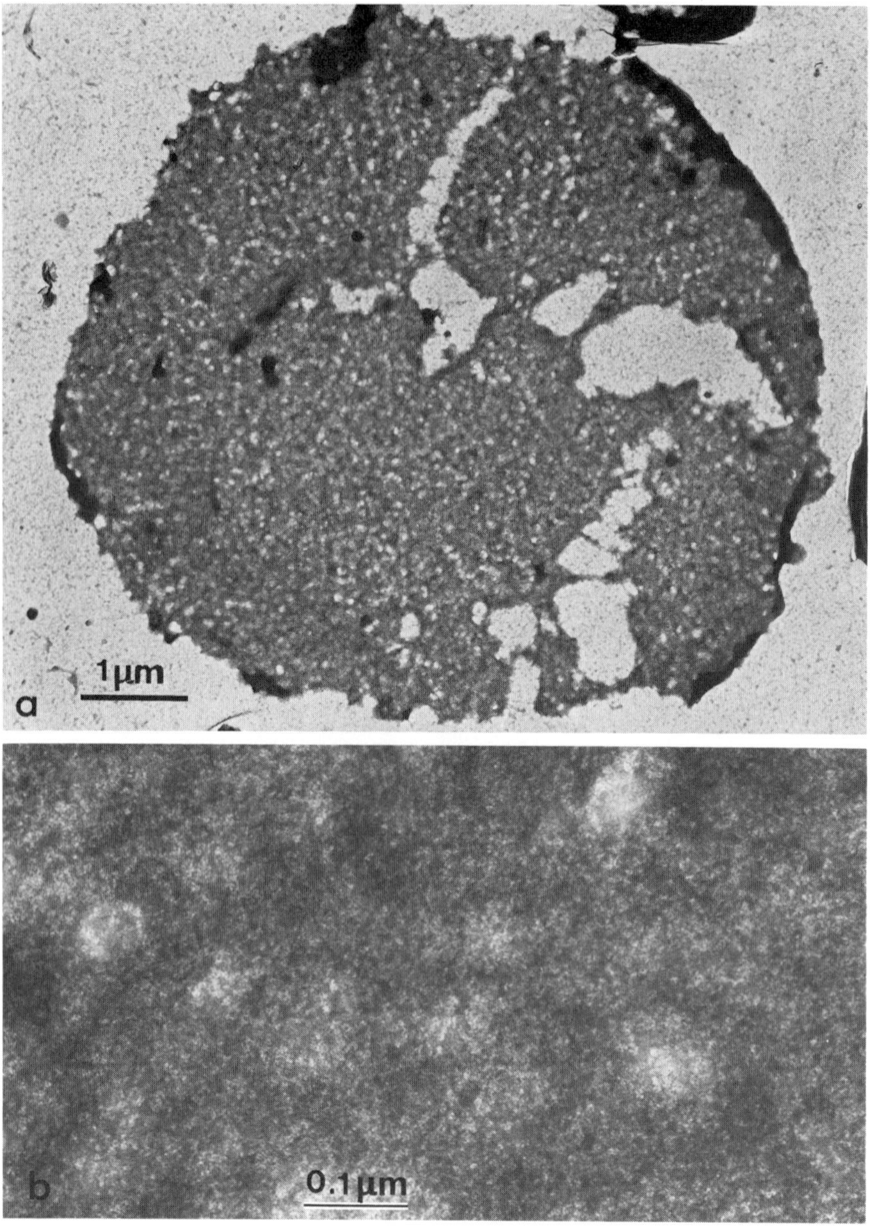

Figure 16. A single layer of human erythrocyte membrane negatively stained with uranyl acetate by the immobilized cell negative staining–carbon film procedure (*Protocol 4*). The external surface of the cell is adsorbed to the carbon film and the negative stain has access only to the cytoplasmic surface. In (**a**) the circular outline of the membrane torn directly from an intact erythrocyte at the lysis stage is shown. In (**b**) the complex membrane skeleton network of spectrin and actin is revealed.

5. Conclusions

It is now more than 40 years since the introduction of negative staining as a working procedure for the electron microscopist. Numerous technical variants have been described, but the information presented in this chapter has been restricted to four protocols, three of which are soundly tried and tested and one indicates the way ahead to exciting possibilities. Despite the limited resolution achievable from negatively stained specimens (*c.* 2 nm), much valuable biological, virological, and biochemical information has been obtained throughout this period of time and the technique continues to be widely used. It provides a bridge between the low-resolution tissue-based studies and the high-resolution crystallographic electron microscopy. A more extensive presentation of negative staining techniques and their applications has recently been published (37). The future of negative staining as a routine technique is undoubtedly secure, since it provides valid data which correlates well with that obtained from the other technical approaches for TEM specimen preparation, many of which are described elsewhere in this book.

References

1. Johnson, M. W. and Horne, R. W. (1970). *Journal of Microscopy,* **91,** 197.
2. Speiss, E., Zimmerman, H.-P., and Lundsdorf, H. (1987). In *Electron microscopy in molecular biology* (ed. J. Sommerville and U. Scheer), pp. 147–66, IRL Press at Oxford University Press, Oxford.
3. Horne, R. W. and Wildy, P. (1979). *Journal of Microscopy,* **117,** 103.
4. Hall, C. F. (1955). *Journal of Biophysics and Biochemical Cytology,* **1,** 1.
5. Huxley, H. E. (1957). *Proceedings of the Stockholm Conference Electron Microscopy,* p. 260. Almqvist & Wiksell, Stockholm.
6. Horne, R. W. and Brenner, S. (1960). *Fourth International Congress on Electron Microscopy, Berlin, 1958,* Vol. 2, p. 625, Springer-Verlag, Berlin.
7. Brenner, S. and Horne, R. W. (1959). *Biochimica et Biophysica Acta,* **34,** 103.
8. Bangham A. and Horne, R. W. (1964). *Journal of Molecular Biology,* **8,** 660.
9. van Bruggen, E. F. J. (1960). *Journal of Molecular Biology,* **2,** 81.
10. Horne, R. W. and Waterson, A. P. (1960). *Journal of Molecular Biology,* **2,** 75.
11. Benedetti, E. and Emmelot, P. (1965). *Journal of Cell Biology,* **26,** 299.
12. Horne, R. W. (1973). *Journal of Microscopy,* **98,** 286.
13. Oliver, R. M. (1973). In *Methods in enzymology,* Vol. 27 (ed. D. H. W. Hirs and S. Timasheff), pp. 616–72. Academic Press, New York and London.
14. Harris, J. R. (1982). In *Electron microscopy of proteins* (ed. J. R. Harris), Vol. 2, pp. 49–103, Academic Press, London.
15. Kiselev, N. A., Sherman, M. B., and Tsuprun, V. L. (1990). *Electron Microscopy Reviews,* **3,** 43.
16. Muscatello, U. and Horne, R. W. (1968). *Journal of Ultrastructure Research,* **25,** 73.
17. Munn, E. A. (1968). *Journal of Ultrastructure Research,* **25,** 363.
18. Shaffer, S. (1980). *Journal of Ultrastructure Research,* **70,** 153.

19. Unwin, P. N. T. (1974). *Journal of Molecular Biology,* **87,** 657.
20. Massover, W. H. and Cowley, J. M. (1975). In *Proteins of iron storage and transport in biochemistry and medicine*, (ed. R. R. Crighton), pp. 237–44, North-Holland, Amsterdam.
21. Harris, J. R. (1969). *Journal of Molecular Biology,* **46,** 329.
22. Fernandez-Morán, H., Marchalonis, J. J., and Edelman, G. M. (1968). *Journal of Molecular Biology,* **342,** 467.
23. Malech, H. I. and Albert, J. P. (1979). *Journal of Ultrastructure Research,* **69,** 191.
24. Nermut, M. V. and Perkins, W. J. (1979). *Micron,* **10,** 247.
25. Smith, C. A. and Seegan, G. W. (1984). *Journal of Ultrastructure Research,* **88,** 111.
26. Frank, J. (1989). *Electron Microscopy Reviews,* **2,** 53–74.
27. Huxley, H. E. and Zubay, G. (1960). *Journal of Molecular Biology,* **2,** 10.
28. Gregory, D. W. and Pirie, B. J. S. (1973). *Journal of Microscopy,* **99,** 261.
29. Harris, J. R. and Agutter, P. S. (1970). *Journal of Ultrastructure Research,* **33,** 219.
30. Harris, J. R. (1986). *Micron Microscopica Acta,* **17,** 175.
31. Horne, R. W. and Pasquali-Ronchetti, I. P. (1974). *Journal of Ultrastructure Research,* **47,** 361.
32. Harris, J. R. (1982). *Micron,* **13,** 147.
33. Horne, R. W., Pasquali-Ronchetti, I., and Hobart, J. M. (1975). *Journal of Ultrastructure Research,* **51,** 233.
34. Ghiretti-Magaldi, A., Zanotti, G., Tognon, G. and Mezzasalma, V. (1985). *Biochimica et Biophysica Acta,* **829,** 144.
35. Harris, J. R. and Holzenburg, A. (1989). *Micron Microscopica Acta,* **20,** 223.
36. Harris, J. R. (1991). *Journal of Electron Microscopy Technique*, **18,** 269–76.
37. Hayat, M. A. and Miller, S. E. (1990). Negative Staining. McGraw Hill Publishing Co., New York.

10

Transmission electron microscopy of vitrified biological macromolecular assemblies

MURRAY STEWART

1. Introduction

The production of frozen-hydrated biological specimens in thin films of vitrified water enables unstained material to be examined by transmission electron microscopy in conditions that closely resemble those in living cells. This exciting prospect offers the chance of eliminating many of the preparative artefacts associated with dehydration and heavy metal staining that have long bedevilled the structural investigation of some macromolecular assemblies. Moreover, because freezing takes place very rapidly, these methods offer the chance to examine dynamic systems with high time-resolution and also to examine the effects of solvent composition on the structure and interactions of cellular components and macromolecules. For these reasons, the examination of vitrified specimens has become an important adjunct to other methods, such as negative staining, shadowing, and sectioning, for the examination of biological macromolecular structure and assembly. Particularly striking results have been obtained from objects containing membranes, such as virus particles and two-dimensional crystals of membrane proteins such as gap junctions and acetylcholine receptors.

There have been a number of recent reviews of the transmission electron microscopy of frozen-hydrated biological material (1–5) to which the reader is directed for a comprehensive discussion of many aspects of the method. This chapter provides an introduction to the methods and description of the freezing techniques and electron microscopy procedures that are employed. It first discusses the theoretical background to freezing before progressing to a description of the methods used to prepare thin vitrified films containing biological macromolecules. Methods for examining this material by transmission electron microscopy are then described, together with a description of the specialized equipment required, before discussing some of the difficulties encountered in interpreting images of unstained vitrified biological specimens. Finally some applications are discussed and particularly favourable types of specimen identified.

2. Basic theory of freezing

The usual form taken by water at temperatures below its freezing point, namely crystalline ice, is generally extremely harmful to biological material. Ice-crystal damage of tissues usually results in the perforation of membranes and loss of viability. Moreover, in solutions containing salts and other soluble molecules, eutectics are often formed at the boundaries of growing ice-crystals and these can produce substantial changes in the local composition of the solution. When freezing biological material it is usual to attempt to avoid, or at least reduce these problems by retarding ice crystal formation by using a variety of cryoprotectants. However, these agents are often not completely benign and may result in considerable alteration to the structure of the material under investigation. Substances such as glycerol and DMSO, for example, often alter membranes, particularly with respect to permeability. However, another way in which the problems associated with ice-crystal formation can be circumvented is to freeze so quickly that the water, in which the material is suspended, is trapped in a vitreous glass-like state (1, 6, 7). The actual rate of cooling that is necessary to achieve this vitrification is extremely high and cannot be attained simply by plunging material into liquid nitrogen. However, sufficiently high rates of cooling can be achieved by using supercooled ethane or propane or by slam freezing. These methods produce vitrified water and thereby prevent damage from ice-crystal formation and associated changes in solvent composition due to the formation of eutectics (8). Dubochet and his colleagues (1) have reviewed at length the theory of freezing and vitrification of aqueous solutions.

3. Preparation of vitrified films containing biological macromolecules

Although cryosections of biological specimens have also been examined (reviewed in ref. 1; and see Chapter 3), the freezing of bulk material and subsequent production of cryosections is often difficult and requires considerable skill and experience. Consequently, most investigations have concentrated on macromolecules and their assemblies vitrified in thin films. Samples in vitrified films are in some ways analogous to negatively stained material, albeit that the degree of preservation is often superior. However, each method uses broadly similar samples, such as suspensions of biological macromolecules, membranes, and aggregates such as thin crystals and virus particles.

Thin vitrified films are most easily produced directly on electron microscope grids. Generally this is done using 400-mesh grids that have been coated with a fenestrated film (9) and glow-discharged in the presence of an aliphatic amine (10, 11). Aebi and Pollard (12) describe simple apparatus in which grids can be glow discharged. A thin aqueous film of the suspension is then produced and vitrified by plunging into liquid ethane or propane[a] (1, 2, 13) at temperatures below −130°C. This can be done using very simple apparatus, originally described by Dubochet and his co-workers, along the lines described in *Figure 1*.

Grids are held in a pair of fine tweezers that are attached to the end of a steel rod suspended above a small bath of supercooled liquid ethane or propane. The tips of the tweezers should be treated with Teflon or some other suitable material to facilitate removal of the grid after vitrification. The liquid hydrocarbon is held in a small metal vessel surrounded by a liquid nitrogen bath and may be stirred with a magnetic bar (NB: Use a water-operated stirrer, not an electrical one) or manually with a metal rod. *Protocol 1* describes the production of vitrified films.

[a] **Warning:** Before attempting to vitrify material in this way you must become familiar with the hazards associated with the use of liquid ethane and propane. Consult your laboratory safety officer to ascertain precisely what precautions may be required (often by law) to guard against fire and explosion. Mixtures of air and ethane or propane are highly inflammable and have a very low flash-point. Remember in particular that 1 ml of liquid will produce about a litre of gas. At the very least, manipulations should be carried out in a fume hood with an explosion shield and care should be taken to ensure that there is no electrical equipment in the vicinity which could produce a spark. Note also that there is a much higher risk of frost-bite from splashing with liquid hydrocarbons than with liquid nitrogen. In particular, the use of safety glasses to protect the eyes is strongly recommended, especially when filling the apparatus from a gas cylinder.

Protocol 1. Vitrifying suspensions on grids

You will need:

- vitrification apparatus described in *Figure 1*, preferably located in a fume hood
- a cylinder of ethane or propane and appropriate control valves
- liquid nitrogen
- fenestrated 400-mesh EM grids (9) glow-discharged in alkylamine (10, 11)
- a suspension of your sample at comparatively high concentration (the optimum concentration is found empirically)

1. Fill the outer container with liquid nitrogen.
2. Feed in a thin stream of hydrocarbon (propane or ethane) from a cylinder. Initially this will be gaseous, but it will soon begin to liquefy in the cooled vessel. (NB: Take care about splashing of liquid hydrocarbon during filling.) The hydrocarbon will eventually solidify when cooled by liquid nitrogen and it is helpful initially to do this to see what happens. After a little practice one can usually judge fairly easily when the liquid hydrocarbon is about to freeze.

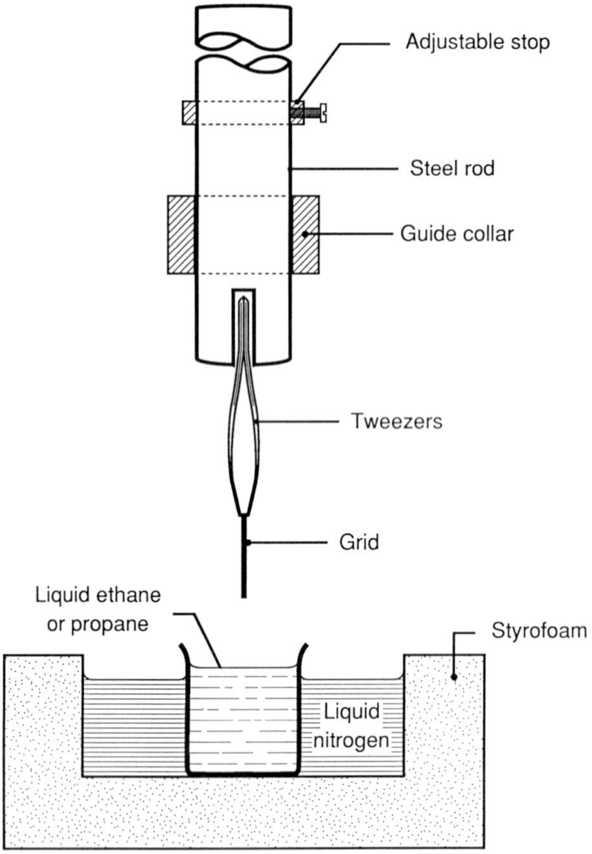

Figure 1. Schematic representation of the sort of apparatus that can be employed to vitrify thin films of suspensions of biological macromolecules. (After Dubochet, ref. 1.) Note that this illustration is not to scale and that the precise dimensions of the components are not very critical.

Liquid hydrocarbon (ethane or propane) is held in a small metal cup (about 2 cm diameter and about 4 cm long) surrounded by a bath of liquid nitrogen, held in a container made from styrofoam or some other suitable insulator. The liquid hydrocarbon can be stirred using either a magnetic stirrer bar which should preferably be driven by a water powered stirrer (to avoid the hazard associated with sparks from an electrical stirrer) or manually using a metal rod. The sample is applied to an EM grid held in tweezers that are themselves clamped into the end of a steel rod about 2 cm diameter and about 10–15 cm long. The steel rod runs through a cylindrical guide and is held so that the grid is located about 5–10 cm above the surface of the liquid hydrocarbon. An adjustable stop collar on the rod ensures that, when it is released, the grid is plunged below the surface of the liquid hydrocarbon but does not hit the bottom of the container. Generally, the weight of the rod ensures that the grid is plunged sufficiently quickly into the liquid hydrocarbon, although some workers use rubber bands to increase the speed with which the rod falls. A small metal screw cap from a vial serves as a suitable container for vitrified grids when they are transferred to the liquid nitrogen in the outer container.

Protocol 1. *Continued*

3. When the hydrocarbon is supercooled (and is starting to freeze so that it forms a slush), apply a small aliquot (about 2–5 μl) of sample to the grid and then blot it off. The aim is to produce a layer of water about 100–200 nm thick. Blotting both sides of the grid by pressing firmly with filter paper for about 1–2 sec is often adequate, but some experimentation is usually required to find the optimum conditions for each specimen.[a] When large specimens are being employed, there are advantages in blotting only from the side opposite to that to which the specimen was applied (14).
4. Release the steel rod, so that the grid plunges into the liquid hydrocarbon.
5. Transfer the frozen grid to the liquid nitrogen in the outer container,[b] draining off excess liquid hydrocarbon above the inner container while taking care that the specimen is not warmed. Specimens are stored under liquid nitrogen until examination in the electron microscope.

[a] Some evaporation usually takes place between blotting and vitrification. This can increase the concentration of salts and buffers in the film and so can cause considerable difficulties with salt-sensitive specimens. Often best results are obtained by working in distilled water with the minimum of added salt and buffer. If it is necessary to use appreciable concentrations of salt, the extent of evaporation can be reduced by working in a high relative humidity atmosphere. However, this can produce problems due to condensation of water into the liquid hydrocarbon bath.

[b] A razor blade can be helpful when removing the grid from the tweezers.

4. Electron microscopy: equipment

You will need:

- a high-vacuum electron microscope, preferably with a low-dose kit
- a cryo-specimen holder
- an efficient anticontamination device

In addition, you will need to know how to:

- record images under low-dose conditions
- obtain high-angle electron diffraction patterns

To examine vitrified specimens by transmission electron microscopy it is necessary to have a specimen stage that will operate at below −140°C and to have a sufficiently clean microscope vacuum and efficient cold trap to prevent contamination of the specimen. Several microscope manufacturers sell cryo-stages, and many workers in the field have found that the specialist stages manufactured by Gatan Inc. are easy to use and work well in practice. Often some modification to the electron microscope cold trap may be necessary to prevent contamination of the specimen. Some designs for improved cold traps have been published (see, for example, ref. 15) and again many workers have obtained satisfactory results using the special cold traps manufactured by Gatan Inc. It is probably unrealistic, however, to attempt the examination of vitrified specimens in microscopes that do not have turbomolecular or ion-getter pumps and which cannot reliably attain column vacuum better than 10^{-6} Torr. In addition to contamination, instrumental problems can arise with respect to stage drift and vibration of the specimen and sometimes it can be difficult to attain extremely high resolution (say better than about 1 nm) particularly with side-entry stages. However, resolutions better than 2 nm can usually be attained without difficulty.

5. Examination of specimens

Grids with vitrified films are loaded into the microscope specimen holder under liquid nitrogen and transferred rapidly into the electron microscope. It is important to maintain the sample temperature below about −140°C, to prevent crystallization of the vitrified water (11, 16). Some time is usually required for the film of hydrocarbon that usually coats the grid to evaporate and for the stage to reach thermal equilibrium. Times of the order of 15–30 min are common. During this time, and also when examining the specimen in the microscope, it is important to check that the sample has not become dehydrated by freeze-drying and that the vitreous film has not crystallized. Freeze-drying is usually easily detected by the increase in specimen contrast it produces. However, with a new specimen (and even with some well-characterized ones) it is prudent, at the end of a run, to warm up the stage in the microscope in order to freeze-dry the specimen. One can then compare this known freeze-dried specimen with the images recorded earlier. Ice-crystallization (and also the presence of a vitrified water film) is best checked by electron diffraction. Electron microscope manuals usually describe how to obtain electron diffraction patterns and usually have a special diffraction condition that can be selected. *Protocol 2* gives a general method for obtaining these patterns.

Protocol 2. Electron diffraction

Check that your electron microscope has the following:

- a diffraction aperture holder (fitted with a range of apertures between 10 and 100 μm)
- a selected area diffraction function control

1. Obtain a normal brightfield image on the screen at about 20 000×.
2. Focus on an adjacent area. Reduce illumination by substantially underfocusing the second condenser lens.
3. Insert and centre a diffraction aperture that roughly covers the field of interest.

4. Select the diffraction function of the microscope and adjust for a camera length of 100–250 mm.
5. Overfocus the second condenser lens and then focus the central spot of the diffraction pattern. Mask out the central spot using the beam stop if fitted.
6. Record a series of exposures of between 1 and 30 seconds (this must be done manually as exposure meters are not reliable for electron diffraction patterns; the precise time depends on the microscope and illumination conditions).

In electron diffraction patterns, vitrified water produces a ring at a spacing of about 0.37 nm, whereas crystalline ice shows distinct sharp spots at 0.366 nm and at either 0.343 nm and 0.389 nm (for hexagonal ice) or 0.224 nm (for cubic icc). It is not uncommon to find a few small crystals of hexagonal ice contaminating areas of otherwise vitrified water. These crystals are easily recognized by their sharp edges and dark contrast and usually arise from the condensation of water during transfer of the vitrified specimen (see *Figure 2*).

Once the stage has stabilized, one can start searching for suitable areas. The thickness of the vitrified film tends to vary substantially over the grid, usually in some places it is too thick and electron-opaque to see through, whereas in other places it will be too thin and will have broken. Areas of suitable thickness are found by scanning at low magnification (several thousand times). Unstained vitrified specimens show very little contrast and one can usually do little more than detect the presence of material by direct observation on the microscope's viewing screen, particularly since the radiation sensitivity makes the use of low levels of illumination obligatory.

5.1 Radiation damage

Generally, severe radiation damage (bubbling) is seen after doses of 2000–5000 electrons per square nanometre (1, 17, 18), which is about an order of magnitude lower than that usually employed to record micrographs of biological material. The radiation sensitivity of vitrified specimens makes it vital to record images using low dose conditions and often makes it difficult to record tilt series to produce three-dimensional reconstructions. Problems due to radiation damage can be circumvented by (*i*) recording images at the lowest magnification practicable, (*ii*) minimizing the electron dose received by the specimen before the image is recorded, and (*iii*) maximizing the contrast and density of the exposure by developing in full-strength D-19 for up to 20 min. Generally, most photographic films used for electron microscopy can record information with high efficiency provided that the finest spatial frequencies (highest resolution features) present in the image occur at spacings of about 25 μm on the film. So, if one is aiming for a resolution of

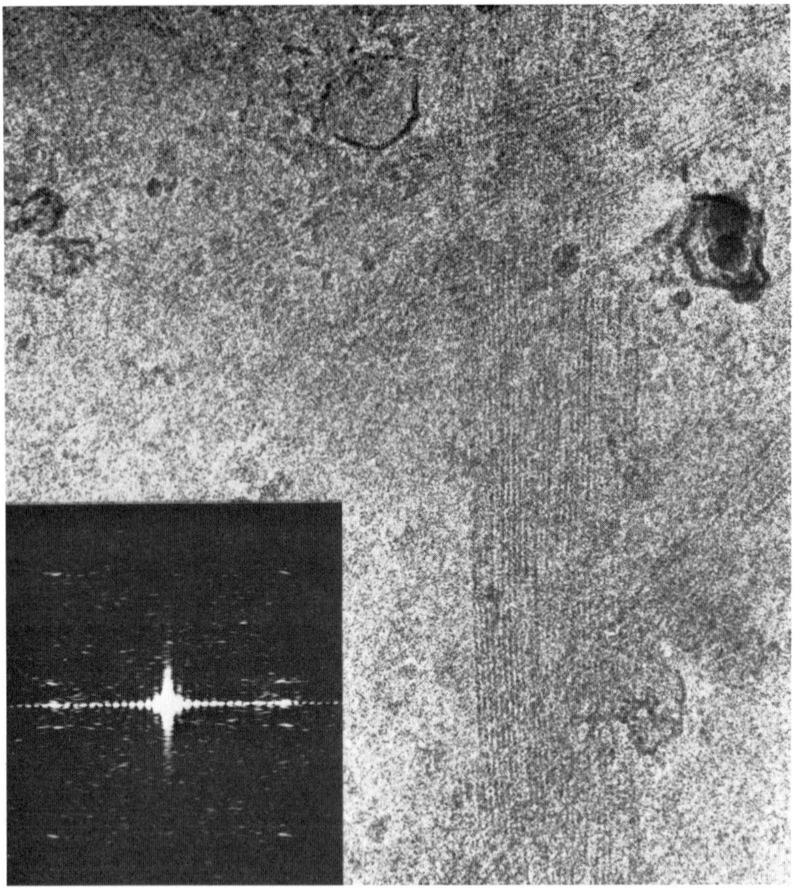

Figure 2. Transmission electron micrograph of vitrified actin paracrystals, which appear dark as protein is more dense than vitrified water. The small, dark objects (arrows) are ice-crystals. Although little detail can be seen in these paracrystals by eye, optical diffraction (inset) shows that their structure has been well preserved. (Reproduced from ref. 31.)

2 nm, an electron microscope magnification of about 12 000× should be adequate. Note that the exposure required to produce a given density on the film increases with the square of the magnification.

Many electron microscopes have a special kit that facilitates the recording of low-dose images. These kits usually enable the adjustment of focus and astigmatism on an area adjacent to the area to be photographed and so ensure that the exposure received by the desired area before the actual recording of the image is small compared with the exposure used to actually record the image. When a low-dose kit is not available, *Protocol 3* can be used to record low-dose images.

Protocol 3. Recording low-dose images without a low-dose kit

1. Adjust filament excitation so that the exposure meter indicates an exposure time of about 8 sec when the beam is expanded to just cover the area of the viewing screen that corresponds to the size of the film being used. Set the exposure manually to 2 s longer than indicated by the meter.
2. Expand the beam by two steps of coarsest condenser control. Generally this should expand the beam by at least a factor of 10.
3. Scan the sample using this very low level of illumination (if necessary reducing the magnification) until a suitable area is found.
4. Rapidly move the suitable area off the main viewing screen and refocus the condenser by two steps to reach the condition produced in step 1.
5. Adjust focus and astigmatism (often easiest using the carbon of the support film or the Fresnel fringes surrounding a piece of dirt or a small ice-crystal).
6. Re-expand the beam by two steps.
7. Rapidly bring the desired area to the centre of the stage and initiate an exposure.
8. Soon (about 1–2 s) after the film is transported and the shutter opened, focus the beam by two steps.

5.2 Contrast and focus

Contrast in unstained vitrified specimens derives from the intrinsic density difference between vitrified water and biological material. Therefore, because it is denser, protein appears dark, whereas in negatively stained material it is white. However, the contrast in unstained vitrified specimens is substantially lower than with negative staining and so images are usually recorded several μm underfocus (about an order of magnitude greater than is usually employed for biological material) to increase phase contrast (1, 19) as illustrated in *Figure 3*.

The low-dose conditions used to record images means that they are sometimes not so well defined statistically as those obtained with negatively stained specimens. This, combined with their generally low contrast can make it difficult to make out fine structural details. Computer image processing is therefore an important adjunct to the electron microscopy of vitrified specimens. Averaging methods, usually based on Fourier transforms and correlation averaging (20–22), are particularly effective with crystalline specimens or with objects that have a high degree of symmetry such as virus particles or nuclear pores, and it would be difficult to undertake a detailed investigation of a biological structure without access to these methods.

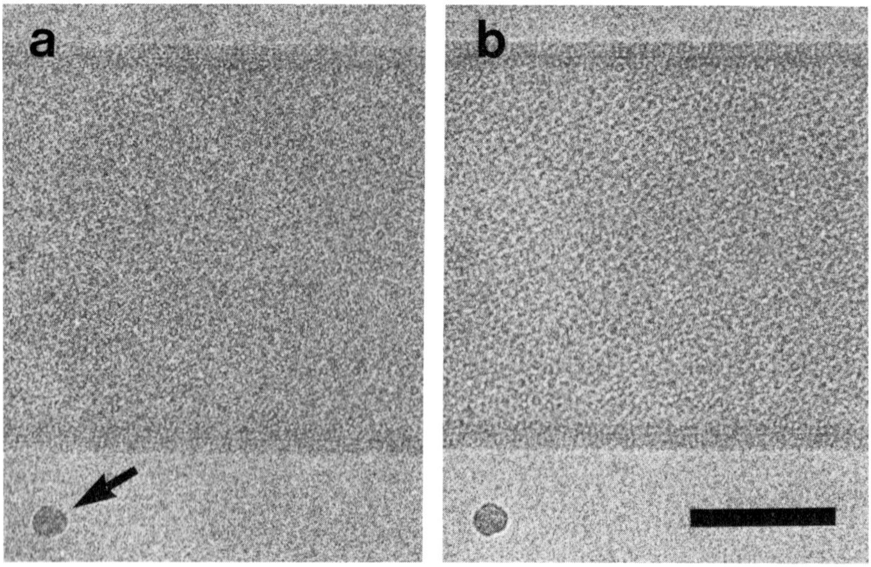

Figure 3. Transmission electron micrographs of vitrified acetylcholine receptors at different defocus values. (a) 800 nm under focus; (b) 2 μm under focus. Note how much clearer the pattern is in the highly defocused image (b). There is a small ice crystal at the bottom left (arrow in a) and the white Fresnel fringe (particularly clear in b) is a useful guide to defocus. Protein is dark because contrast reflects the density difference between sample and vitrified water. Bar is 100 nm. (Micrographs courtesy of Dr Nigel Unwin.)

6. Image interpretation

Micrographs of vitrified material are usually recorded at high defocus values and so some care is needed in interpreting these images. Moreover, images of these specimens are formed almost entirely by phase contrast (1, 23, 24), whereas images from stained material usually also contain substantial contributions from amplitude contrast. The predominately phase nature of the images obtained with vitrified specimens results in their relationship to the original object being somewhat different to that encountered in negatively-stained material and this can also sometimes complicate detailed interpretation. The theory of image formation in thin biological objects and the influence of defocus on amplitude and phase contrast have been discussed in detail elsewhere (25, 26), but it is important to appreciate their implications for unstained vitrified specimens. It is most convenient to discuss these problems in terms of the microscope contrast transfer function (CTF), which is analogous to the frequency response spectrum for Hi-Fi equipment and describes the fidelity with which each spatial frequency present in the object is represented in the image. Broadly, in a negatively stained object at optimal defocus (about 500 nm), low frequency amplitude contrast components complement higher frequency phase contrast components and so most spatial frequencies to about 2 nm are present in the image with approximately equal weight (4, 25). Consequently, the CTF for these objects is approximately linear. But in unstained objects, not all spatial frequencies are represented in the final image with equal fidelity and only a band of frequencies is represented strongly, whereas frequencies above and below this band are represented weakly. Which frequencies are represented in the band of maximum contrast transfer depends on the defocus and, for values commonly employed for vitrified specimens, usually correspond to resolutions of about 3 to 6 nm (see ref. 4). Moreover, because the CTF oscillates at high spatial frequencies (see ref. 25), some higher frequency components of the image may be included with reversed contrast. Consequently, considerable caution is sometimes needed in interpreting higher resolution features of these images, particularly with respect to defining particle boundaries and the relative weights of different image components. These problems are usually less pronounced in crystalline objects, but bounded objects, such as viruses, helices, and small cytoplasmic components, may be substantially altered. These problems are least severe in tube-like objects in which most material is present at about the same radius (22, 27, 28). Attempts have been made to correct for these problems (see, for example, refs 27 and 29), but the inherently small signals at low spatial frequencies often make accurate correction difficult. Because of these difficulties of interpretation, vitrified specimens are usually most effective in providing information about subunit location and changes in structure (either as a result of conformational change or labelling) rather than precise details of molecular outlines.

7. Indications for using vitrified specimens

A wide range of biological specimens have been examined by electron microscopy while held in thin vitrified films and *Figures 2–4* show some illustrative examples. Provided adequate precautions are taken, micrographs of vitrified material can be recorded to resolutions comparable to those obtained by more conventional methods, such as negative staining. Moreover, because the environment of the sample remains close to that of its native milieu, fine structural details are often preserved more faithfully. However, in many specimens, there is little difference between the results obtained by negative staining and those obtained in vitrified films, and so it is important to choose specimens carefully in order to maximise the benefit of examining them in this way. Generally, most impressive results are obtained when specimens are preserved poorly by conventional methods, or when advantage can be taken of the ability to manipulate the composition of the aqueous phase to introduce structural transformations. Because of their weak contrast and the phase nature of the images obtained with frozen hydrated material, there can often be problems in interpreting micrographs of bounded objects, and so crystalline objects or at least particles with high internal symmetry are generally advantageous.

Frozen-hydrated specimens are particularly useful for examining the structure of proteins in membranes, because they enable the whole protein to be examined and not just that portion which projects outside the membrane. Impressive results have been obtained with two-dimensional crystals of gap junctions (30), acetylcholine receptors (1, 32), *Figure 3*, and the F_1-ATPase (33), where three-dimensional models have been produced by combining information from tilted views. Viruses containing membranes have similarly been fertile objects and have given an unprecedented quantity of detail about their internal structure (34–36).

A powerful use of vitrified specimens is in detecting small structural changes produced by labelling or adding or subtracting subunits. Single mercury labels attached to tropomyosin paracrystals have been observed in this way (37), whereas material being translocated through nuclear pores has been observed (38) together with a wealth of detail about the internal structure of

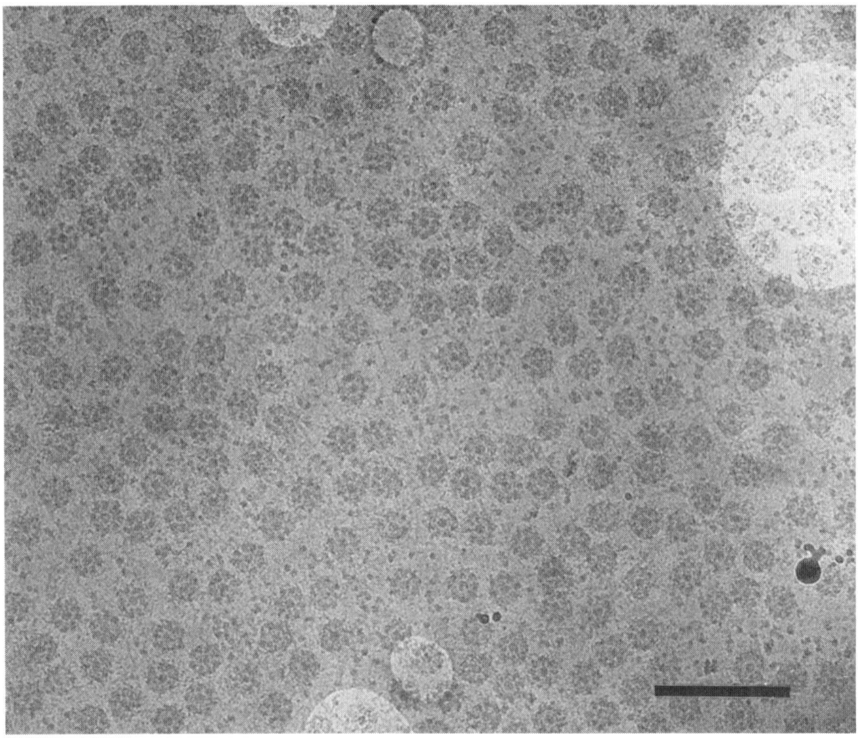

Figure 4. Vitrified *Xenopus* oocyte nuclear pores at high defocus to illustrate the eightfold internal structure of these objects. Bar is 0.5 μm. (Micrograph courtesy of Dr Chris Akey.)

the pores (*Figure 4*) and the conformational changes proposed to be associated with its central 'transporter' granule (38). By analysing images of decorated muscle thin filaments with and without tropomyosin the location of this molecule and its relationship to bound myosin heads has been established with high precision (39), while rapid vitrification has enabled intermediates in microtubule assembly to be observed. Insights into gap junction (30) and acetylcholine receptor function (32) have been obtained by obtaining images corresponding to different physiological states by manipulating solvent composition while an unexpected effect of ATP concentration on muscle thick filament structure has also been observed in this way (40). Vitrification is also a particularly powerful method for examining delicate material such as chromosomes (41), or material that is distorted by the forces involved in drying conventional specimens, such as coated vesicles (42).

Acknowledgements

I am most grateful to my colleagues in Cambridge, and particularly to Nigel Unwin, for their many helpful comments, criticisms, and suggestions.

References

1. Dubochet, J., Adrian, M., Chang, J.-J., Homo, J.-C., Lepault, J., McDowall, A. W., and Schultz, P. (1988). *Quarterly Reviews in Biophysics*, **21,** 129.
2. Lepault, J. and Dubochet, J. (1987). *Methods in enzymology*, Vol. 127 (ed. L. Packer) 719–730.
3. Chiu, W. (1986). *Annual Reviews in Biophysics and Biophysical Chemistry,* **15,** 237.
4. Stewart, M. and Vigers, G. (1986). *Nature,* **319,** 631.
5. Stewart, M. (1989). *Electron Microscopy Reviews,* **2,** 117.
6. Bruggeller, P. and Mayer, E. (1980). *Nature,* **288,** 569.
7. Angell, C. A. and Choi, Y (1986) *Journal of Microscopy*, **141,** 251.
8. Külber, C. (1988). *Quarterly Reviews in Biophysics,* **21,** 229.
9. Murray, J. M. and Ward, R. (1987). *Journal of Electron Microscopy Technique,* **5,** 285–290.
10. Murray, J. M. and Ward, R. (1987). *Journal of Electron Microscopy Technique,* **5,** 279–284.
11. Dubochet, J., Lepault, J., Freeman, R., Berriman, J. A., and Homo, J.-C. (1982). *Journal of Microscopy,* **128,** 219.
12. Aebi, U. and Pollard, T. D. (1987). *Journal of Electron Microscopy Technique,* **7,** 29–33.
13. Ward, R. and Murray, J. (1987). *Journal of Electron Microscopy Technique,* **5,** 275–277.
14. Toyoshima, C. (1989). *Ultramicroscopy,* **30,** 439–444.
15. Homo, J.-C., Booy, F., Labousse, P., Lepault, J., and Dubochet, J. (1984). *Journal of Microscopy,* **136,** 337.

16. Heide, H.-G. and Zeitler, E. (1985). *Ultramicroscopy,* **16,** 151.
17. Zeitler, E. (1982). *Cryomicroscopy and radiation damage.* North-Holland, Amsterdam.
18. Talmon, Y., Adrian, M., and Dubochet, J. (1986). *Journal of Microscopy,* **141,** 375.
19. Adrian, M., Dubochet, J., Lepault, J., and McDowall, A. W. (1984). *Nature,* **308,** 32.
20. Stewart, M. (1988). *Journal of Electron Microscopy Technique,* **9,** 301.
21. Stewart, M. (1988). *Journal of Electron Microscopy Technique,* **9,** 325.
22. Vogel, R. H., Provencher, S. W., von Bunsdorff, C.-H., Adrian, M., and Dubochet, J. (1986). *Nature,* **320,** 533.
23. Eusemann, R., Rose, H., and Dubochet, J. (1982). *Journal of Microscopy,* **128,** 239.
24. Toyoshima, C. and Unwin, N. (1988). *Ultramicroscopy,* **25,** 279.
25. Erickson, H. P. and Klug, A. (1971). *Philosophical Transactions of the Royal Society of London*, Ser. B, **261,** 105.
26. Glaeser, R. M. (1985). *Annual Reviews in Physical Chemistry,* **36,** 243.
27. Mandelkow, E.-M. and Mandelkow, E. (1985). *Journal of Molecular Biology,* **181,** 123.
28. Toyoshima, C. and Unwin, N. (1988). *Nature,* **336,** 247.
29. Lepault, J. and Leonard, K. (1985). *Journal of Molecular Biology,* **182,** 431.
30. Unwin, P. N. T. and Ennis, P. D. (1984). *Nature*, **307,** 609.
31. Stewart, M. (1990). In *Modern microscopies: Techniques and applications* (ed. P. Duke and A. Michette), pp. 9–39, Plenum Press, London.
32. Unwin, N., Toyoshima, C., and Kubalek, E. (1988). *Journal of Cell Biology,* **107,** 1123.
33. Gogol, E. P., Lucken, U., Bork, T., and Capaldi, R. (1989). *Biochemistry,* **28,** 4709.
34. Baker, T., Drak, J. and Bina, M. (1988). *Proceedings of the National Academy of Sciences, USA,* **85,** 422.
35. Prasad, B. V. V., Wang, G. J., Clerx, J. P. M., and Chiu, W. (1988). *Journal of Molecular Biology,* **199,** 269.
36. Fuller, S. D. (1987). *Cell,* **48,** 923.
37. Stewart, M. and Lepault, J. (1985). *Journal of Microscopy,* **138,** 53.
38. Akey, C. A. (1989). *Journal of Cell Biology,* **109,** 955.
39. Milligan, R. A. and Flicker, P. (1987). *Journal of Cell Biology,* **105,** 29–39.
40. Menetret, J.-F., Hofmann, W., and Lepault, J. (1988). *Journal of Molecular Biology,* **202,** 175.
41. Dubochet, J., Adrian, M., Schultz, P., and Oudet, P. (1986). *EMBO Journal,* **5,** 519.
42. Vigers, G. P. A., Crowther, R. A., and Pearse, B. F. M. (1986). *EMBO Journal,* **5,** 529.

11

Quantitative biological X-ray microanalysis

ALICE WARLEY and BRIJ L. GUPTA

1. Introduction

When a specimen is examined in the electron microscope the interaction of the electron beam with the atoms of the specimen results in the generation of X-rays: these originate from two different sources. The interaction of the electron beam with orbiting electrons in atoms of the specimen causes the emission of characteristic X-rays, whose energy depends on the atomic number of the element in which the transition is occurring and the electron shells involved. X-rays are also produced when the electron beam is slowed by interaction with the atomic nuclei of the specimen. In the X-ray energy spectrum this emission forms a continuous band of radiation beneath the characteristic elemental peaks which is variously called continuum, white radiation, or *Bremsstrahlung* (braking radiation). The generation and collection of X-rays (*Figure 1*) forms the basics of the technique of X-ray microanalysis (1, 2), more properly called electron probe X-ray microanalysis.

The development of X-ray microanalysis began in the 1950s and since its earliest days the technique has been applied to the study of biological specimens (3, 4). With a beryllium windowed X-ray energy dispersive Si(Li) spectrometer (EDS) the method allows quantitative analysis of all elements of atomic number >11 (Na) at concentrations as low as 1 mmol/kg wet weight (5). Recently, the development of thin-windowed and windowless EDX detectors allows analysis of elements down to atomic number 5. It is important to realize two points about X-ray microanalysis: the technique measures the total amount of an element present in the irradiated area, and therefore it does not discriminate between free and bound forms; the elements are measured in the specimen *as it exists at the time of analysis*. Since biological materials invariably undergo some form of manipulation before examination in the electron microscope, the method of specimen preparation has a great influence on the results from X-ray microanalysis.

The two major routes used for specimen preparation begin with either conventional (aldehyde) fixation, or cryofixation. In this chapter we will

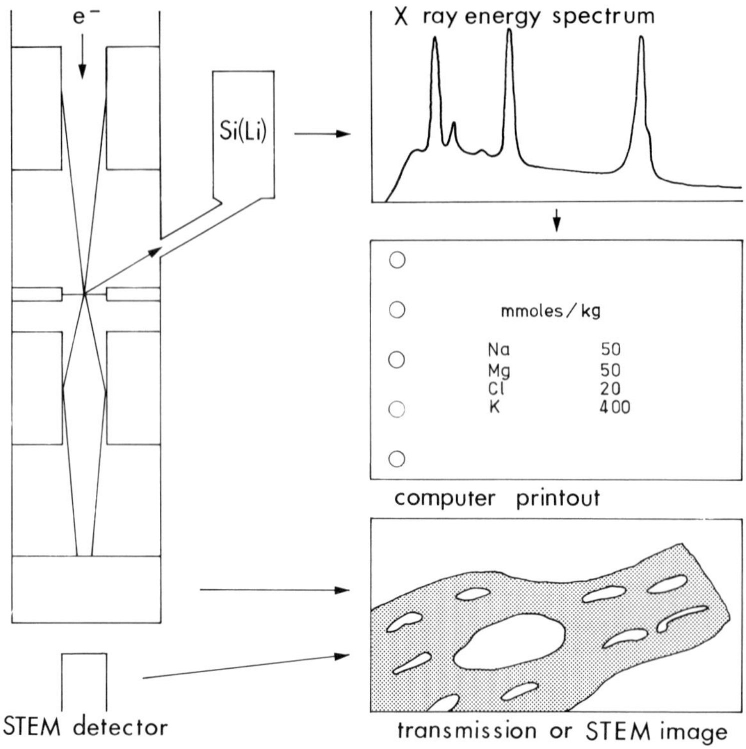

Figure 1. Basics of the technique of X-ray microanalysis. In the electron microscope an area of interest in the specimen is selected visually. The electron beam is focused down on to the chosen area; the emitted X-rays are collected by an X-ray spectrometer. The X-ray signals are processed electronically to produce and display a spectrum from which quantitative information is derived.

concentrate on the use of cryotechniques for the analysis of the diffusible elements Na, K, Mg, Ca, and Cl. Other techniques will be mentioned only briefly.

2. General precautions for X-ray microanalytical studies

It is easy for any specimen to become contaminated with elements from the environment during preparation and analysis, so rigorous attention to detail is necessary at all times. General sources of contamination are:

- dust, including chemical dust
- dirty glassware
- sweat (Na and Cl)

- vacuum sources, evaporators, and microscope (Cl, S, Si)
- desiccants, e.g. silica gel, P_2O_5, $CaCl_2$, $CuSO_4$ (numerous)
- talcum powders, from the body and from surgical gloves (Al, Ca, Mg, Si, Zn)
- adhesive tapes (Cl)

To avoid contamination from these sources:

- make sure surfaces are clean before beginning work
- ensure glassware is cleaned thoroughly and stored in a dust-free environment
- do not handle small items with fingers, use clean tools
- use a vacuum evaporator with a cold-trap and use Si-free oils
- use vacuum desiccators (contamination can still occur from O-rings)
- do not use talcum powders, powdered surgical gloves, or adhesive tapes
- do not store chemicals or make solutions in the microprobe laboratory
- do not use the specimen preparation area for photographic processing

3. Preparation of the specimen

3.1 Conventional fixation and embedding

Conventional methods of specimen preparation are used for the study of accumulated material in pathological samples (6). It is generally assumed that the fixation, even with ion-precipitating agents such as oxalate present, and embedding do not cause substantial movement of the heavy elements. Details about the use of fixatives and embedding techniques can be found in Chapters 1 and 2 in this volume and in other technical publications about electron microscopy (7). Thus, practical details will not be considered further here. It is important to note that conventional fixation causes the cell membranes to become permeable, so that mobile elements are lost from the cell, or may be transferred from one site to another. Consequently, these methods cannot be used for the study of diffusible elements.

3.2 Cryopreparative techniques

The aim of cryopreparation for X-ray microanalysis is to produce a specimen for analysis in which the elemental composition is not measurably changed from the living state. This is achieved by cooling the specimen as rapidly as possible to low temperatures (<-150°C), in order to preserve morphological detail in the cell, and to arrest the diffusible elements at their subcellular locations. In order to prevent movement of elements the specimen is then maintained at low temperatures throughout any further steps, until it is either stabilized by freeze-drying, or analysed in the frozen-hydrated state. It should be understood that, for X-ray microanalysis, the preparation of the specimen does not begin with cryofixation but with the preparation of the specimen for freezing. The way in which the specimen is handled before cryofixation is vitally important, as movement of elements can easily be induced, making the specimen unreliable for further study.

3.2.1 Preparation of the specimen for cryofixation

The general causes of elemental redistribution during the preparation of the specimen for cryofixation are:

- mechanical damage
- anoxia
- lowering of temperature (e.g. a drop from 37°C for homeotherms)
- desiccation due to exposure to a dry atmosphere (8)
- use of a non-compatible medium (e,g. cryoprotectant)

For both cell cultures and tissues, lowering of temperature and desiccation can be avoided by working in a humidity chamber (5, 9). Precautions for handling cells and tissues are given below and more detailed protocols for each are given in the next section.

i. Cells

In addition to the above, cells in culture and isolated cells, suffer from the fact that ice-crystals readily form in an excess of external medium, making sectioning difficult and presenting the danger that the tonicity of the medium may be altered, causing elemental redistribution across the membrane. Useful precautions when handling cells are:

(a) Do not remove cells grown as monolayers from their substrate.

(b) Concentrate cells in suspension by use of a microcentrifuge (10).

(c) Remove excess medium from cell-monolayers by gentle blotting (11).

(d) Use a non-penetrating cryoprotectant (12) to minimize ice-crystal formation in the medium. **Caution:** make sure that the cryoprotectant does not alter intracellular elemental concentrations.

(e) Protect cell-suspensions from the atmosphere during plunge-freezing; for example, by enclosure in suitable thin metal foils (ref. 13; *Figure 2*), or freeze-etch supports. **Caution:** exposure to metals, however brief, could cause elemental redistribution.

ii. Tissue

Small tissues, usually tubular, from a range of invertebrates, are often much easier to handle than those from vertebrates. For larger animals, simple cubing or slicing to produce a small piece of tissue from a larger tissue or an organ (as is often suggested for morphological studies) alters intracellular

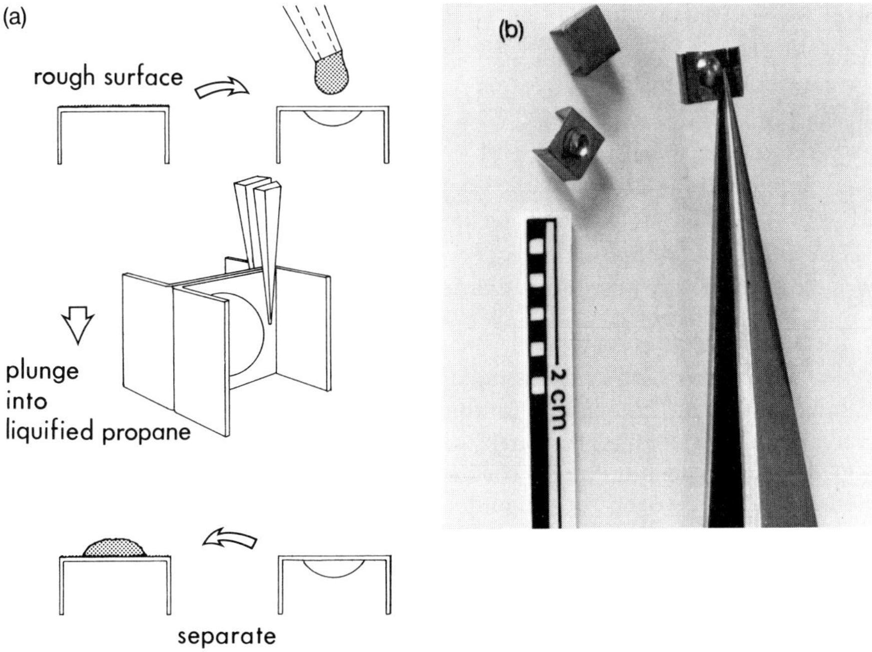

Figure 2. Thin metal foils for protecting cell suspensions during cryofixation. In (a), the use of the foils is shown diagrammatically. In (b), the size of the foils is shown. (The diagram was redrawn from ref. 13 by kind permission of Dr K. Zierold, Dortmund, Germany, and Academic Press.)

elemental concentrations. There are several ways by which the tissue can be handled to maintain elemental concentrations as near as possible to the living state. These are:

(a) Small, thin organs such as the pancreas can be isolated and perfused with oxygenated physiological saline.
(b) Small, intact, functioning pieces of tissue such as papillary muscle or sections of transporting epithelia can be removed and incubated in an aerated physiological medium and frozen under controlled conditions.
(c) Some tissues can be cryofixed *in situ*.
(d) For some complex tissues (e.g. kidney), either isolated single cells or multi-cellular preparations can be made. **Caution:** the effects of isolation on intracellular elemental concentrations need to be taken into account.

3.2.2 Cryofixation

Detailed descriptions about the theory and practice of cryofixation can be found in refs 14 and 15. However, for biological X-ray microanalysis is the following points should be borne in mind:

(a) Use of pre-fixation is not permitted: it causes elemental redistribution.
(b) Use of penetrating cryoprotectants is not permitted: they also cause element redistribution.
(c) The quality of cryofixation is not as critical as in some other methods (cf. Chapter 10). The formation of small ice-crystals in the specimen can be tolerated provided that the ice-crystals are smaller than the probe-size to be used for analysis.

The following methods are most commonly used for the cryofixation of the specimens for X-ray microanalysis.

i. Immersion cryofixation

Immersion cryofixation achieves rapid freezing by plunging the specimen into a liquid cryogen. Equipment for immersion cryofixation is commercially available (Reichert-Jung) although many laboratories have constructed their own apparatus which is equally successful (*Figure 3*). Different liquids which can be used as cryogens are: liquid/solid nitrogen, halocarbons (Freons 22 and 13), propane, ethane, isopentane, and mixtures of propane and isopentane. The following practical points should be borne in mind.

(a) **Rigorous standards of safety must be maintained** (16, 17) **when handling cryogens**.
(b) The specimen must be protected even from slight desiccation before and during immersion in the cryogen.
(c) Cryogens such as Freons can be subcooled without freezing by vigorous stirring (*Figure 3*).
(d) A layer of cold cryogenic gas formed above the surface of the cryogenic vessel may be the most important factor in determining the rate of cooling of the specimen. Either this layer should be eliminated in the design of the freezing apparatus or the specimen must be moved through this layer as quickly as possible.
(e) Mechanical propulsion of the specimen into the cryogen rather than dipping by hand will give better and more reproducible freezing.
(f) A large volume of cryogen in a deep vessel (50 to 100 ml) should be used.
(g) Residual cryogens may contaminate the specimen (e.g. Freons with Cl, Br, or F).
(h) A layer of liquid propane (B.P. −40 C) remaining on the specimen makes cryosectioning difficult.

ii. Immersion cryofixation of a cell suspension

A method for immersion cryofixation of a cell suspension using a Reichert KF80 device is given in *Protocol 1*. For other freezing equipment follow relevant instructions. Details of the foils holding the cells are shown in *Figure 2*, and a micrograph of the material prepared in this way is shown in *Figure 4*.

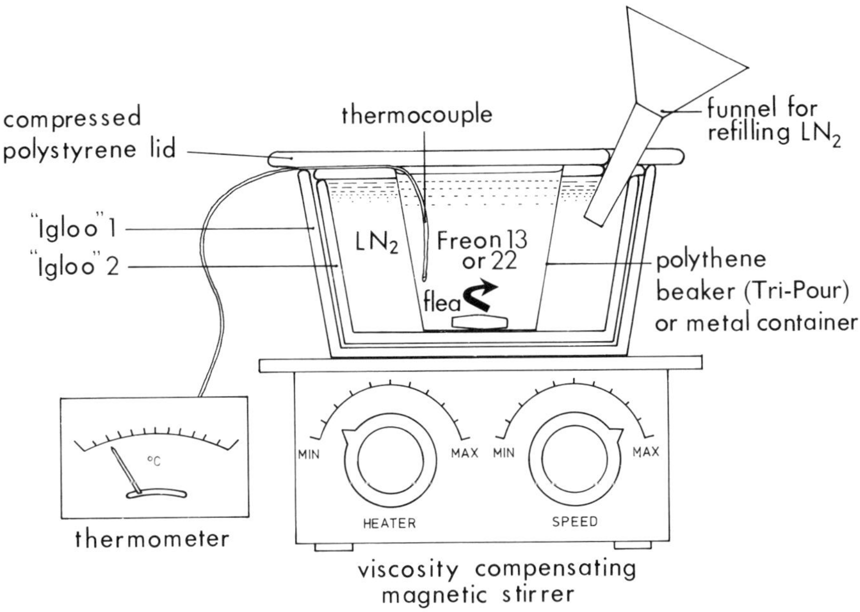

Figure 3. A simple apparatus for freezing specimens using supercooled cryogens (either Freon 13 or Freon 22). The apparatus is made from: two compressed polystyrene 'Igloos' (Jencon, Hemel Hempstead, Cat No. 193/3) or 'Thermobox 215 mm', (Elpas, England); 'Tri-Pour' polypropylene beaker, 50 ml; a viscosity-compensating, magnetic stirrer (Cornings, PC-351) used with a 2- or 3-cm 'flea'; an electronic thermometer with a suitable thermocouple extension (Matronix, Mat 400T). (*i*) Clean and thoroughly dry all components. (*ii*) Assemble the 'Igloo' without the 'Tri-Pour' beaker, fill-up with LN_2 and allow to settle. (*iii*) Insert the thermocouple into LN_2, cover tightly and calibrate the thermometer (−196°C). (*iv*) Add the 'flea' to the 'Tri-Pour' beaker in a fume cupboard and fill it up to the brim with clean, fresh Freon. Keep the beaker covered in order to avoid any condensation of moisture in the cryogen. (*v*) Keeping it tightly covered, transfer the beaker to the LN_2 'Igloo', insert the thermocouple, cover tightly, and stir vigorously. Keep the LN_2 level topped up to the brim through the side funnel. In a few minutes Freon 13 should cool down to about −190°C without freezing. **Caution.** (a) Do not use this apparatus for propane or ethane. (b) The failure to supercool is usually due to dirt or frost contaminating the cryogen. If the cryogen freezes, one should start all over again.

Protocol 1. Immersion cryofixation of a cell-suspension

1. Cool immersion fixation apparatus, fill with the cryogenic liquid, and allow to cool to the minimum temperature above freezing.
2. Remove a sample of the cell suspension from the culture vessel.
3. Concentrate the specimen by centrifugation.
4. Remove the supernatant and rapidly transfer a small droplet of the pellet to the cavity of a suitable smoothly polished metal sheet. Cover the polished sheet with a flat sheet having a roughened surface to form a sandwich (*Figure 2*).

Figure 4. Electron micrograph of a frozen-dried cryosection of isolated cardiac myocytes cryofixed according to *Protocol 1*. The quality of the freezing is good; subcellular detail can be clearly seen.

Protocol 1. *Continued*

5. Grip the sandwich with a pair of fine-pointed, self-closing forceps.
6. Mount in the plunging device and plunge immediately into the cooled cryogen.
7. In the chamber of the plunge-freezing apparatus separate the sandwich. The cell suspension should freeze on to the roughened surface of the non-dimpled plate.
8. Clearly label a storage tube, and cool the tube thoroughly by filling with liquid nitrogen. Place the cooled tube in the chamber, place specimen in the cooled tube, cap, and transfer to a storage Dewar under liquid nitrogen.

iii. *Immersion fixation of a tissue*

Protocol 2. Immersion cryofixation of a complex insect-tissue (cockroach salivary glands)

1. Carefully dissect out the salivary gland with associated secretory ducts, etc., under a high humidity environment
2. Transfer the gland to a dextran–Ringer's solution (Section 5.3.4) in a shallow glass vessel to allow microscopic observation. Keep in a high humidity environment and agitate to mix.

3. In order to avoid the formation of a dextran-free layer of Ringer's solution around the tissue (unstirred layer), gently pull the tissue through the solution several times.
4. Cut the tissue into small (0.5–1.0 mm) units.
5. In the meantime prepare the cryogen for freezing the specimens (*Figure 3*).
6. Carefully mount individual pieces of the tissue on to metal pins or any other mount suitable for the cryoultramicrotome. Scooping out the tissue using the tip of a fine, blunt glass or plastic needle is better than lifting with forceps.
7. Carefully suck off extra medium so as to reduce the size of the drop to a minimum without damaging the tissue.
8. Carefully freeze the sample, maintaining a humid environment around the specimen (4).
9. Retrieve the specimen, rapidly shake off extra cryogen, and place in liquid nitrogen.

The quality of freezing of a specimen can be checked immediately after cryofixation by use of an observation 'Igloo' (*Figure 5*). Place the specimen under nitrogen, transfer to the 'Igloo' and place the specimen in the stabilized liquid nitrogen in the metal cup. Any noticeable boiling of liquid nitrogen indicates that the specimen has warmed up during transit. Examine each specimen carefully using the stereomicroscope. A well-frozen specimen should have a lifelike appearance. A frosty 'sugar lump' appearance is a definite sign of ice-crystal formation. Such frosty specimens can only be used for studies which require measurements in macrocompartments with low resolution (a few microns). After examination the specimens should be transferred to the storage Dewar if not required immediately.

iv. *Cryofixation against a cold metal block*

In general, metals have a high thermal conductivity so the polished surface of a cooled metal is an ideal medium for cryofixation. This has led to the development of a number of quite complex impact–cryofixation devices. Caution needs to be exercised when using such an automatic device: during the time that elapses between removal of a tissue, mounting in the apparatus, and cryofixation, redistribution of elements will occur. In some laboratories such apparatus has been successfully modified so that cryofixation is achieved in a physiologically well-defined state (18).

A much simpler metal cryofixation device can be constructed from a pair of electronics pliers to which polished copper jaws with a mirror-smooth surface have been added (*Figure 6*).

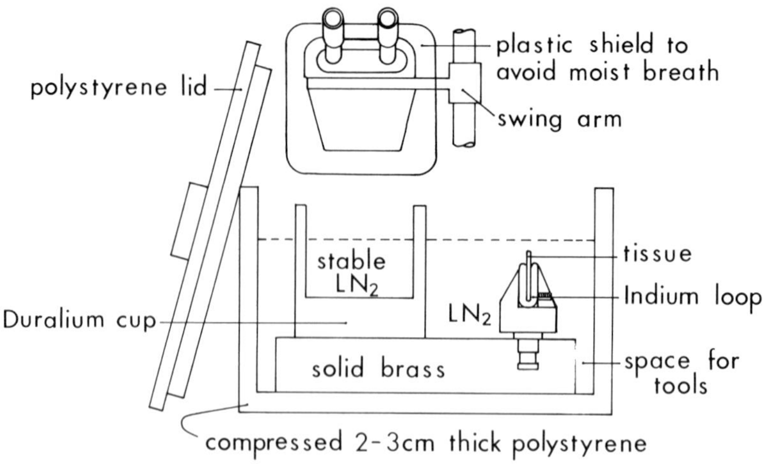

Figure 5. Observation 'Igloo' made from two compressed polystyrene containers and various metal components. For use, (*i*) clean and thoroughly dry all the components. (*ii*) Set up all the components, and fill-up with LN_2; keep covered between filling. (*iii*) When all the metal components have equilibrated and LN_2 has settled down, place all the instruments for handling the frozen specimens in place and allow to equilibrate. Keep the level of LN_2 in the metal cup slightly lower than in the container. Keep covered until needed. The LN_2 in the metal cup should be as clear as water, allowing observation through the stereomicroscope, provided that heat-absorbing filters are used in the illuminating system (high-intensity fibre-optics lamps work best).

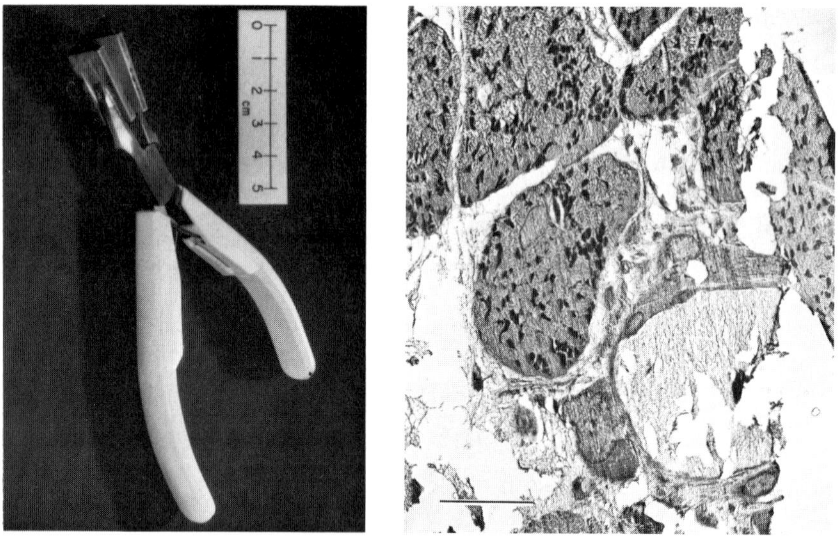

Figure 6. Freezing of tissue *in situ*. (a) A pair of copper-clad pliers in which the jaws have been replaced by copper blocks polished to a mirror-smooth surface. (b) An electron micrograph of a frozen-dried section of cardiac tissue cryofixed according to *Protocol 3*. Marker = 10 μm.

Protocol 3. Freezing of tissue using copper-clad pliers

1. Cool the pliers by immersing the blades in liquid nitrogen.
2. Shake off excess nitrogen; place the cooled pliers in liquefied propane; close the jaws of the pliers under the surface of the liquid propane.
3. Remove the pliers from the propane; keeping the jaws of the pliers closed until the very last moment, rapidly grab and squeeze the specimen between the jaws of the pliers.
4. Keeping the jaws closed round the specimen place the specimen into the observation 'Igloo' and then release the specimen.
5. Fracture the specimen into smaller pieces under the surface of the liquid nitrogen in the metal cup of the observation 'Igloo'.
6. Place in storage vial (see below) and transfer to storage system.

Cooled pliers are suitable for freezing layers of epithelia in an Ussing-type chamber, monolayers of cell cultures, or organs and tissues *in situ* (19). Other devices which have been described for freezing of tissues *in situ* are solid Freon clamps (20) and a chilled cryobiopsy needle (21). A specimen frozen in this manner is shown in *Figure 6*.

3.2.3 Storage and transfer of frozen specimens

Frozen tissue must be maintained at low temperatures (<150°C) to prevent ice-crystal growth. Liquid-nitrogen storage systems designed for tissue culture are suitable for this purpose. Specimens should be stored such that their location is known, and they can be retrieved without disturbance of other specimens. All specimens must be maintained at low temperatures during transfer between apparatus such as the freezing device to the storage Dewar and the storage Dewar to the cryoultramicrotome. Many well-frozen specimens are inadvertently damaged during such manipulations. A very practical and useful storage system for liquid nitrogen and liquid-nitrogen Dewars is shown in *Figure 7*.

3.2.4 Cryofixation followed by embedding techniques

Cryofixation followed by low temperature embedding or freeze-drying and embedding can be used for preparing specimens for X-ray microanalysis. It has not yet been shown conclusively that these procedures do not result in elemental redistribution. In addition, there is considerable doubt about the interpretation of results from these methods (22). For these reasons, such embedding techniques will not be discussed here. Further details can be found in Chapter 3 of this volume and in refs 23 and 24.

3.2.5 Cryosectioning

Cryostats developed for histology, which operate at −30°C can be used for

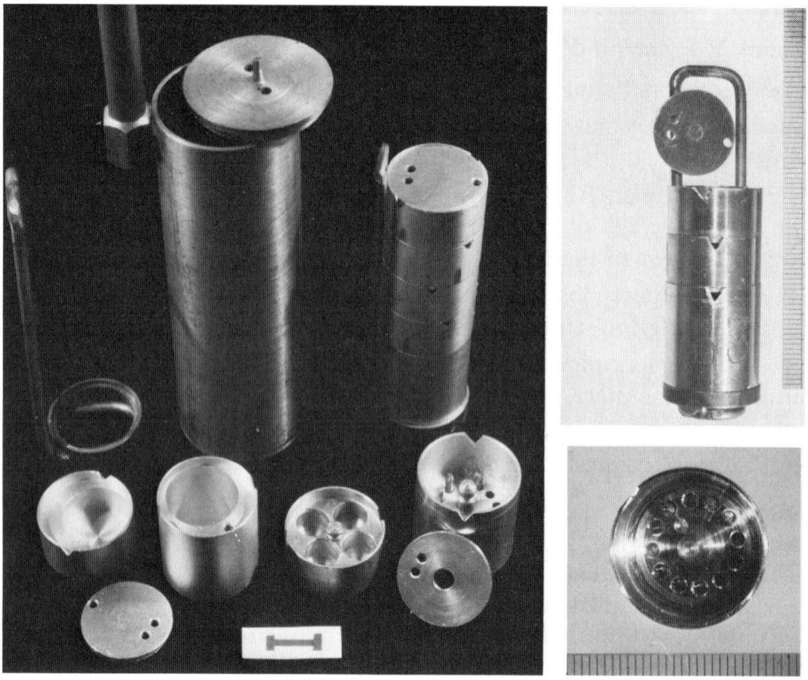

Figure 7. A home-made 'Tiffin' assembly of duralium pots for storing either pin-mounted or loose pieces of frozen tissues. When used with the observation 'Igloo' (*Figure 5*) the specimens in the pots can be examined under a stereomicroscope without the need to handle them.

cutting thicker sections (>1 μm) in work where high spatial resolution is not required (23). However, the relatively high temperatures at which these machines operate lead to the formation of large ice-crystals and may cause redistribution of diffusible elements. In most laboratories cryoultramicrotomes cooled by liquid nitrogen, operating at much lower temperatures (< −80 °C) are used. The temperature used for cryosectioning depends on the sample to be sectioned and must be determined experimentally for each application. In addition to the usual gadgets required for sectioning, the following points are important when cryosectioning for X-ray microanalysis.

(a) All tools must be cooled and kept frost-free before use.

(b) A diamond knife can be used for sectioning, but glass knives are in common usage. A good glass knife, preferably prepared by the balanced break method (25), should be used. The quality of glass knives can be improved by coating with tungsten (26). Details of a jig for the tungsten coating of glass knives is given in ref. 15. Advantages of tungsten coating are: the knife remains sharp longer; the knife is less prone to the development of static electricity; the knives can be stored (in a clean atmosphere). For prolonged storage a vacuum desiccator should be used in order to prevent the oxidation of the cutting edge which blunts the knives.

For cutting 1–2-μm-thick cryosections, low bevel-angle carbon-steel knives are better than glass knives (27, 28); 1-cm-wide carbon steel knives are now supplied by some manufacturers (Reichert-Jung).

(c) For X-ray microanalysis, EM grids made from low-atomic-number materials, such as beryllium and carbon, are often recommended. Such grids have two drawbacks. First, the grids produce no characteristic signal in the EDX spectrum and therefore the contribution of the grid to the continuum measurement cannot be estimated (Section 5.2.1). Second, they are usually made of a closely woven mesh and therefore are more likely to intercept scattered electrons and produce excess continuum. Grids made from copper, nickel, or titanium have the advantage that the characteristic X-ray peak can be used for adjusting the drift in the X-ray peaks in the EDX spectrum (Section 5.1). Grids with thin bars and a mesh size of 100 or lower, single slotted grids, or multi-slotted grids must be used to minimize the contribution of the grid to the EDX spectrum. The use of multislotted grids on an SEM holder is illustrated in *Figure 8*. Preparation of such a holder is given in *Protocol 6*.

Figure 8. 1-cm, low-background holders for use with STEM stages in an SEM. The holder on the right has been heavily nickel-plated and has a nickel multi-slot grid attached: it is also covered with a formvar–carbon film.

(d) The grids need to be coated with a support film. Suitable film materials are pure carbon, pioloform, collodion, or formvar. Methods for coating grids with these different support films can be found in ref. 29.

(e) Sections must be cut dry and transferred without the use of fluids to avoid the redistribution of elements. Therefore the use of sucrose droplets, as used for transferring cryosections in immunocytochemistry, is not permitted. Cutting, collecting, and pressing of cryosections is shown diagrammatically in *Figure 9*.

(f) If static electricity is a problem it can be neutralized by use of an anti-static gun.

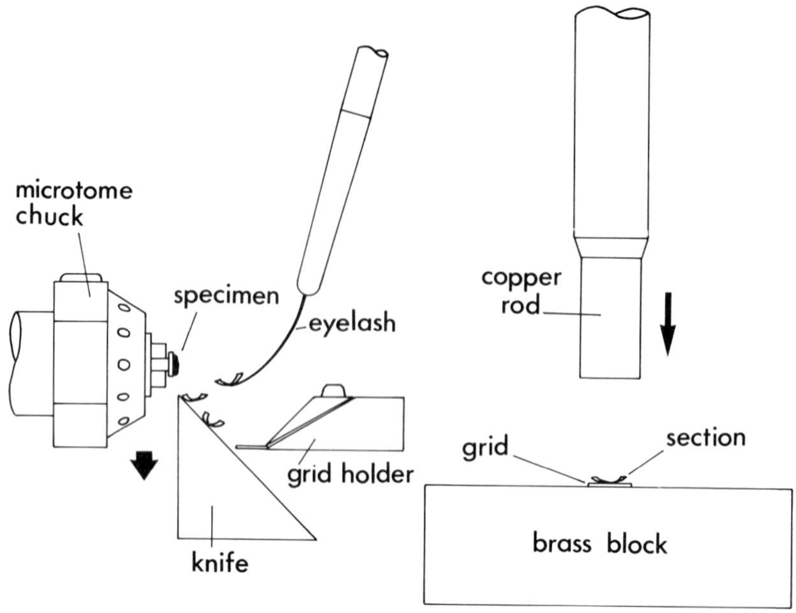

Figure 9. Dry method of transferring cryosectons from the edge of the knife to the grid. Grids are held close to the edge of the knife in a proprietary grid-holder. Cut sections are transferred on to the grid, using an eyelash. The grids are then transferred to a clean, flat surface, and the sections pressed on to the support film by use of a cooled, polished copper rod.

In general, more detailed instructions for cryosectioning are given in the manufacturer's instructions for a particular cryomicrotome. Further detail can also be found in ref. 30.

3.2.6 Transfer and storage of cryosections

Subsequent handling of cryosections depends on whether the sections are to be frozen-dried in an external freeze-drier, or whether the cryosections are to be transferred directly to the electron microscope.

i. External freeze-drying

When external freeze-drying is used care must be taken to ensure that the cold sections are not exposed to the atmosphere, allowing water vapour to condense on the section thus causing redistribution of elements. A procedure for external freeze-drying is given in *Protocol 4*.

Protocol 4. External freeze-drying

1. Pre-cool a transfer device (*Figure 10*) in the chamber of the microtome.
2. Place grids bearing sections inside the transfer device in the chamber of the cryoultramicrotome.

Figure 10. Cryotransfer pot for transferring grids bearing sections to an external freeze-drier. The pot made of brass is similar to one shown diagrammatically in ref. 15. The pot is cooled in liquid nitrogen, and liquid nitrogen is placed in the lower chamber of the pot. The grids are placed on the small tray which is covered by a thin brass weight, and transferred to sit on the ledge inside the pot. The pot is covered with its cooled lid and immediately removed to the chamber of the freeze-drier.

3. Cover the transfer device with the lid and transfer either to the chamber of a freeze-drier, or to a carbon-coating unit. **Caution:** do not place the device directly on the metal base of the coater, position insulating material between the two, so that the transfer device warms up slowly.
4. To evacuate the chamber use high vacuum ($< 10^{-3}$ Torr if possible).
5. Freeze-dry the sections overnight to allow adequate time for the sections to dry and warm up to the room temp.

At this stage the frozen-dried cryosections are hygroscopic and vulnerable to moisture from the atmosphere, the sections need to be coated with a thin layer of carbon, which protects the cryosections and also increases their conductivity. If the cryosections are frozen-dried in a carbon-coater, then, if possible, coating should take place without breaking the vacuum. This means using a transfer device from which the cover can be removed remotely. If the vacuum in the freeze-drying unit must be broken, admit dry nitrogen rather than room air. Every precaution should be taken to ensure that the cryosections are protected from the atmospheric moisture. Sensible working procedures include:

- handling cryosections in a glove box with the sections exposed to a flow of dry nitrogen
- use of Anglepoise lamps in the working area to create a warm atmosphere locally

- warming the sections to a few degrees above room temperature to prevent condensation of moisture

Transport grids to the carbon coater in a desiccator which has been warmed slightly.

ii. Transfer of cryosections to the electron microscope
Details of the procedures to be followed for transferring ultra-thin sections to the EM column can be found in Chapter 10, Section 5 in this volume. Details for transfer of thick cryosections and the procedure for freeze-drying, which is the same for both thick and thin sections, can be found in Section 5.3.5.

3.2.7 Storage

If the cryosections are not required immediately, then it is possible to store them. The main problem when storing cryosections is to ensure that they are not exposed to conditions which add contamination to the specimen and degrade the EDX spectrum. Cryosections which have been frozen-dried can be stored in an efficient vacuum desiccator. **Caution:** see Section 2 on contaminants. If long-term storage is envisaged, the conditions used must be checked thoroughly.

Frozen-hydrated cryosections mounted on grids alone can be stored successfully under liquid nitrogen (31).

4. Analytical equipment

4.1 Electron microscopes

Transmission electron microscopes (TEM) are used for obtaining analytical data at high resolution from thin sections of a specimen, whereas scanning electron microscopes (SEM) are used for analysing thicker (0.5–2 μm) sections at a lower resolution. In general, qualitative X-ray microanalysis can be carried out by using any electron microscope fitted with an X-ray spectrometer. However, if reliable quantitative analysis is to be carried out, some additional modifications and more stringent working conditions are required. The more important of these features in both SEM and TEM are outlined below.

(a) The minimum diameter of the focused electron beam in the conventional TEM is about one micron, therefore for high-resolution studies TEM equippped with either an accessory lens or with STEM is needed.

Caution: when a reduced raster or a static probe is located on a full screen STEM image, (in either a SEM or STEM) a shift in registration can occur. This leads to anomalous results. This problem can be coped with either by increasing the magnification so that the area of interest occupies most of the screen, or by locating the feature by its electron-density characteristics of the image on the wave form monitor.

(b) Knowledge about the exact probe-current is needed for accurate quantitative analysis. The microscope should, therefore, be equipped with a Faraday cage. Methods for continuously monitoring the beam current at the C2 aperture (32) or in the imaging chamber of the TEM (33) have been described.

(c) A good clean vacuum is essential to prevent contamination which affects quantification in the following ways: absorption of the X-rays from light elements leads to their underestimation; adds mass to the specimen affecting accurate quantitation of all elements; adds spurious elements to the X-ray spectrum. Another problem with contamination is that it deposits on the nosepiece of the EDS spectrometer and degrades its performance.

In order to avoid contamination, a liquid-nitrogen-cooled anti-contaminator must be used, and regularly charged, during the analytical run. A device consisting of cold plates providing a large surface area situated close to the specimen is more effective than a cold collar (31).

(d) Production of extraneous X-rays from the column can be minimized by having a well-aligned, well-collimated electron beam; by using thick apertures to prevent X-rays generated higher in the column from reaching the specimen area; and by removal of the objective aperture during analysis.

(e) For the analysis of frozen-hydrated sections a stable cold stage, with accurate monitoring and good temperature control (see Section 5.3.5) is essential. When quantitative analysis of frozen-dried sections is being undertaken, the use of a cold stage is also an advantage (but see ref. 34). **Caution:** when in use, the temperature of the stage must be maintained higher than that of the anticontaminator to prevent condensation of contamination on the specimen.

(f) In the TEM a specially designed and constructed specimen holder is an important factor in analysis. As for grids, microanalytical specimen holders shielded with carbon or beryllium are used to minimize the production of X-rays. Contribution to the X-ray spectra from such soft elements causes the same problems in correction procedures as with the grids (Section 3.2.5). The design of the specimen stage is also important, as it should not obstruct the pathway of X-rays between the specimen and the X-ray detector, or degrade instrumental performance in any other way.

The following optional extras are also useful for the analysis of frozen-hydrated sections:

- a beam blanking or deflecting facility in order to avoid irradiation of the specimen when not being analysed (continuous switching of the beam on and off introduces beam instability and affects critical alignment)
- an image waveform monitor
- on the SEM a retractable optical reflecting microscope for inspecting the specimen stage without the use of the electron beam (in order to minimize beam damage)

4.2 X-ray spectrometry

i. X-ray spectrometers

For biological analysis EDS have now largely replaced wavelength dispersive spectrometers (WDS), although WDS can prove useful for studying elements present in low concentrations in thicker cryosections (Section 5.3.5). The advantages of EDS are their ability to analyse a large number of elements simultaneously, and their effectiveness at much lower probe currents than are needed for WDS. Detailed discussion about workings of EDS spectrometers can be found in ref. 35.

ii. Computer software

Qualitative analysis is easily carried out by noting the position of a peak in the X-ray spectrum and assigning an element to it. **Caution:** commercial software for automatic identification of peaks is written by non-biologists and will often mis-identify peaks.

Quantitative analysis requires a reliable 'deconvolution' of the EDX spectrum (that is, the ability to separate the characteristic X-ray peak of each element present in the X-ray spectrum from the underlying X-ray continuum—sometimes loosely called background); the ability to determine the total number of counts in the peak minus the continuum beneath the peak (net peak integral); and the ability to separate overlapping peaks. This processing of the spectrum is carried out by software supplied with the detector system and the details of the different approaches used are usually included with each system.

Quantification software based (sometimes loosely) on the continuum normalization process (Hall) is usually provided by the manufacturer of the spectrometer. **Caution:** The operator must familiarize himself with details of quantification theory and understand what any particular system offers; commercial programs differ considerably; if a computer program is not available it is still possible to carry out the procedures given in Section 5.2. The calculations can be performed manually or by use of a spreadsheet on a personal computer.

5. Quantitative analysis

Quantitative analysis of bulk inorganic specimens, or heavily mineralized biological specimens using corrections for the mean atomic number (Z), absorption of x-rays (A), and secondary fluorescence (F) within the irradiated volume (ZAF corrections) is well-understood and carried out routinely (36). In soft biological specimens the elements which give detectable X-ray peaks are present in a low concentration (often much less than 5% of the dry mass) in a matrix which consists of the elements carbon, hydrogen oxygen, and nitrogen which do not normally produce detectable peaks. In addition, the composition of the matrix may differ between different cells and between different components in cells. The organic matrix composition of the soft biological specimen affects their microanalysis in several ways:

(a) ZAF corrections for the organic matrix are not yet well-understood, consequently analysis of bulk organic tissue is carried out in only a few laboratories (37). Quantitative analysis of biological material is most often carried out on thin sections in which ZAF corrections can be ignored.

(b) Because of the high and unknown matrix composition the quantitative analysis of soft biological materials cannot be carried out without the use of standards.

(c) The organic matrix is sensitive to irradiation by the electron beam, which causes disruption of bonds leading to a loss of mass from the specimen. This loss of mass is not uniform and can differ substantially between different compounds (for example, protein and carbohydrate). As mass is lost focally during the course of analysis the concentration of the unaffected elements per unit mass of the specimen under the beam actually increases. There is still much to be learnt about the mass-loss. The usual practical approaches for coping with it are: to use a cold stage when undertaking analysis (Section 4.1) to reduce mass-loss, and to choose standards of similar composition to the specimen, making the assumption that the loss of mass will be equivalent between the specimen and the standard.

(d) In the living state the soft biological materials consist mainly of water. This water is removed if the specimen is analysed in the frozen-dried state, but some estimation of the original water content of the specimen should be made, for interpreting the results in physiological terms.

5.1 Analysis of frozen-dried ultra-thin cryosections

Cryoprepared biological specimens can be analysed either in a frozen-hydrated or in a frozen-dried state. Since the majority of studies are carried out on frozen-dried specimens, this will be considered first.

The advantages of using frozen-dried cryosections over frozen-hydrated cryosections are:

- handling of the cryosections is easier
- the morphological detail of the specimen in the electron image is better
- the spatial resolution of analysis is better
- X-ray peak to continuum signals are better
- frozen-dried sections are stable and can be stored

The disadvantages are:

- data are usually obtained in mmol/kg dry weight
- water contents in such sections cannot be analysed directly
- matrix free fluid compartments cannot be analysed
- fields at, or close to, the cell surface cannot be analysed reliably

Practical steps for undertaking analysis of a frozen-dried section are given in *Protocol 5*.

Protocol 5. Obtaining microanalytical data from frozen-dried cryosections for use with the continuum normalization method for quantification

1. Insert a calibration specimen, e.g. evaporated Al film on a Ni or Cu grid; obtain a spectrum; check the position of the peaks in the spectrum; adjust the gain if necessary.
2. Insert the empty specimen holder; measure the probe current; direct the beam on to the rim of the holder; collect and store several spectra.
3. Insert a bare grid without a film; measure the probe current; direct the beam on to or immediately adjacent to the grid bar; collect and store several spectra. **Caution:** adjust the probe current in order to keep the total count rate in the EDX spectrum <2000 c.p.s. This is to avoid overloading the detector, which will cause pulse pile-up and degrade deconvolution of the X-ray spectrum.
4. Insert a suitable standard; obtain an image; measure the probe current; analyse areas of the standard and of the film adjacent to the standard using identical probe current and X-ray counting times; collect and store several spectra from the film and the standard; check the spectra for any unexpected peaks.
5. Insert the section-bearing grid; obtain an image; survey the specimen; move beam away from the specimen and measure the probe current; select areas of interest; adjust the probe size; and analyse; collect and store the spectra; view the complete spectrum (there may be something there that you do not expect). **Caution:** choose areas in the centre of the grid opening and in the central region of the grid, do not analyse close to grid bars or near the edge of the holder.
6. Collect and store spectra from the adjacent film.
7. Photograph the specimen and record the areas analysed.

5.2 Quantification

5.2.1 Continuum normalization (Hall) method

The continuum normalization method was developed by Hall (36, 38)] and is the most frequently used method for determining elemental concentrations in thin biological specimens. When continuum normalization is used, the concentration of a given element x is expressed in terms of mass x/total mass in the area analysed. The mass of x in the area of analysis is proportional to the number of counts in its characteristic peak (I_x) and the total mass in the area of analysis is proportional to the continuum radiation (W) generated by the specimen itself.

The basic equation is:

$$C_{x,\mathrm{sp}} = \frac{(I_x/W)_{\mathrm{sp}}}{(I_x/W)_{\mathrm{std}}} \cdot \frac{\overline{Z^2/A}_{\mathrm{sp}}}{\overline{Z^2/A}_{\mathrm{std}}} \cdot C_{x,\mathrm{std}} \qquad [1]$$

where C is the concentration in mmol/kg, the subscripts sp and std refer to specimen and standard respectively, and the value of $\overline{Z^2/A}$ is the mean value of atomic number squared divided by the atomic weight for all constituent elements of the sample.

i. Estimating the correct continuum

The continuum reading is taken by setting a window in a given region of the EDX spectrum which is free from characteristic peaks. The major point about the continuum normalization method is that, to obtain the correct concentration, the counts in the continuum region must be those emitted solely from the specimen (38). The total continuum in the spectrum consists of the continuum generated by the specimen itself, and the continuum generated from extraneous sources (*Figure 11c*). Accurate estimation of the specimen continuum must therefore include corrections for all sources and not just the film. Hall (38) has shown that corrections for the holder and the grid can be carried out in a simple manner provided that each of the different sources generates a discrete characteristic peak. The measurements made in steps 2 and 3 of *Protocol 5* provide this data. The procedure for making these corrections is the same for the holder, the grid, and (if required) the specimen stage. A worked-out example will be described here in detail using the experimental data shown in *Table 1*. *Figure 11* shows diagrammatically what is happening at each step. For simplicity, this example assumes that there is only one element associated with the grid. For more comprehensive correction one should examine a complete spectrum in each of the steps 1–5, and identify any unexpected peaks. Corrections for all such peaks can be carried out in the same manner if required. The column for W_t *Table 1* is the total counts recorded in the continuum region (set at 4.2–6.2 keV) for an individual spectrum and I(Ni) is the net peak integral for Ni (the grid material) in the spectrum.

(a) Data generated from the analysis of a grid (*Protocol 3,* step 3) is given in *Table 1*, spectrum 1. From this the ratio of counts in the continuum to counts in the Ni peak (R_g) is determined.

$$R_g = W_t/I(\mathrm{Ni}) = 5588/58342 = 0.095. \qquad [2]$$

This ratio R_g (sometimes called the grid correction factor) is used to calculate the number of counts in the continuum contributed by the grid-peak in subsequent spectra.

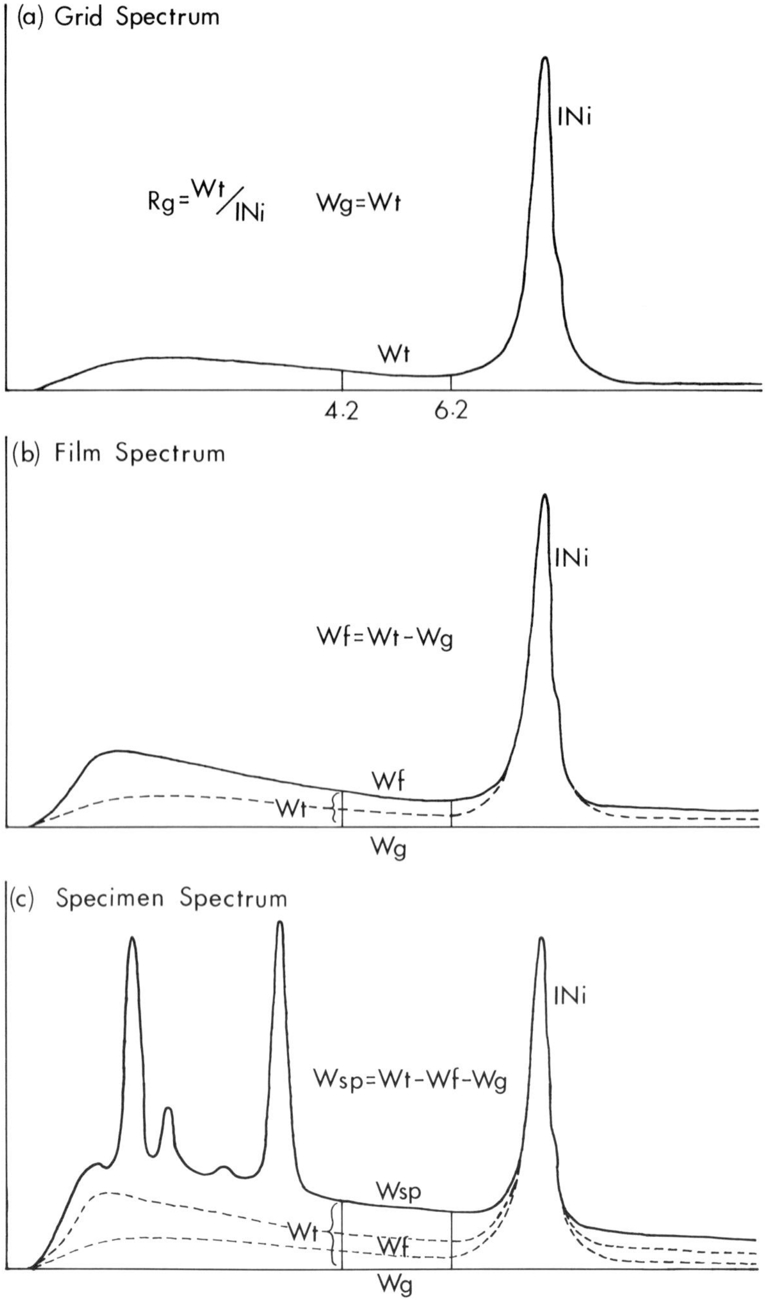

Figure 11. Corrections to the continuum. Spectra from a grid (a), a film (b), and a specimen (c) are shown diagrammatically. Total continuum in the specimen spectrum is contributed by the grid and the film as well as the specimen.

Table 1. X-ray data obtained from analysis of a cardiac myocyte

Spectrum	Na	Mg	Cl	K	*I* (Ni)	W_t	(4.2–6.2)
1 grid					58342	5588	
2 film					2706	1144	
3 specimen	75	178	218	4242	5625	4021	
Calibration constants from analysis of standards							
	2607	1044	373	330			
Calculated concentrations using equation 1 (mmol/kg dry weight)							
	73	71	31	539			

(b) The next step is to determine the continuum counts (W_f) generated from the supporting film (*Figure 11b*). A spectrum is collected from the film support adjacent to the specimen (*Protocol 5*, step 6). From this data (*Table 1*, spectrum 2) W_f is determined.

$$W_f = W_t - (R_g \times I(N_i)) \qquad [2]$$
$$= 1144 - (0.095 \times 2706) = 887.$$

(c) Finally the continuum counts generated by the specimen itself (W_{sp}) are determined (*Figure 11c, Table 1*, spectrum 3).

$$W_{sp} = W_f - (W_t + R_g \times I\ \mathrm{Ni}) \qquad [3]$$
$$= 4021 - (887 + 5625 \times 0.095) = 2600.$$

ii. Estimation of Z^2/A

For the use of equation 1 the value of $\overline{Z^2/A}$ (sometimes called the G-factor) needs to be known. This can be calculated for a standard of known composition, but has to be estimated for the unknown specimen. *Table 2*, taken from refs 2 and 38, shows some estimates for $\overline{Z^2/A}$ values. If a standard of similar composition to the specimen is used, then the term $\overline{Z^2/A}_{sp}/\overline{Z^2/A}_{std}$ cancels out. Hall and Gupta (39) have shown that small errors in the estimation of Z^2/A are unlikely to cause large errors in quantification.

iii. Standards

The term $C_{x,\mathrm{std}}/(I_x/W)_{\mathrm{std}}\cdot\overline{Z^2/A}$ in equation 1 is a constant. This is determined from the analysis of a standard of known composition. A detailed review of the standards which have been used for X-ray microanalysis of biological specimens has been published recently (40). For each element, a number of standards of different composition should be analysed. A plot of $(I_x/W)_{\mathrm{std}}$ against the concentration of x should give a straight line. The mean value of $C_{x,\mathrm{std}}/(I_x/W)_{\mathrm{std}}\cdot\overline{Z^2/A}_{\mathrm{std}}$ is the constant. The constant is calculated separately for each element and is used in equation 1 to determine elemental concentrations when unknown spectra are analysed (*Table 1*). Further details about standards for X-ray microanalysis can be found in refs 1 and 2.

Table 2. Values for $\overline{Z^2/A}$ for different constituents of biological materials (taken from refs 2 and 38)

Material	$\overline{Z^2/A}$
Water	3.67
Triglyceride	2.98
Glucose	3.40
Protein (with S)	3.28
Protein (no S)	3.20
Freeze-dried tissue (liver)	3.19
Frozen-hydrated tissue (liver)	3.53

iv. Advantages of continuum normalization

Continuum normalization depends on the measurement of the ratio I_x/W. This has the advantage that it is independent of changes in the specimen thickness and fluctuations in the probe current both of which affect I_x and W to the same extent. Measurement of the probe current is necessary for correct estimation of the continuum which is needed for the film correction and for the measurement of water by using only changes in the specimen mass (41). It is not necessary to make measurement of standards every time that analysis is carried out, although it is as well to measure one standard at the beginning of an analytical run in order to check the system.

5.2.2 Other methods of quantification

There are other methods of quantification used mainly by the research group in the physiological laboratory in Munich (28, 42) which do not require the measurement of continuum. The methods were used initially for analysis of 1-μm-thick cryosections in a frozen-dried state (28) but has recently been adapted for analysis of thinner (~0.5 μm) sections (43). This method is discussed in ref. 39 in comparison with the continuum method (see also Section 5.3.5).

5.3 Microanalysis of 1-μm-thick sections in a SEM

For reasons discussed by Hall (44), such analysis can be performed more successfully in an SEM type microanalyser using either 0.5- to 2-μm-thick cryosections or bulk samples. Ultra-thin hydrated cryosections are too unstable under the high beam currents used for X-ray microanalysis (44, 45). Here we illustrate a procedure which has been successfully used by Gupta, Hall, and co-workers (46) for the microanalysis of the compound racemose innervated salivary gland of the cockroach, *Periplaneta americana*. The procedure can be readily adapted for any other cell or tissue preparation.

The advantages of frozen-hydrated cryosections over freeze-dried cryosections are:

- matrix-free fluid spaces can be analysed
- concentration values are obtained directly both in mmol/kg wet weight and in mmol/kg dry weight for every area analysed
- local concentrations of water-fractions can be readily obtained
- distinction is made between free (solvent) and total water fractions
- by assuming that all the measured elements are free, concentration values in mmol/litre water can be calculated
- direct estimates of concentrations in mmol/litre of specimen areas analysed can be obtained by using alternative methods of data conversion

The disadvantages are:

- a lot of additional equipment is required
- methods are more tedious and need extra precautions in specimen preparation, handling, and analysis
- specimens are much less stable and readily suffer beam damage
- hydrated cryosections cannot be stored for long periods (months) and once loaded in the SEM cannot be retrieved without risk of damage
- images are poor in contrast and detail compared with the images from frozen dried specimens
- analytical spatial resolution is poorer (0.5–1 μm)

5.3.2 Basic approach

Rapidly freeze living tissue samples, cut cryosections, analyse cryosections first in the frozen-hydrated state, obtain quantitative data in mmol/kg wet weight, freeze-dry sections within the SEM column at about −80°C, re-analyse frozen-dried sections and obtain data in mmol/kg dry weight (optional: again dry sections within SEM at −40°C, re-analyse fully dry sections to obtain second set of data in mmol/kg dry weight. This data is more comparable to the conventional chemical analysis by spectrophotometry).

5.3.3 Equipment

For the cryopreparation of specimens, storage, cryosectioning, etc., the essential requirements are the same as given for ultra-thin cryosections, multi-slot (Ni) grids should be used. For thicker cryosections analysed in an SEM the following additional facilities are required

- low-background specimen holders, heavily electroplated with nickel suitable for the SEM stage (*Figure 8*)
- facilities for making the carbon/plastic films on the cryosection holders hydrophilic by glow discharge (15, 29)

- a cryotransfer system for retrieving the cryosection bearing holders from the microtome to the transit storage Dewar and for loading holders from the latter on to the cold stage of the SEM (*Figures 12, 13*)

  Note: commercially available cryotransfer systems for the SEM are designed for handling bulk specimens and often do not have provision for handling cryosections. However, it should not be difficult to adapt these for use with cryosections.
- in the EDX detection system a sensitive count rate meter fed from a selected band in the continuum region is also useful

---

Protocol 6. Preparation of cryosection holders

1. Cast a formvar (or other) film on glass and allow to dry.
2. Thoroughly clean and dry nickel-plated collars using procedures for cleaning delicate EM components.
3. Thoroughly clean and dry multi-slot nickel grids.
4. Lightly polish the face of the metal-holder on vellin paper.
5. Under a magnifying system, attach a grid on the top of the hole of the holder, using either tiny amounts of carbon-dag or rapid-setting resin polymers. Press the grid down with a clean metal disc or a large diameter glass rod so that the grid does not stand proud of the holder surface.
6. Using a suitable trough (29) mount the plastic film on to the grid-mounted holders.
7. Dry the coated holders in a clean, dry atmosphere.
8. Make triangular nicks in the plastic film at the outer rim of the holder face using a clean toothpick. This allows the coating film (next step) to make conducting contact with the metal. Stabilize the plastic film by evaporating a layer of carbon ($\sim$10 nm).
9. Store the holders in a vacuum desiccator until needed.
10. Alternatively, the grids may be coated with plastic first and then carefully attached to the holder before carbon coating. **Caution:** do not damage the film.

5.3.4 Peripheral medium

In the study of frozen-hydrated cryosections Gupta and Hall we have often used a physiological Ringer's solution containing 10–20% (w/v) high-molecular-weight dextran. Such dextran–Ringer's is used as the suspending medium during initial monitoring of the physiological functions of the tissue *in vitro*. The tissues are then frozen in a small drop of this medium which is sectioned with the specimen and acts as a peripheral standard for use with the Rick method of quantification (Section 5.2.2). More importantly, analysis of the peripheral medium as a specimen, using the continuum normalization procedure (using previously determined standard constants), provides a check for the validity of the analytical procedures and for the exact level of hydration in the cryosections as analysed. If such a medium is not compatible with the physiological function, or when the tissues need to be analysed directly after removal from the organism, a thin layer of the Dextran medium may be added on the surface immediately before freezing. The preparation of such a medium for use with microprobe analysis is *critical*. Physiological solutions used in most laboratories are often not made accurately enough. For microanalysis the solutions should be made either gravimetrically (w/w) or volumetrically (w/v). In microanalysis one measures the mass of an element per unit mass of the specimen under the beam, therefore gravimetric values for the salt solutions are often more relevant. As shown in the example below, the differences can be quite significant.

Protocol 7. Preparation of dextran–Ringer's solution as used by Gupta and Hall (46)

1. Prepare a Ringer's solution containing mmol/litre water: Mannitol = 60; Trehalose = 5: $KHCO_3 = 1$; $K_2HPO_4 = 9$; $KH_2PO_4 = 0.15$; KCL = 7; NaCl = 120; $MgCl_2 = 2$; $CaCl_2 = 2$; pH = 7.2. Total mmol = 336
2. Take 25 ml of Ringer's in a small, graduated glass vessel.
3. Weigh the vessel and record the weight.
4. Slowly dissolve 5 g of dextran in the Ringer's. Add the dextran in small quantities and dissolve by sonicating the capped vessel between each addition. When all the dextran has been dissolved weigh the vessel to check the exact amount of dextran dissolved. Record the weight and the final volume.
5. Store the tightly sealed vessels at −20°C if not required immediately.
6. For use thaw the vessels in a dry atmosphere. Mix by sonication before use. Shaking the vessels or stirring the solution generates too much froth. Clear any froth by sonication. The final composition of this medium is shown in *Table 3*.

   Caution: when mixing in a sonicator take care not to overheat the solutions. Do not use powerful sonicators which will fragment the dextrans.

When used as a reference standard *in situ*, for the continuum normalization method, the relevant values would then be mmol/kg wet weight for frozen-hydrated sections, mmol/kg dry weight for fully frozen-dried sections, and mmol/litre H_2O when the data are converted for comparison with the values measured by other methods such as ion selective microelectrodes. For the Rick method of quantification the relevant values would be mmol/litre of the specimen.

Table 3. Final concentrations of elements in the dextran–Ringer's solution, prepared as in *Protocol 7*

	mmol/litre H_2O	**mmol/kg Wet wt**	**mmol/kg Dry wt**	**mmol/litre Total[a]**
Na	120	100	556	118
Cl	135	112	625	132
K	10	8	46	9
P	~2	<2	~10	<2
Mg	~2	<2	~10	<2
Ca	~2	<2	~10	<2

Total dry wt fraction of the total wet wt:

Dextran	0.167
Others	0.013
Net	0.180

The dry wt is therefore 18% of the wet wt, assuming that there is no residual bound water, as would be the case if the solution were to be dried in an oven over a long period.

[a] 20% dextran (w/v) adds about 1% to the volume of the solution and is therefore negligible. Other macromolecular additives may change the final volume more significantly.

5.3.5 Analysis of 1-μm-thick frozen-hydrated sections

For specimen preparation and sectioning see Section 3; for additional steps in analysis see *Protocol 5*.

(a) Have all the transfer equipment and transit/storage Dewars ready for handling the cryosection holders.

(b) Cut cryosections 1 μm thick; transfer to the cooled section-holder and press under a copper weight with a mirror-smooth surface, for a few minutes.

(c) Retrieve the specimen-holder using the sliding rod and metal sheath system (*Figures 12 and 13*) or any alternative methods for your system, and transfer to the transit/storage Dewar.

Caution: in order to maintain the fidelity of cryosections for analysis in a frozen-hydrated state these steps are much more critical. When the sections are analysed freeze-dried, neither frost contamination nor partial dehydration

Figure 12. A flow-diagram to show the cryosectioning, mounting and retrieval of about 1-μm-thick frozen-hydrated sections for analysis in a SEM. (a) Cryosections are cut either in a cryostat at about −80°C or in a cryoultratome at about −120°C, transferred to a pre-cooled STEM holder (*Figure 8*) and pressed under a copper weight for a few minutes. (b) A brass tube and sliding rod assembly (total length about 40 cm) is cooled in a Dewar as shown in (c). The Dewar should be cooled down to LN_2 temperature first and the excess nitrogen discarded. It is important to dip only the copper bit, insulated from the

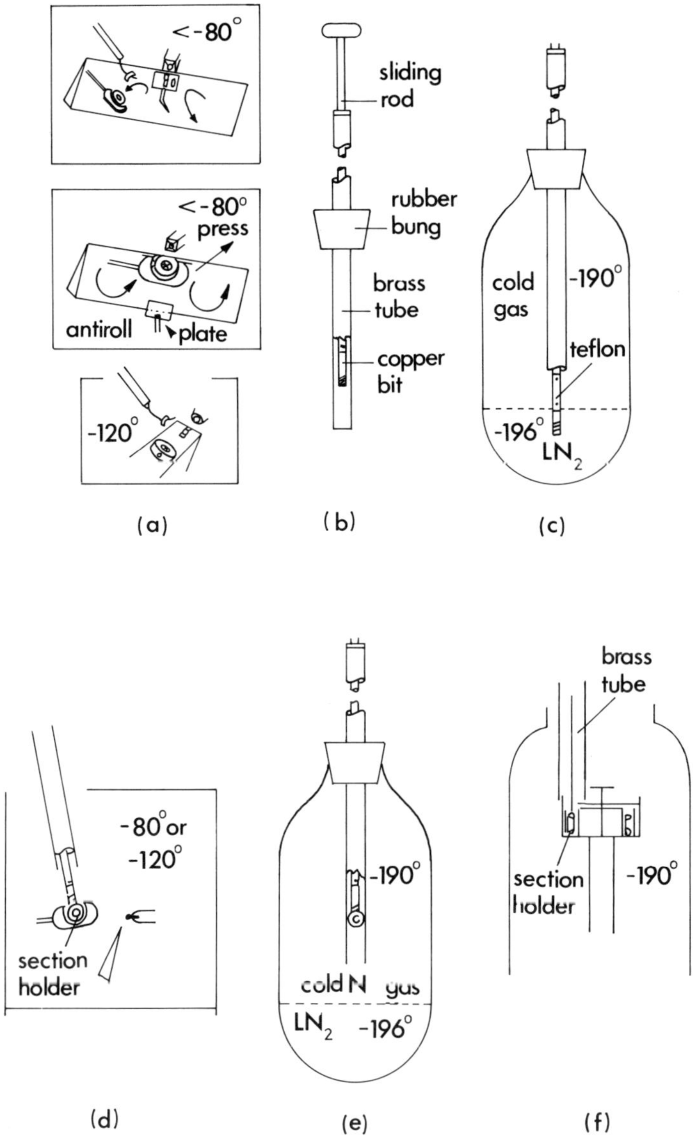

rest of the rod, in liquid nitrogen and allow it to equilibrate completely. Do not dip the brass tube in LN_2; the tube gets filled with the cold nitrogen gas and is vented at the top. The cold tip is screwed lightly into the holder with the cryosections (d) and immediately transferred to the LN_2 Dewar as shown in (e). The cryosection-holder is then transferred to the transit/storage Dewar. (f). It is important to thaw and completely dry the tube and rod assembly before using it for another holder. It is best to keep more than one such assembly in order not to interrupt the cryosectioning session.

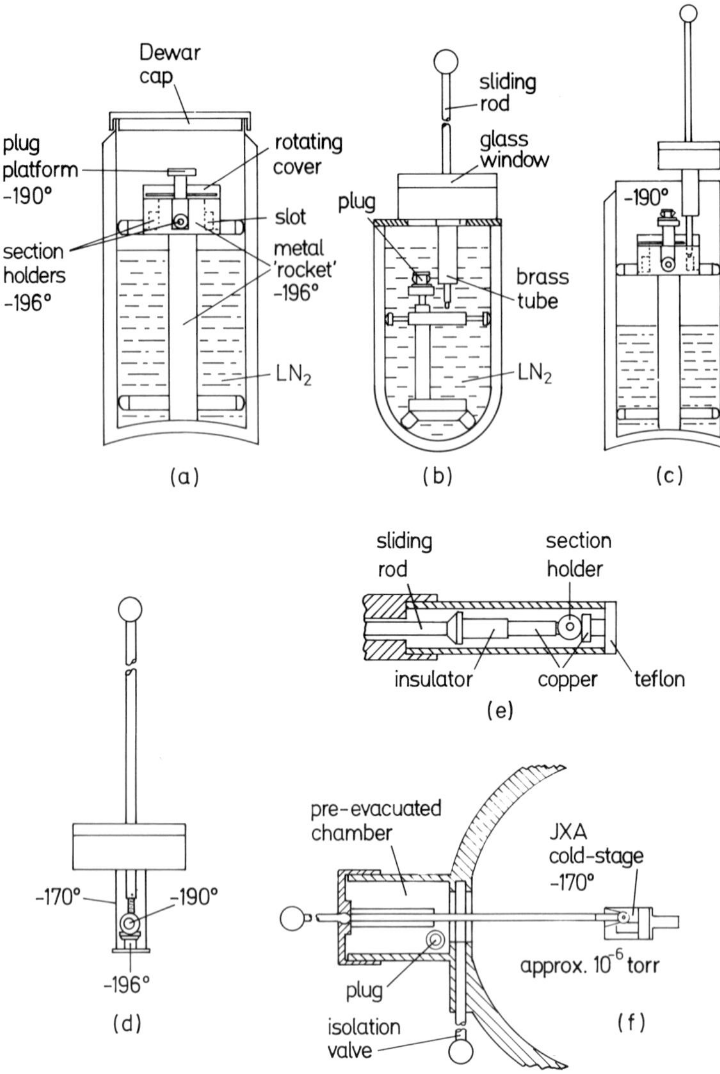

Figure 13. A flow-diagram to show the loading of frozen-hydrated sections on to the cold-stage of a SEM (here JEOL JXA-50A). The transit/storage Dewar with the cryosection-holders in position is shown in (a). For loading a holder, the Dewar is positioned close to SEM. (When initially cooling the Dewar, care must be taken to allow the metal 'satellite' to equilibrate completely. The LN_2 level should then be adjusted to slightly below the housing for the holders. It is important that liquid nitrogen is not allowed to enter the slots for the cryosection-holders at any stage of the operation: that is, at the start, when refilling with LN_2 during storage, or when moving the Dewar to and from the SEM. Splashing with liquid nitrogen usually results in the loss of holder-films with the cryosections). A Dewar-arrangement for pre-cooling a JXA-50A specimen-loading rod adapted for frozen-hydrated cryosections is shown in (b). The copper bit and the protective plug are all

will affect the results from analysis. If the section-holder is at a temperature of <−120°C moisture will condense on to the holder as well as the cryosections in the form of amorphous ice and not as visible frost. Such ice adds to the hydrated mass of the specimen, producing anomalous results (5, 27).

Instrumental conditions used for the example analyses were: JEOL JX-50A operating at 45 kV, Kevex-Link EDS with Quantem FLS software, horizontal specimen stage, take-off angle 35 degrees, distance 20 mm, detector area 30 mm resolution 155 eV FWMH, spectra collected with a 0.4 nA probe current for 100 s live time, continuum band 4.5 to 18 keV stage temperature −170°C, anti-contamination plate −190°C. The complete step-by-step procedure is given in *Protocol 8*.

Protocol 8. Analysis of 1-μm-thick frozen-hydrated sections

1. Fully prepare the analytical instruments and all ancillary gadgets. **Caution:** a frozen hydrated section once loaded into the SEM cannot be readily retrieved without some damage to it.
2. Load a blank holder assembly; carry out calibration; collect spectra for holder and grid corrections as outlined in *Protocol 5*, steps 1–3.
3. Remove blank holder from the column; check the temperature of the stage and of the anti-contamination plate; fill up the cooling Dewars if necessary; again check all other instrumental parameters.
4. Isolate the electron beam and load a frozen-hydrated specimen from the transit/storage Dewar (*Figure 13*).
5. Observe the specimen either through the optical microscope, if available, or in a STEM image obtained with a very low current and a low (×500) magnification (*Figure 14*).
6. If the sections look satisfactory, isolate the beam and dry the sections by raising the stage temperature to between −100°C and −80°C for about 10 min
7. Re-cool the stage to normal operating conditions (−170°C). Obtain an image as in step 5. (We used partial brightfield/darkfield image.)

---

allowed to equilibrate with the LN_2. The rod is withdrawn into the sheath and the latter is capped [as in (d) and (e) but without the holder]. For retrieving the cryosection-holder (c), the plug is stationed on the platform, a holder withdrawn into the brass-sheath, the plug securely replaced (d, e) and the assembly immediately attached to the pre-evacuated air-lock of the SEM. The air-lock is rough-pumped, the plug gently pushed out (f), the isolation gate-valve opened, and the cryosection-holder loaded on to the cold stage. The rod is withdrawn and the gate valve securely closed. With due care and practice, the whole operation takes 2–3 min, during which time the temperature of the cryosection holder does not rise above −140°C as measured in BML, Cambridge, by Gupta and Hall periodically over several years.

Protocol 8. *Continued*

8. Keeping the electron dose low, observe the specimen at different magnifications; select large fields that you wish to analyse; take polaroid pictures; use these for identifying the features of interest in the poor STEM images from the frozen-hydrated samples as below.
9. Set the probe current to the desirable level for analysis; direct the beam on to the peripheral medium close to the tissue; adjust the raster size/magnification to irradiate an area of about 1 × 1 μm or larger; collect a spectrum.

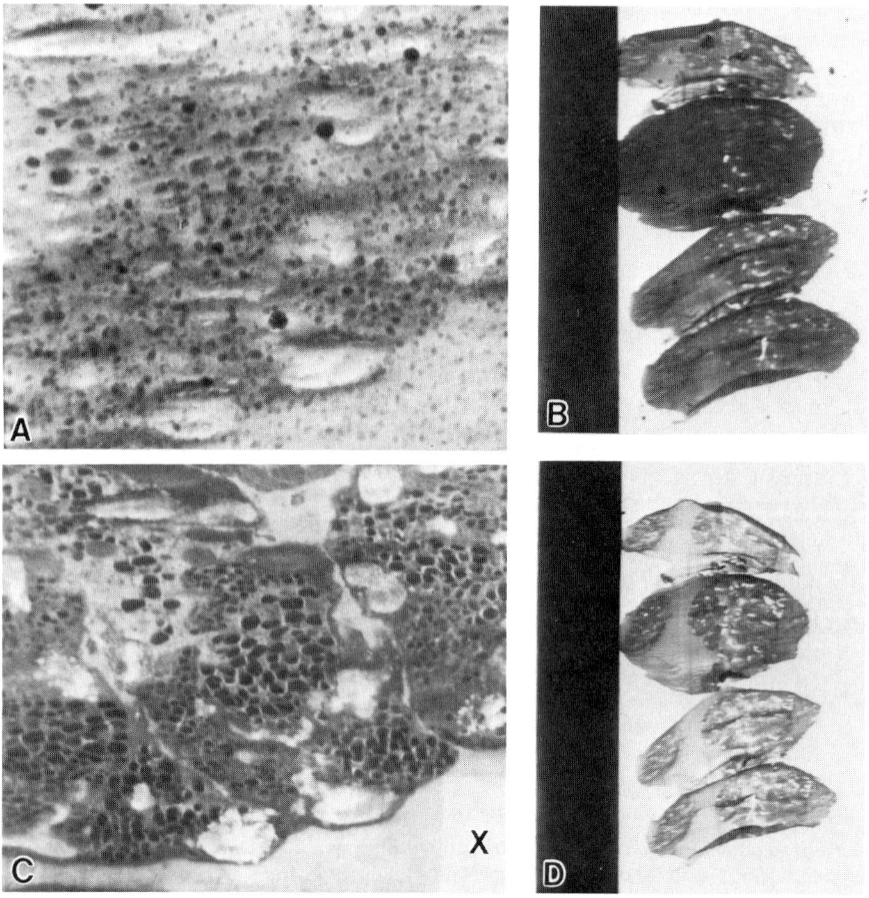

Figure 14. STEM images from 1-μm-thick cryosections of cockroach salivary glands in a frozen-hydrated (**A, B**) state and after freeze-drying (**C, D**) within the STEM column. Note the 'raster mark' (×) in (**C**). In the frozen-hydrated state the cryosections (**A, B**) show some frost contamination on the surface. When analysing the film, frost is deliberately included as a part of the X-ray background correction. After freeze-drying the sections have shrunk slightly, the frost contamination has disappeared, and the STEM contrast has vastly improved.

10. Isolate the electron beam; process the spectrum through the computer program and store.

11. Repeat on several fields around the specimen. A typical print-out is shown in *Table 4*. A good correlation between the measured and expected values shows that the analytical system is working well.

Table 4. Abridged data-sheet from one dextran–Ringer's field obtained in BML, Cambridge (date 9 March 1982)

Total continuum	=	5662 cts
Holder + grid correction (Ni)	=	569 cts substracted
Film correction	=	569 cts substracted
Corrected continuum	=	2033 (120)[a]
G-factor (iterated)	=	3.552
Mass thickness (dry)	=	0.228 μm
Matrix elements	=	Glucose

Concentration of elements in mmol/kg (dry wt)

	Na	**Mg**	**S**	**Cl**	**K**	**Ca**	**P**
Measured	558	6	19	613	47	18	11
(SD)[a]	(88)	(32)	(20)	(48)	(11)	(13)	(20)
Known	556	10	–	625	46	10	10

[a] (SD) values for one run reflect the uncertainties from all the numerical factors which enter each estimation.

12. If the specimen is satisfactory, go on to analyse all the tissue compartments of interest; process and store each spectrum as collected.

13. Remove the specimen; load another from the transit/storage Dewar as in steps 3 and 4.

14. Keeping the beam current as low as possible, obtain a STEM image at low magnification. If the image has contrast and details comparable to the ones from the frozen-dried sections, then the sections have inadvertently dried. However, sections which are only partially hydrated also produce reasonably good STEM imagfes and are useful for analysis in a frozen-hydrated state.

15. Set the various controls for the STEM detector to obtain an image with maximum density range as in *Figure 14*; adjust the position of the transmitted beam on the STEM detector if necessary.

16. Use the polaroid pictures of the frozen-dried sections to locate the corresponding fields in the hydrated sections; identify different components. *The secret of success is to be patient and not give up. The longer that you observe the image the more likely you are to identify features of interest.*

Protocol 8. *Continued*

17. Direct the beam on to the peripheral medium; collect the data as in *Step 9*. A high continuum background in the spectrum (*Figure 15*) and reasonable values for the elemental concentrations in the computer print-out provide a check for the hydration of the specimen (*Table 5*)

18. Direct the beam on to a whole cell or a nucleus and adjust the irradiated area within the field of interest (to cover the whole CRT screen).

19. Collect a spectrum; isolate the electron beam; process the spectrum and store. *Figure 15* shows some EDX spectra from a good hydrated cryosection.

20. Analyse all the features of interest. **Caution:** when analysing matrix free compartments use as large a field as possible for analysis in order to reduce the electron dose. Such spaces are more sensitive to beam damage.

21. Isolate the beam; dry the sections by raising the stage to between −100°C and −80°C for >10 min.

22. Re-cool the stage; set the beam current for analysis as before; obtain a STEM image; identify the fields analysed in the hydrated state by the residual raster marks caused by latent mass-loss (*Figure 14*).

23. Direct the beam on to the fields adjacent to the raster marks and re-analyse all the fields to obtain data in mmol/kg dry weight.

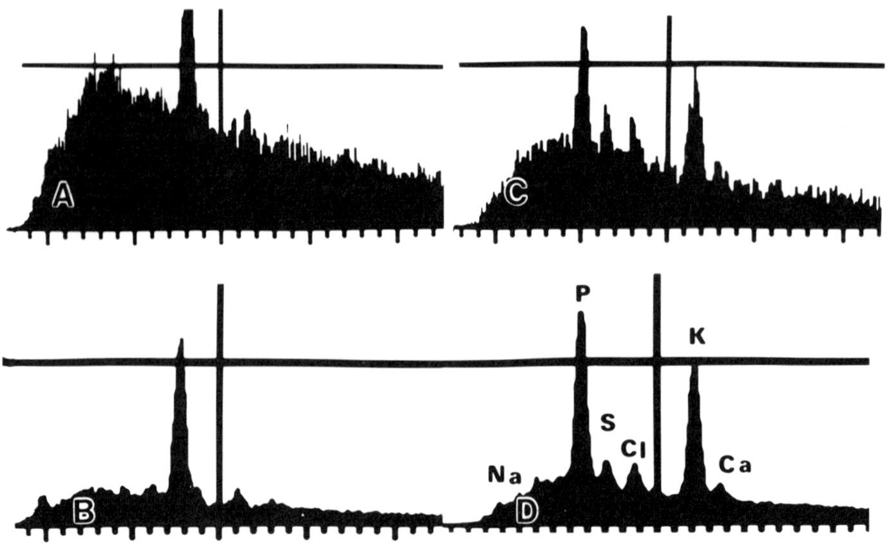

Figure 15. EDX spectra from 1-μm-thick cryosections of cockroach salivary glands. (A) frozen-hydrated dextran–Ringer's solution; (B) frozen-dry dextran–Ringer's solution; (C) frozen-hydrated C-cell cytoplasm, (D) frozen-dry C-cell cytoplasm.

Table 5. Abridged data-sheets for dextran–Ringer's solution from two frozen-hydrated sections in the same series as in *Table 4*

Section 1

Total continuum	=	20725 cts
Holder + grid correction (Ni)	=	5790 cts substracted
Film correction	=	1561 cts substracted
Corrected continuum	=	13464 (231)
Mass thickness (hydrated)	=	1.47 μm
G-factor (iterated)	=	3.64
Matrix element	=	0.2 Dex/0.8 water

Elemental concentrations in mmol/kg wet wt

	Na	Mg	S	Cl	K	Ca	P
Measured	100	−6	3	114	8	2	14
(SD)	(29)	(39)	(7)	(8)	(5)	(5)	(8)

Section 2

Total continuum	=	13637 cts
Holder + grid correction	=	3807 cts substracted
Film correction	=	1561 cts substracted
Corrected continuum	=	8269 (178)
Mass thickness (hydrated)	=	0.93 μm

Elemental concentrations in mmol/kg wet wt

	Na	Mg	S	Cl	K	Ca	P
Measured	102	5	−0.4	115	7	8	−3
(SD)	(37)	(14)	(8)	(10)	(2)	(6)	(9)
Known values	100	2	–	113	8	2	2

Characteristic X-ray quanta in the two examples above were:

Continuum	Na	Mg	S	Cl	K	Ca	P
8269	163	25	− 4	1425	81	109	− 26
13464	227	−39	54	2094	128	45	198

Note: These examples show (a) the variation in the thickness of cryosections following each other, and (b) the efficacy of Hall's continuum method in compensating for such variations.

5.3.6 Loss of mass during analysis

During the analysis of frozen-dried sections a loss of mass often occurs from the individual fields depending on the electron dose and on the stability of the components. When continuum normalization is used this loss of mass can be compensated for in the following way.

(a) Using the count rate meter, record the continuum counts from the selected band during the first 5 s of every run. Hall (46) has demonstrated that in 1-μm-thick sections the loss of mass does not seem to occur in the first few seconds, irrespective of the electron dose.

(b) Calculate the c.p.s. from these initial counts.

(c) Calculate the total continuum for the run by multiplying by the live time of the run.

(d) Correct for the holder and film to find the actual continuum for the specimen.

(e) Use this initial continuum value to calculate the concentration values by using the characteristic X-ray quanta for the print-out from the run.

5.3.7 Partially hydrated sections

It is not usually possible to identify the structural features of interest, except the major compartments in the poor STEM images from frozen-hydrated cryosections.

Partially dry the sections by heating the cryostage. Monitor the level of hydration: by the improvement in the low magnification STEM image; by monitoring the continuum counts from the selected band on the count rate meter. As soon as the count rate drops by 80% of the starting rate, isolate the beam and rapidly re-cool the stage. When normal conditions are restored, analyse the specimens. The exact level of hydration can be estimated by comparing the measured elemental concentrations in the peripheral medium with their known values. For further information see refs 5, 39, and 47.

5.3.8 Estimation of water fraction

Dry-weight fraction (F_d) can be estimated directly from the elemental concentration (C_{Ah}) as measured first in the frozen hydrated state and subsequently when the same sections have been dried (C_d). The simple formula is

$$F_d = C_{Ah}/C_d. \quad [5]$$

The aqueous fraction (F_A) is simply

$$F_A = 1 - F_d. \quad [6]$$

Some results are illustrated in *Tables 6* and *7* below.

Table 6. Pooled data for hydrated dextran–Ringer's solution from 25 different measurements in the study of cockroach salivary glands: BML, Cambridge (9 and 10 March 1982)

	Concentrations in mmol/kg						
		Na	**Mg**	**Cl**	**K**	**Ca**	**P**
Dextran–Ringers							
Hydrated	13864	112	6	134	10	6	9
(SE)	(848)	(3)	(3)	(4)	(2)	(3)	(4)
Dry	4632	508	14	662	46	15	20
(SE)	(433)	(23)	(3)	(25)	(3)	(2)	(3)
Known values							
Wet weight	–	100	<2	113	8	<2	<2
Dry weight	–	556	~10	625	46	~10	~10

	Na	**Cl**	**K**
F_d	112/508 = 0.22	134/662 = 0.20	10/43 = 0.22

Average = 0.213 + 0.011

Residual hydration as % of full: 0.18 (known)/0.213 (measured) × 100 = 85%

	Corrected concentrations mmol/kg wet wt (C_{Ah} × 0.85)					
	Na	**Mg**	**Cl**	**K**	**Ca**	**P**
Dextra–Ringer hydrated	95	5	113	8.5	5	7.7

Table 7. Data for the zymogen cells (C-cells) of cockroach salivary glands from 'hydrated' cryosections after correction for partial hydration as in *Table 6*

	Corrected concentrations in mmol/kg wet wt						
	Na	**Mg**	**S**	**Cl**	**K**	**Ca**	**P**
Hydrated							
Nuclei	5	5	26	27	164	1.1	188
Cytoplasm	9	8	41	45	126	2.3	120
Sec. gran.	16	4	111	26	76	17	23
Dry							
Nuclei	24	25	146	178	846	6.7	881
Cytoplasm	43	25	174	212	484	12	547
Sec. gran.	33	13	241	61	181	35	43
F_d							
Nuclei	0.21	0.20	0.18	0.15	0.19	0.16	0.21
Cytoplasm	0.21	0.29	0.24	0.21	0.26	0.19	0.22
Sec. gran.	0.48	0.31	0.46	0.43	0.42	0.49	0.53

F_d Mean (SD)
- Nuclei = 0.19 (0.024)
- Cytoplasm = 0.23 (0.030)
- Sec granules = 0.46 (0.070)

Acknowledgements

A. W. is supported by the Garfield Weston Foundation. The authors would like to thank Mr L. Mack for the line drawings and Dr K. Zierold (Dortmund, Germany) for donating the metal foils.

References

1. Morgan, A. J. (1985). *X-ray microanalysis in electron microscopy for biologists*. Royal Microscopical Society Handbook No. 5. Oxford University Press, Oxford.
2. Roomans, G. M. (1988). *Journal of Electron Microscopy Technique,* **9,** 19.
3. Hall, T. A. (1986). *Micron Microscopica Acta,* **98,** 91.
4. Gupta, B. L. (1991). *Scanning Microscopy,* **5.** (In press.)
5. Gupta, B. L. and Hall, T. A. (1982). In *Techniques in cellular physiology*, Vol. 128 (ed. P. F. baker), pp. 1–52. Elsevier/North-Holland County Clare, Ireland.
6. Shelburne, J. D., Tucker, J. A., Roggli, V. L., Linton, R. W., and Ingram, P. (1989). In *Electron probe microanalysis applications in biology and medicine* (ed. K. Zierold and H. K. Hagler), Springer Series in Biophysics, Vol. 4, pp. 305–16. Springer-Verlag, Berlin and Heidelberg.
7. Glauert, A. M. (1974). *Fixation, dehydration and embedding of biological specimens*. Elsevier, Amsterdam.
8. Hall, T. A. and Gupta, B. L. (1982). *Journal of Microscopy,* **126,** 333.
9. Somlyo, A. V., Shuman, H., and Somlyo, A. P. (1989). In *Methods in enzymology*, Vol. 172 (ed. S. Fleischer and B. Fleischer) pp. 203–29. Academic Press, London.
10. Warley, A. (1986). *Journal of Microscopy,* **144,** 183.
11. Buja, L. M., Hagler, H. K., Parsons, D., Chien, K., Reynolds, R. C., and Willerson, J. T. (1985). *Laboratory Investigations*, **53,** 397.
12. Barnard, T., Gupta, B. L., and Hall, T. A. (1984). *Cryobiology,* **23,** 559.
13. Zierold, K. (1988). In *Methods in microbiology,* Vol. 20 (ed. F. Mayer) pp. 91–111. Academic Press, London.
14. Robards, A. W. and Sleytr, U. (1985). *Low temperature methods in biological electron microscopy*. Vol. 10 in the series *Practical methods in electron microscopy* (ed. A. M. Glauert). Elsevier, Amsterdam.
15. Roos, N. and Morgan, A. J. (1990). *Cryopreparation of thin biological specimens for electron microscopy: Methods and applications*. Royal Microscopical Society Handbook No 21. Oxford University Press, Oxford.
16. Sitte, H., Neumann, K., and Edelmann, L. (1987). In *Cryotechniques in biological electron microscopy* (ed. R. A. Steinbrecht and K. Zierold), pp. 285–9. Springer-Verlag, Berlin.
17. Ryan, K. P., and Liddicoat, M. I. (1987). *Journal of Microscopy*, **147,** 337.
18. Nassar, R., Wallace, N. R., Taylor, I., and Sommer, J. R. (1986). *Scanning Electron Microscopy*, 1986/I, 309.
19. Warley, A. (1989). *Scanning Microscopy,* **3,** 1247.
20. Somlyo, A. V., Bond, M., Silcox, J. C., and Somlyo, A. P. (1985). In *Proceedings of the 43rd Annual Meeting of EMSA* (ed. G. W. Bailey), pp. 10–13. San Francisco Press, San Francisco, California.

21. von Zglinicki, T., Rimmler, M., and Purz, H-J. (1986). *Journal of Microscopy,* **141,** 79.
22. Hall, T. A. (1989). *Scanning Microscopy,* **3,** 461.
23. Wroblewski, J., Wroblewski, R., and Roomans, G. M. (1988). *Journal of Electron Microscopy Technique,* **9,** 83.
24. Elder, H. Y., Bovell, D. L., Pediani, J. D., Wilson, S. M., McWilliams, S. A., and Jenkinson, D. M. (1988). *Institute of Physics Conference Series,* **93,** 575.
25. Griffiths, G., McDowell, A., Back, R., and Dubochet, J. (1984). *Journal of Ultrastructure Research,* **89,** 65.
26. Roberts, I. M. (1975). *Journal of Microscopy,* **103,** 113.
27. Gupta, B. L. and Hall, T. A. (1981). *Tissue Cell,* **13,** 623.
28. Rick, R., Dorge, A., and Thurau, K. (1982). *Journal of Microscopy,* **125,** 239.
29. Sommerville, J. and Scheer, U. (1987). *Electron microscopy in molecular biology: A practical approach*, IRL Press at Oxford University Press, Oxford.
30. Roomans, G. M., Wei, X., and Seveus, L. (1982). *Ultrastructural Pathology,* **3,** 65.
31. Dubochet, J., Adrian, M., Chang, J. J., Homo, J. C., Lepault, J., McDowell, A., and Schultz, J. (1988). *Quarterly Review in Biophysics,* **21,** 129.
32. Nicholson, W. A. P. and Schreiber, J. (1975). *Journal de Microscopie et Biologie Cellulaire,* **22,** 169–176.
33. Nicholson, W. A. P. (1981). *Journal of Microscopy,* **121,** 141.
34. von Zglinicki, T. and Uhrik, B. (1988). *Journal of Microscopy,* **151,** 43.
35. Statham, P. J. (1981). *Journal of Microscopy*, **123,** 1.
36. Hall, T. A. (1971). In *Physical techniques in biological research* (ed. G. Oster), pp. 157–275. Academic Press, New York.
37. Marshall, A. T. (1987). In *Cryotechniques in biological electron microscopy* (ed. R. A. Steinbrecht and K. Zierold), pp. 240–57. Springer-Verlag, Berlin.
38. Hall, T. A. (1979). In *Microbeam analysis in biology* (ed. C. P. Lechene and R. R. Warner), pp. 185–203. Academic Press, New York.
39. Hall, T. A. and Gupta, B. L. (1986). In *Principles of analytical electron microscopy* (ed. D. C. Joy, A. D. Romig, Jr., and J. I. Goldstein), pp. 219–48. Plenum Press, New York.
40. Warley, A. (1990). *Journal of Microscopy,* **157,** 135.
41. Saubermann, A. J., Beeuwkes, R., and Peters, P. D. (1981). *Journal of Cell Biology,* **88,** 268.
42. Rick, R., Dorge, A., Bauer, R., Gehring, K., and Thurau, K. (1979). *Scanning Electron Microscopy,* 1979/II 619.
43. Rick, R. and Schratt, W. (1989). In *Electron probe microanalysis: Applications in biology and medicine* (ed. K. Zierold and H. K. Hagler). Springer Series in Biophysics, Vol. 4, pp. 213–21. Springer-Verlag, Berlin and Heidelberg.
44. Hall, T. A. (1986). *Journal of Microscopy,* **141,** 319.
45. Zierold, K. (1983). *Scanning Electron Microscopy,* 1983/II, 809.
46. Gupta, B. L. and Hall, T. A. (1983). *American Journal of Physiology,* **244,** R176.
47. Hall, T. A. and Gupta, B. L. (1983). *Quarterly Review in Biophysics,* **16,** 279.

12

Three-dimensional electron microscopy of tissues and cells

NIGEL JAMES

1. Introduction

The three-dimensional (3D) visualization and analysis of structures is of such considerable importance to investigators in many biomedical and other disciplines that it requires no justification. A fundamental difficulty in many subjects is that the 3D structures of interest are either opaque or too small to be seen without microscopical facilities. The smallest of structures requires the use of the electron microscope, which immediately places several important constraints upon the investigator. First, only extremely small portions of a structure can be analysed. Second, only dead and biased representations of the biological structures can be studied. Third, the most likely specimen preparation to be of use is a section of sufficient thinness to allow either light or a beam of electrons to penetrate, to yield a visible image or generate a photographic negative. Such a section is variously either relatively 'thick' or 'thin' according to the dimensions of the test objects under study, which is either limited in the total amount of information it contains or possesses inherent artefacts limiting adequate visualization when examined by *two-dimensional* (2D) microscopical techniques in which all images are essentially *projections*.

The limitation of information may be due to a variety of features, such as the relative size of the component parts of test objects under study compared with the section thickness or, even more important frequently, insufficient space between the test objects being represented in the section, which precludes any inferences on 3D spatial structure being drawn fom the 2D sections. The spatial relationship between test objects, i.e. the properties of the space between test objects (the matrix), is a fundamental element of the *topology* of cells and tissues, and so important in their function. What tissue has a random arrangement of its cells? What cell has a random arrangement of its organelles? Each component has an *exact* relation to its functional relatives. The investigator may not yet know, or be able to determine, the nature of those relations, but current developments in cell biology and the medical sciences, stress more readily the importance of topology and the need for its quantification.

Reconstruction in a research project may be an absolute requirement for acquiring knowledge, either subjectively merely for an elementary understanding of the nature of a structure, or objectively, using sophisticated quantitative morphometric techniques which are increasingly more important in biomedical studies.

Reconstruction is difficult to define accurately and comprehensively. Reconstructions are easy to recognize when based on the classical process of stacking images of serial sections taken from a test object to yield a 3D image, but more sophisticated and widespread definitions and approaches may be required, particularly with the advent of modern computer technology, advances in stereological methods, and newer imaging techniques.

In the present author's view, the realization by most investigators of the need for providing quantitative data as an end-result of a reconstruction does not necessarily require that the classical image generating reconstruction actually ever needs to be visualized. For example, if an investigator needs to know the 3D size distribution of a system of biological test objects approximating to spheres when only a series of 2D thin sections containing spherical profiles is available, then some reconstruction technique is clearly an essential requirement. But does the investigator actually need to see the *images* of the 3D spheres? In this case, the reconstructive process would best be the stereological process of *unfolding*. Three-dimensional reconstruction techniques are sometimes described in standard textbooks of stereology (1) in addition to independent texts of the subject (2) and the two subjects both overlap and are complementary.

Three-dimensional aspects of cell and tissue structure are sufficiently important and so many varied techniques exist for such a study that it is possible even for whole symposia on 3D studies of cells and tissue to be published (3) without any reference to those techniques discussed in the present chapter. Included within the repertoire of electron microscopical procedures, for example, is that of corrosion casting, in which 3D capillary networks are infiltrated with a synthetic resin which is then polymerized and which is then followed by the dissolution of the surrounding tissue. Subsequent study of the reconstructed capillary network in three dimensions is carried out by scanning electron microscopy.

Ultimately, all reconstructive work should be of a quality allowing full numerical analysis of the reconstructions, for two clear reasons. First, the quality necessary for quantification is a safeguard for reconstructive standards and for ensuring, as part of the necessary statistical design, that adequate sampling procedures and representative selection of test objects are thereby ensured. Second, since the aim of a reconstruction is to obtain information on the 3D nature of test objects, then much more objective quantitative numerical data than subjective data is contained in almost all specimens. Quantification is at least an important but more often an essential component of any reconstructive study. The acquisition of quantitative data means that an investigator knows more about a system and has a superior knowledge of it in consequence. Quantification enables comparison of control and experimental material to be more readily and more validly compared, and the making of empirical models for modelling, testing hypotheses, and the making of predictions, all of which are the essential features of science.

The extraction of the quantitative information in reconstructed specimens is often the easiest but most neglected aspect of 3D reconstruction. As a powerful technique for the quantitative reconstruction of tissue structures, stereology is of prime importance. Stereology can be defined as the science whereby data on structure having a higher dimension can be obtained from specimens of a lower dimension. In electron microscopy, this normally (i.e. most commonly) consists of obtaining 3D information from images which are in effect 2D projections of sections which, for theoretical purposes, are often regarded as having zero thickness. Stereology and reconstruction techniques clearly share a common basis for some problems.

Other reconstruction problems at electron microscopical level may relate to objects which have sizes smaller than the thickness of a typical thin section (<70 nm) or may be test objects lying within relatively thick sections (e.g. >25 μm). Electron microscopists have several available techniques for examining single sections to yield spatial and other types of reconstructed data. These include the use of stereomicroscopy, often using either intermediate or high-voltage electron microscopy (IVEM or HVEM), or newer methods such as electron microscope tomography (EMT), which is a technique almost analogous to X-ray computerized tomography of radiologists, the laser confocal microscopy of optical microscopists (4), or X-ray microscope tomography (5), in which techniques images at varying levels are obtained within a single test object at the appropriate magnification. It is clear, however, that each test structure necessitates its own appropriate reconstruction procedure.

The first practical aspect of reconstruction by any investigator is the careful design of experiment, having determined the nature of the information sought. Some information, in practice, may either be difficult or even impossible to obtain. Other categories of data may require extremely expensive forms of electron microscope together with appreciable amounts of computer hard- and software not currently available at the investigator's laboratory. However, a significant and substantial amount of data in biological and medical investigations is obtainable from 'lower level' facilities. These consist of a conventional transmission electron microscope, commonplace in a department of cell biology or pathology, a moderately powerful desk-top computer, a digitizing tablet, and the 'appropriate' computer software.

In reconstruction work, as with many other subjects, there are two corresponding categories of investigator: the high-level investigator having microscopical and computing facilities rarely available to the biologist, and the less experienced investigator attempting to begin reconstruction work and not knowing how or where to begin. It is to the latter type of investigator that this chapter is directed. The practical descriptions are of the main techniques for obtaining 3D information, using either section stacking, stereoscopical or tomographic techniques in microscopy, and the numerical elucidation of 3D quantitative data, stereology. Several small systems which for a moderately competent investigator should be within grant-seeking range are described, whilst the larger systems, outside the usual range and even needing building facilities, are also described to illustrate what might *not* readily be achieved or which might need time-sharing facilities with collaborators in alternative laboratories.

Whatever type of reconstruction work is attempted, however, needs not only the microscopical and computer hardware but the appropriate software and computer graphics support. Such support has either to be provided by computing services support staff or, in the case of larger *de novo* attempts at reconstruction, software writers are needed as part of the reconstruction team. Practical advice is difficult with software since authors seldom publish either their algorithms or even language details (6). The best practical advice on software is that investigators must understand the principles of what the software is attempting to do and how it is achieved.

2. Section stacking reconstructions

Historically, Born (1883) (7) originally developed reconstruction techniques using serial sections for the light microscope which have remained universally popular, though until recently it has proved relatively difficult to obtain comparable reconstruction data at the electron microscopical level. The lack of computer facilities and the availability of microscopical facilities capable of using sections up to 70 nm originally led to the tracing of hundreds of profiles by hand from transmission electron micrographs, followed by the production of plastic 3D models (7, 8), or even models made of wood (9). More recently, typical computer-produced displays of reconstructed biological structures, using superimposed contour lines, have been made (10, 11). Electromechanical digitizing tablets have long been available for 3D reconstruction displays of electron microscope preparations and have been used in specific programs; for example, TROTS (12). In more modern software, the 3D perceptual abilities of the human operator particularly with respect to stereoscopic, abilities, depth perception, and pattern recognition are increasingly being recognized.

Reconstruction techniques using serial sections are gruelling, even for the light microscope, where some authors consider that a useful upper limit to the number of serial sections is about 100; above that, the problem might best be solved by other techniques such as reconstruction by stereology. A typical example of an operationally convenient but elegant reconstruction of a plant dictyosome from a series of only 29 ultra-thin sections has recently been provided by Lockhausen, *et al.* (13). Electron micrographs of 40 nm-thick sections prepared using a Philips EM 420 were read into a VAX 11/780 computer using an Optronics P-1000 optical drum microdensitometer and data stored with 512 × 512 point resolution. Further processing was achieved using Ramtek 9460 VDU. Interactive software encompassing contour profiling and editing, contour alignment by translation, and rotation of individual single plane image, the editing of topological connections, and polygon tiling with final display shading, was used. Details of further processing in the system are available (14). The system would appear to provide a powerful and readily usable and understandable system suitable for many 3D reconstruction problems for test objects capable of being sectioned by relatively small numbers of serial thin sections.

Advances in 3D reconstruction have become possible due to the development and accessibility of computers, digitizing tablets, and technical advances in electron microscopy. At the simplest and least sophisticated level of computer-aided reconstruction is the use of a microcomputer, with or without an interfaced digitizing tablet, to acquire and store profile data which is then analysed using the appropriate program software. It is potentially within the grasp of all biomedical departments possessing a transmission electron microscope. This is the area in which this chapter can best help the potential reconstructor of the future. Those working in the subject will have their own specialist electron microscope, mainframe computer, and their own specialist software unique to their own problems and requirements.

Two-dimensional electron microscopical images can be projected on to grid paper and points on the profile outlines of the selected test object identified uniquely by their x and y coordinates. Serial or parallel serial sections (for example every tenth) can be similarly recorded and the positions of the serial sections numerically identified as the third, orthogonal, z axis. The x, y, and z data points can then be stored in a computer array prior to manipulation. Each z coordinate may have many hundreds of x, y coordinates for each profile and an alternative, though needing larger computer space, is to use a computer-interfaced manual digitizing tablet and computer graphics (15). Typical examples included in publications in the last decade include the Kontron MOP AMO3 and Summa Graphics Bit Pad. Each profile is drawn around with an electronic cursor and the co-ordinates of each point of the profile outline are entered into the computer. Alternatively, the electron micrographs can be placed directly on the bit pad for outlining by cursor in the same manner as the outlining of *camera lucida* projections of optical microscopy. Three-dimensional reconstruction is within the scope of the same laboratory, using an Apple Mackintosh Plus microcomputer interfaced with a Houston Instruments EDT11 Hipad digitizing tablet and appropriate software (16). The reconstruction program uses 3D computer graphics (15) to display the reconstructed structures after rotation, variable distance viewing, or magnification. An alternative microcomputer system for use with either optical or electron micrographs using a desk-top computer has recently been described by Yaegashi *et al.* (17). Their system consists of a Hewlett-Packard 310 (HP 9000 Series 300) desk-top computer with capacity extended to 40 megabytes (HP 9134L), a graphic tablet (HP 46088A), printer (HP 2671G), and colour memory-mapped graphics display with 1024 × 768 pixels (HP 98782A). Their software is written in HP-BASIC 4.0 requiring 2 megabytes of RAM and yields 3D reconstructions with optional surface tiling. Some sub-programs were written and compiled in Pascal to implement more rapid processing. The ready implementation of stereological techniques by the system was again stressed by these authors. For an 'average' investigator not having access to a major image processing laboratory, desk-top hardware should, provided appropriate software support is available, be sufficient to generate 3D reconstructions at contemporary *biological* standards and carry out adequate stereological analyses.

The distance between each point is determined by the resolving ability of the digitizing tablet. High-resolution tablets generate more co-ordinate points than low-resolution tablets and consequently require more computer space for storage and manipulation, and the speed of arithmetic computer operations may become an important consideration.

Many computer programs are available for manipulating the co-ordinate data so that 3D viewing can be achieved from any direction. For example, elementary fully written programs are available in standard textbooks (1) and references to specific programs are available in reviews (6) or in specialist journals of relevant societies such as the International Society for Stereology. Often, however, authors fail to describe their reconstruction programs, or give their algorithms, or even identify the computer language in which they are written. Such 'black box' techniques are not to be condoned and much needs to be done to standardize 3D computer program practice (6, 18). As in scientific investigations generally, the use of procedures not known to the users or not understood by them, followed by their blind interpretation, is fundamentally a dishonest procedure. Investigators using a computer or computerized technique should, as in any other speciality, at least have some knowledge of the basic principles upon which the analysis is based. As an absolute minimum, the specific algorithms should be known to the investigator if interpretation or further use is to be made of the resulting data. A useful example for illustrating the general principles is given by Elias and Hyde (1) as the GETSET program for a 48K desk-top Microcomputer which would yield a fixed viewing position from any perspective. Motion or real-time rotation or dilation or deformations require much larger computers with specialized high-level graphics software. Stereological measures of size can readily be derived from such a program.

For the rotation of co-ordinate axes, the rotation around the z axis, y and z are rotated to new positions y' and z' using the relations:

$$Y' = y\cos\theta_1 + Z\sin\theta_1$$

and

$$Z' = Z\cos\theta_1 - Y\sin\theta_1.$$

For rotation around the Y-axis, both the Y' and the Z' newly rotated axes are rotated to positions X'' and Z'' using the relations:

$$X'' = X'\cos\theta_2 - Z'\sin\theta_2$$

and

$$Z'' = z'\cos\theta_2 + X'\sin\theta_2$$

where the terms θ_1 and θ_2 are the angles of rotation about the X and Y axes respectively.

The calculation of mean caliper diameter (D) is an example of the data which can be derived from a 3D reconstruction. The mean caliper diameter is given by the relation:

$$\bar{D} = \frac{\sum_{j=1}^{m} \sum_{i=1}^{n} D_{(\theta i,\phi j)} \sin(\phi_j)}{\sum_{j=1}^{m} \sum_{i=1}^{n} \sin(\phi_j)}$$

where n is the number of x rotations, m the number of y rotations, $D_{(\theta i,\phi j)}$ is the caliper diameter at any x–y rotation, and ϕ_j is the y-axis rotation.

Alternative procedures exist which allow for the calculation of a measurement such as $\bar{D}$. For example, the investigator may not wish to calculate $\bar{D}$ from a few 3D computer reconstructions, since he may not be certain how typical his necessarily few reconstructions are compared with the sampling of a few sections from a much larger population of test particles. This statistical sampling problem is always a fundamental problem in electing to use 3D reconstructions, though the alternative possible, more accurate stereological techniques capable of yielding values within a prior predetermined statistical limits, fail to yield the spatial information required of 3D reconstructions.

The acquisition of spatial pattern and information on spatial structure is relatively under-achieved. This is partly due to the mathematical complexity of defining other than relatively simple shapes (19), which may not be of relevance to microscopists, but also to the problems of data handling and analysis using computers.

It is sometimes desirable to compare the shapes of two structures for each of which data is retained as a set of co-ordinate points in 3D space. Some analysis is possible even with the use of a microcomputer, and the approach using quaternion algebra (20) is followed here. The work of Barrett and Mackay (20) indicates the value of quaternions by their publishing of the program QUAT, which allows for the comparison of the shape of objects in three dimensions.

A 3D body can always be described in terms of vectors from an arbitrary origin to each identifiable co-ordinate point contained within its boundary. To rotate the structure the vectors are transformed to new vectors corresponding to the new orientation. A quaternion is covenient for representing a rotation through an angle θ about a specified direction. Let a vector $\boldsymbol{r}$ be rotated through an angle θ about an axis having direction cosines l, m, and n to yield a new vector $\boldsymbol{r}'$. The process may be repeated as often as required.

Several procedures are possible, though that followed here is given by Rooney (21) as it is readily amenable for inclusion within a software program. We obtain $\boldsymbol{r}'$ by evaluating the expression $\boldsymbol{r}' = \boldsymbol{Q}^{-1}\boldsymbol{r}\boldsymbol{Q}$ where $\boldsymbol{Q}$ is the quaternion $\cos(\theta/2) + \sin(\theta/2)\mathbf{n}$ and its inverse $\mathbf{Q}^{-1}$ is given by $\cos(\theta/2) - \sin(\theta/2)\boldsymbol{n}$ and where $\boldsymbol{n}$ is the unit vector with cosines l, m, and n. Evaluation yields:

$$\boldsymbol{r}' = [\cos^2(\theta/2) - \sin^2(\theta/2)]\boldsymbol{r} - (\boldsymbol{n} \times \boldsymbol{r})\sin\theta + (\boldsymbol{n}\cdot\boldsymbol{r})\boldsymbol{n}\sin^2(\theta/2),$$

to yield an expression combining scalar and vector products of $\boldsymbol{n}$ and $\boldsymbol{r}$.

The rotation operation is equivalent to multiplying the vector $\boldsymbol{r}$ by the matrix $\boldsymbol{R}$ given by:

$$\begin{bmatrix} ll(1-\cos\theta)+\cos\theta & ml(1-\cos\theta)+n\sin\theta & nl(1-\cos\theta)-m\sin\theta \\ lm(1-\cos\theta)-n\sin\theta & mm(1-\cos\theta)+\cos\theta & nm(1-\cos\theta)+l\sin\theta \\ ln(1-\cos\theta)+m\sin\theta & mn(1-\cos\theta)-l\sin\theta & nn(1-\cos\theta)+\cos\theta \end{bmatrix}$$

where l, m, and n are the direction cosines of the axis of rotation and θ its angle of rotation.

The next stage consists of finding θ and also l, m, and n for the first of the two shapes for comparison. Having found these values for the first 3D shape, the second structure is rotated to the positions for the first shape for comparison, and differences in the positions of the corresponding co-ordinate points are printed by QUAT.

For numerical evaluation we obtain in orthonormal co-ordinates the relation:

$$(x - x')\boldsymbol{i} + (y - y')\boldsymbol{j} + (z - z')\boldsymbol{k} =$$

$$\tan\theta/2 \begin{bmatrix} \boldsymbol{i} & \boldsymbol{j} & \boldsymbol{k} \\ l & m & n \\ (x+x') & (y+y') & (z+z') \end{bmatrix}.$$

Three linear equations are thereby generated for each of N pairs of related co-ordinate points giving a total of $3N$ of the form:

$$\begin{array}{lll} & mt(z + z') - nt(y + y') & = (x' - x) \\ -lt(Z + Z' & +nt(x + X') & = (y' - y) \\ lt(y + Y' - mt(x + x') & & = (z' - z) \end{array}$$

where $t = \tan\theta/2$. These equations are solved using the standard elementary least-squares procedure.

3. Three-dimensional imaging in electron microscopy

In conventional stereometry two micrographs forming a stereo-pair are analysed, using a mirror stereoscope or other method, such as using a dual channel frame-store system in which right- and left-hand images can be studied directly and on-line reconstruction and analysis performed. Stereoscopic measurements from stereo-pair images were first discussed by Boyde (22) and the first computerized system designed to allow 3D measurements to be extracted from stereo-pair images, taken directly from the electron microscope, soon followed (23).

3.1 Electron microscope stereoscopy

With the scanning electron microscope much of its use has been for subjective 3D imaging since for much of its history accurate quantification has not been possible due to its great depth of focus. For example, in optical microscopy heights can be measured using a graduated focus control, and due to the relatively large numerical aperture (NA) the depth of focus is small and very fine discrimination is possible. The same technique, whilst potentially capable of being implemented in SEM, is much less satisfactory. The SEM beam defining aperture of 200 μm at an operating magnification of ×1000 would have a maximum theoretical expectation of ⩾1 μm. Considerable computer control is required for altitude measurement, since much standard 3D stereology technology is not applicable to scanning electron micrographs.

More effective is the technique of obtaining images of multiple stereo-pairs thick section using the tilting stage of an intermediate voltage electron microscope and reconstruction using on-line modern 3D computer graphics facilities (24). Peachy and Heath (24) have refined the earlier approaches of Boyde and colleagues and have used intermediate voltage (400 kV IVEM 4000-EX, Japan Electron Optics Laboratory), electron microscopy, and their own computer graphics software for thick section reconstructions. The instrument possessed integral film and TV camera facilities capable of signal digitization and interfacing with and IBM PC-XT computer for gray level processing functions. Subsequent analyses were carried out on a VAX 11/750 computer (Digital Equipment Company). In addition, the VAX computer was interfaced with a graphics computer (Model ONE/380, Raster Technologies), both of which were specially modified on-site for stereoscopic display of both images and graphic reconstructions. All displays were viewed by a mirror stereoscope via a VAX-linked terminal and digitizing tablet facility.

A set of micrographs is then obtained at many known speciment tilts, and a stereoscopic, computer-generated 3D vector model is constructed. When all possible views are computerized a solid appearing model is created. Peachy and Heath (24) emphasize that stereological measures of size and quantity were then derivable from the data set for the model display. This theme is developed later in the Section 4, below.

The authors developed 3D reconstructions of both whole cells and organelles such as the T-system and sarcoplasmic reticulum of the white (type II) fibres of rat muscle. Clearly, their system is capable of being extended further, not only for modifications to the intrinsic hard- and software such as the ability to rotate, zoom, and translate, but also with other biological methods involving organelle and antigen specific labelling technology.

3.2 Energy filtered STEM tomography

Biological sections of up to 0.5 μm in thickness containing structures of biological interest have been studied using a dedicated VG-HB501 STEM to which an energy loss spectrometer has been attached and which was also interfaced with high-level computer hardware for digital imaging (25). The value of energy filtered images was clearly demonstrated and subsequently, using a Zeiss EM902 at 80 kV, energy filtered images at specifically controlled electron-volt values were obtained for electron microscopy tomography (26).

Some basic understanding of the principles of energy filtered STEM is desirable for appreciation of both instrumental and interpretation considerations.

Four major categories of electrons (denoted by P_u, P_e, P_i, and P_m) which enter the biological specimens can be differentiated. The subscripts refer to unscattered, elastically scattered, inelastically scattered, and mixed category electrons respectively (27). For a section thickness t and where Λ_e and Λ_i are the elastic and inelastic mean free paths respectively, expressions for P_u, P_e, P_i, and P_m, where $\lambda_t^{-1} = \lambda_e^{-1} + \lambda_i^{-1}$, are given by

$$P_u = \exp\left(-\frac{t}{\Lambda_t}\right)$$

$$P_c = \exp\left(-\frac{t}{\Lambda_i}\right)\left[1 - \exp\left(-\frac{t}{\Lambda_e}\right)\right]$$

$$P_i = \exp\left(-\frac{t}{\Lambda_e}\right)\left[1 - \exp\left(-\frac{t}{\Lambda_i}\right)\right]$$

and

$$P_m = \left[1 - \exp\left(-\frac{t}{\Lambda_e}\right)\right]\left[1 - \exp\left(-\frac{t}{\Lambda_i}\right)\right]$$

The large difference in average angular scattering for elastic and inelastic electron events requires two detectors, an ADF detector measuring P_{adf} denoted by the expression:

$$P_{ADF} = P_e + P_m = \left[1 - \exp\left(-\frac{t}{\Lambda_e}\right]\right.,$$

and a BF detector measuring P_n given by the expression: $P_u + P_i = P_n$

$$P_n = \exp\left(-\frac{t}{\Lambda_e}\right)\frac{1}{n!}\left(\frac{t}{\Lambda_i}\right)^n \exp\left(-\frac{t}{\Lambda_i}\right)$$

where P_n represents the probability of measuring losses after penetration of specimen thickness t. For the definition of a term x given by the relation:

$$x = \frac{t}{\Lambda_i}$$

The contrast in the n image, $C(x)$ is defined as

$$C_n(x) = \frac{\frac{\mathrm{d}P_n(x)}{\mathrm{d}x}}{P_n(x)}$$

and the signal-to-noise ratio $[\mathrm{SNR}_n(x)]$ is given by the variance ratio

$$\mathrm{SNR}_n(x) = \frac{\left(\frac{\mathrm{d}P_n(x)}{\mathrm{d}x}\right)}{P_n(x)} = P_n(x)C_n^2(x).$$

In practice, the inelastic mean free paths Λ_i were estimated for unstained specimens:

Λ_i nucleus = 125 nm
Λ_i cytoplasm = 80 nm
Λ_i chromosome = 95 nm
Λ_i nucleolus = 70 nm
Λ_i Epon^R = 115 nm

Olins, *et al.* (26) specifically recommend the SNR to be a maximum as the first choice for the optimal imaging conditions of a thick biological specimen.

These authors investigated the Balbiani ring 2 on a specific chromosome of *Choronomus tentans*, contained in a standard glutaraldehyde fixed, OsO_4 and uranyl acetate stained 0.5 μm Epon section. In addition, colloidal gold was used to provide controlling points for the registration and alignment of the projected images. Operating parameters of the EM902 included the use of a 30 μm objective aperture, the spectrometer slit-width electron window set at a width of 20 eV, and with a primary accelerating voltage of 80 kV. Fifty-one projections at a magnification of ×6000 were found to be sufficient at 2.5 degree section tilt between images. Further examples of electron microscope tomography can readily be found in the literature (28, 29).

3.3 Computer graphics

D'Arcy Thompson (30) introduced a system of plotting the morphology of various living organisms as co-ordinates in two dimensions. At that time computer facilities were not available that would have allowed him to pursue his original intention of transforming the Cartesian co-ordinates to curvilinear co-ordinates, and hand-drawn 2D constructions had to suffice. The principle remains important today, however, in the morphometric analysis of structure, particularly in subjects such as allometry. Modern computer-graphics systems allow ready transformation from one set of co-ordinates to another, and representation obtained on a visual display unit (VDU) using either 2D or 3D graphics. A particularly interesting and possibly most powerful development potentially, which can be superimposed on computerized 3D reconstruction, is the manner in which the 2D co-ordinate plots can be subjected to subsequent manipulative procedures (31). Such procedures, for example using the well-known Gompertzian growth equation, can be used to mimic growth or other structure-deforming processes superimposed on the 3D reconstruction.

In common with other subjects, many of the techniques relevant to the analysis of optical micrographs may often be applicable to electron micrographs, and the application of computer graphical techniques to modify shapes or to obtain procrustean-type analysis is an extremely useful by-product.

4. Three-dimensional reconstruction by stereology

Stereology is a branch of morphometry which, for microscopists, usually consists of obtaining information on the 3D nature of test objects using only projected 2D images of sections. An essential principle of stereology is the dimensional reduction in the preparation of 2D test images from 3D structures. Measurements are then carried out on 2D images and 3D data acquired. The dimensional alteration from 2D measurements to 3D data is also an essential feature of stereology. In addition to the sectioning techniques, other dimensional techniques include projection stereology (32). Numerous standard textbooks are available on morphometry and stereology (1, 32–35).

Biological investigations have, until relatively recently, been non-numerate.

The basis of modern practice in morphometry and stereology is that of geometrical probability. An understanding of the principles of stereological relations and their derivation from the basics of geometrical probability is vital.

Many types of measure of size or quantity can be estimated using stereological techniques. For example, the volume fraction (V_V) and numerical density (N_V) of organelles such as nuclei and mitochondria; unfolding of the size distribution of populations of regular particles from 2D section profiles; membrane surface densities (S_V), the lengths of capillaries per unit volume of tissue (L_V), and quantitative estimates of their degree of orientation (Ω and K). Standard textbooks on morphometry list all the commonly used stereological symbols.

Practical measurements are either carried out on the 2D section images by using a randomly placed overlying transparent test screen bearing relevant patterns of points or test lines (probes) or by the use of computer interfaced digitizing tablets.

An essential practical point is that often, the application of transparent test screens are superior, due to their ready availability, their simplicity, and cheapness. Their use permits much greater flexibility of analysis in that new screens for new investigations can be constructed in only minutes by drawing new test grid patterns and xeroxing them on to plastic sheets.

The large amount of data collected by expensive instruments, for example, the Quantimet 720, is often entirely unnecessary in statistical sampling theory, tends to give spurious confidence to the numerical values of the excessively large samples obtained, and often cannot even differentiate between the images of two apposed objects from a single larger one.

Stereological data encompasses many different categories of information. Simple values could include either measuring the proportion of test points in a test lattice which fall upon the profiles of interest (P_P) or counting the number of intersections per unit length of line that test lines make with profile boundaries (I_L). The numerical values are then used to evaluate measures of size using expressions of the type

$$P_P = A_A = V_V$$
$$S_V = 2I_L.$$

Derivatives of these relations are to be found in several standard textbooks of stereology.

An elegant and concise way of describing the stereological principles is to use the standard notation of set theory. Consider a 3D specimen Y intersected by a section T. Part of the specimen is contained in section T as the intersection of Y with T and is represented symbolically by the relation

$$\mathrm{Y} \cap \mathrm{T}$$

where $\cap$ is the symbol for '. . . the intersection of . . .'. If the specimen Y contains a test structure, X, of particular interest, then a profile of X may appear in T. In this case the intersection of X and T is denoted by the relation

$$\mathrm{X} \cap \mathrm{T}.$$

Inferior number notation (subscript) is used to denote the dimensions of test structures and probes (points, lines, and sections) so that Y_3 is a 3D specimen containing X_3 and T_2 is the 2D test section. The area (A) of specimen intersection between Y_3 and T_2 is given by

$$A(\mathrm{Y}_3 \cap \mathrm{T}_2)$$

while the area of intersection of X_3 and T_2 can be written as

$$A(\mathrm{X}_3 \cap \mathrm{T}_2).$$

Clearly, the relation of the two intersection areas is an areal fraction, A_A, which is expressed by the relation

$$A_A = \frac{A(X_3 \cap T_2)}{A(Y_3 \cap T_2)}.$$

However, there is a finite probability that the probe T does not intersect with X_3, with the consequence that an empty set, $\emptyset$, is generated. This is symbolized by the relation

$$T \cap X = \emptyset.$$

The probability that T hits the specimen Y or its contained test structure X is expressed by the symbol $\uparrow$, and is the probability that a set, $\emptyset$, is *not* generated. Hence the probability of T intersecting the test structure X is given by

$$Pr\{T \uparrow X\}.$$

More complicated statements such as

$$Pr\{T \uparrow X | T \uparrow Y\}$$

can also be constructed using the standard conditional statistical conditional symbol '|' for conditional probabilities. This statement can be read as 'the probability of the probe T hitting the test structure X given that T hits the specimen Y'. Additional symbols include T_3 for a thick tissue section, T_2 for a section plane, T_1 for a unidimensional line in a plane, and T_0, a lattice of dimensionless points. Clearly, various types of intersection between T and X can be expressed according to experimental conditions; for example

$$X_2 \cap T_2 = R_1$$

expresses the intersection of a 2D plane X_2 with a 2D section which results (R) in a one-dimensional line, R_1. The expression

$$X_1 \cap T_2 = R_0$$

denotes a one-dimensional line structure, X_1 (such as a capillary network), intersecting a 2D section, T_2, resulting in dimensionless points, R_0. An areal fraction, A_A, of a test structure, X_3, seen in a section, T_2, of a whole specimen, Y_3, is expressed by

$$A_A = \frac{A(X_3 \cap T_2)}{A(Y_3 \cap T_2)}.$$

Using similar notation for the probabilities of intersection using $\uparrow$, the probability that a test structure X_3 is hit by dimensionless points, T_0, given that the specimen Y_3 intersects with the section, T_2, is given by

$$Pr\{T_0 \uparrow X_3 | T_0 \uparrow Y_3\} = \frac{|A|_X}{|A|_Y} = V_V,$$

which is the volume fraction of X in reference volume Y. Clearly, $|A|_X$ and $|A|_Y$ are the absolute areal values obtained from $A(X_3 \cap T_2)$ and $A(Y_3 \cap T_2)$, respectively, measured in sections.

For the investigator considering using analysing the 3D nature of biological structures (36, 37) the review articles on the practical aspects of stereology by Gundersen *et al.* are highly recommended. The latest stereological techniques are clearly described and include the disector, nucleator, selector, and fractionator.

4.1 Stereology on thick sections

Whilst most stereology is carried out on sections assumed to have zero thickness, the theory of obtaining stereological data from measurements obtained from projected images of thick sections has long been available (32, 38). Given that specimen preparation conditions are adequate and that the basic assumptions of the method are not violated, two types of technique are possible. As an example of the first, single sections of known thickness, t, are prepared and areal fraction and boundary density counts could be obtained from projected images. In the second approach, due to the pioneer work of Miles (38), two sections of differing but *unknown* thickness are used and additional tangent count measurements are required.

Area fraction and boundary density measurements, denoted by A'_A and B'_A respectively are made on the projected images or micrographs for both methods, whilst surface tangent counts per unit area T'_A made at test profile boundaries by sweeping a test line across the images in a constant orientation are also needed for the technique of Miles.

For single-section thickness, t, some simple relations for estimating line lengths (L), surfaces (S), surface densities (S_V), volume fractions (V_V), and numerical densities (N_V) in 3D space can readily be obtained from the various measurements L', A', and N'_A on projected images. The notation (′) denotes measurements on projected images. For line lengths the following derivations are obtained:

$$L = 4/\pi . L'; L_V = \frac{4/\pi . L'_A}{t}.$$

For surfaces we obtain the well-known Cauchy result for convex surfaces without re-entrant (invaginating) surfaces $S = 4A'$ whilst for re-entrant convex surfaces the inequality $S \geqslant 4A'$ is applicable. For surface densities we obtain the corresponding relations for non re-entrant and re-entrant surfaces

$$S_V = \frac{4A'_A}{t}; \; S_V \geq \frac{2A'_A}{t}.$$

In order for the above equality to hold, the additional conditions of no truncation and no test object overlapping must be met. For numerical densities,

$N_V = N'_A/t$, but for volume fractions, the measure of size most commonly and readily used in thin-section studies, the estimating equation is more complex and requires independent measurements of the mean random intercept length (l_3) unless the test-object shape can be specified accurately, e.g. spheres. The estimation equation is given by

$$V_V = \frac{A'_A L_3}{t}.$$

The mean random intercept length is related to test volume and surface by the well-known equation of Tomkieff

$$L_3 = 4\frac{V}{S},$$

from which the volume fraction can be derived by substitution.

The alternative approach of Miles (38) of obtaining estimates of V_V and S_V from two differing but *unknown* thicknesses requires a completely different set of estimating equations and also an additional surface tangent density count (T'_A) on the projected image, although measurements of projected areal fractions A'_A and projected profile boundary densities B'_A are made as for the previous technique. The estimating equations are given by

$$V_V = 1 - e^{-u};\ S_V = 4 v a_{21} e^{-u}$$

where values for u and v are obtained from the relations

$$u = \frac{\{a_{21}(c_1 a_2 - c_2 a_1) - b_{21}(a_1 b_2 - a_2 b_1)\}}{b_{21}^2 - a_{21} c_{21}}$$

and

$$v = \frac{c_1 b_2 - c_2 b_1}{b_{21}^2 - a_{21} c_{21}}.$$

Values for a_i, b_i, and c_i ($i = 1, 2$) are obtained from the two unknown section thicknesses t_1 and t_2 and are calculated from

$$a_i = -\ln(1 - A'_A)$$

$$b_i = \frac{B'_A}{\pi(1 - A'_A)}$$

$$c_i = \frac{T' - A}{2\pi(1 - A'A)} + \frac{\pi b_i}{4}$$

5. Reconstruction adequacy

In statistical experimental design, samples must be representative of the parent population from which the sample is drawn. In addition, an assumption vital to all experimental investigations is that specimens must be identically and independently obtained. These assumptions must never be violated and are inherent in all procedures including those procedures which are misleadingly called 'assumption-free'. No procedure can possibly be assumption-free. In practice, the investigator has to choose between taking great care over reconstructing a few specimens or taking less effort on each specimen but studying a larger number to allow for biological variation. How does the investigator choose? As a general principle, it is important to maintain as large a number of animals and structures as possible. Surprisingly, the number of sections points per test structure, beyond a basic minimum, has relatively little effect on the reconstruction and estimating procedure compared with the numbers of the animals or structures used. This can be illustrated by a general variance, $[V(x)]$, relation used in experimental design:

$$V(x) = \frac{S_1^2}{n_1} + \frac{S_2^2}{n_1 \cdot n_2} + \ldots . + \frac{S_4^2}{n_1 \cdot n_2 \cdot n_3 \cdot n_4},$$

where S_i^2 is the variance of values obtained from animals, test structures for 3D analysis and section planes respectively, and n_i the specific numbers of test items in each category i. The most variable component is normally the variation between animals (n_1), less so, the number of reconstructed objects (n_2) and least of all the number of sections (n_3). For stereology measurements subsequently, the number of digitized points per section (n_4) is even less important.

References

1. Elias, H. and Hyde, D. M. (1983). *A guide to practical stereology*. S. Karger, Basel, Switzerland.
2. Gaunt, P. N. and Gaunt, W. A. (1978). *Three dimensional reconstructions in biology*. Pitman Publishing, Tunbridge Wells, Kent, UK.
3. Motta, P. M. (1989). *Cells and tissues, A three-dimensional approach by modern techniques in microscopy*. In *Progress in clinical and biological research*, Vol. 295, pp. 1–650. Alan R. Liss, New York.
4. Van der Voort, H. T. M., Brakenhof, G. J., and Janssen, G. C. A. M. (1988). *Optik,* **78,** 48.
5. Dover, S. D., Elliot, J. C., Boakes, R., and Bowen, D. K. (1989). *Journal of Microscopy,* **153,** 187.
6. Huijsmans, D. P., Lamers, W. H., Los, J. A., and Strackee, J. (1986). *Anatomical Record,* **216,** 449.
7. Sjöstrand, F. S. (1958). *Journal of Ultrastructure Research*, **2,** 122.
8. Sjöstrand, F. S. (1974). *Journal of Ultrastructure Research,* **49,** 60.
9. Pellegrini, M. (1980). *Journal of Cell Science,* **43,** 137.

10. Perkins, W. J. and Green, R. J. (1982). *Journal of Biomedical Engineering,* **4,** 37.
11. Nierzwicki-Bauer, S. A., Balkwill, D. L. and Stephens, S. E. Jr. (1983). *Journal of Ultrastructure Research,* **84,** 73.
12. Veen, A. and Peachey, L. D. (1977). *Computer Graphics,* **2,** 135.
13. Lockhausen, J., Kirsten, U., Menhardt, W. and Dallas, W. J. (1990). *Journal of Microscopy,* **158,** 197.
14. Menhardt, W., Lockhausen, J., Dallas, W. J., and Kristen, Y. (1986). *Micron Microscopica Acta,* **17,** 349.
15. Newman, W. and Sproull, R. (1978). *Principles of interactive computer graphics,* 2nd edn. McGraw-Hill, New York.
16. Wharton, D. A. and Murray, D. S. (1989). *Journal of Microscopy,* **154,** 157.
17. Yaegashi, H., Takahashi, T., and Kawasaki, M. (1987). *Journal of Microscopy,* **146,** 55.
18. Huijsmans, D. P. (1989). *European Journal of Cell Biology,* **48,** Suppl. 25, 57.
19. Kemdall, D. G. (1989). *Statistics in Science,* **4,** 87.
20. Barrett, A. N. and Mackay, A. L. (1987). *Spatial Structure and the microcomputer: Selected mathematical techniques*, Macmillan, London.
21. Rooney, J. (1977). *Environmental Planning,* **B4,** 185.
22. Boyde, A. (1975). *Photogrammetric Record,* **8,** 424.
23. Holburn, D. M. (1979). Computer aided techniques for topographical analysis in the SEM. Ph.D. thesis, University of Cambridge.
24. Peachey, L. D. and Heath, J. P. (1989). *Journal of Microscopy,* **153,** 193.
25. Colliex, C., Mory, C., Olins, A. L., Olins, D. E. and Tencé, (1989). *Journal of Microscopy,* **153,** 1.
26. Olins, D. E., Olins, A. L., Levy, H., Durfee, R. C., Margle, S. M., Tinnel, E. P., and Durfee, R. C. (1989). *Journal of Microscopy,* **154,** 257.
27. Colliex, C. and Mory, C. (1984). In *Quantitative electron microscopy* (ed. J. N. Chapman and A. J. Craven), pp. 149–216, 25th Scottish University Summer School in Physics.
28. Olins, A. L., Olins, D. E., Levy, H. A., Durfee, R. C., Margle, S. M. and Tinnel, E. P. (1986). *European Journal of Cell Biology,* **40,** 105.
29. McEwan, B. F., Radermacher, M., Rieder, C. L., and Frank, J. (1986). *Proceedings of the National Academy of Sciences, USA,* **83,** 9049.
30. D'Arcy Thompson, W. (1917). *On growth and form.* Cambridge University Press, Cambridge.
31. Fujita, S. (1990). *Journal of Microscopy,* **157,** 259.
32. Underwood, E. E. (1970). *Quantitative serology.* Addison-Wesley, Reading, Massachussetts.
33. DeHoff, R. T. and Rhjines, F. N. (1968). *Quantitative microscopy*. McGraw-Hill, New York.
34. Weibel, E. (1979). *Practical stereology*, Vol. 1. Academic Press, London.
35. Weibel, E. (1980). *Practical stereology,* Vol. 2. Academic Press, London.
36. Gundersen, H. J. G., Bendtsen, T. F., Koprbo, L., Marcussen, N., Moller, A., Nielsen, K., Nyengaard, J. R., Pakkenberg, B., Sorenson, F. B., Vesterby, A., and West, M. J. (1988). *Acta Pathologica, Microbiologica et Immunologica, Scandinavica,* **96,** 379.
37. Gundersen, H. J. G., Bagger, P., Bendtsen, T. F., Evans, S. M., Korbo, L.,

Marcussen, A., Moller, A., Nielsen, K., Nyengaard, J. R., Pakkenberg, B., Sorensen, F. B., Vesterby, A., and West, M. J. (1988). *Acta Pathologica, Microbiologica et Immunologica, Scandinavica,* **96,** 857.

38. Miles, R. E. (1976). In *Proceedings of the Fourth International Congress for Stereology*, National Bureau of Standards Special Publication 431, pp. 3–12. US Government Printing Office, Washington, DC.

Appendix

Principal Suppliers of EM Specimen Preparation Materials and EM Accessories.

Agar Scientific Ltd, 66a Cambridge Road, Stanstead, Essex CM24 8DA, UK
  Agent: Ted. Pella Inc., 4595 Mountain Lakes Boulevard, Redding, CA 96003, USA
  Agent: W. Plannet GmbH, Elektronmikroskopie, Marburger Strasse 90, D-6550 Marburg 7, Germany

Bachem, Bubendorf, Switzerland

Balzers-BAL-TEC GmbH, im Grohenstuck 2, Walluf, Germany

Balzers High Vacuum Ltd, Northbridge Road, Berkhamsted, Herts, HP4 EN, UK

Balzers-BAL-TEC Products Inc., 984 Southford Road, P.O. Box 1221, Middleburry, CT 06762, USA

Bio-Rad, Microscience Division, Bio-Rad House, Maylands Avenue, Hemel Hempstead, Hertfordshire, HP2 7TD, UK

Bio-Rad Laboratories Inc., 19 Blackstone Street, Cambridge, MA 02139, USA

Bio-Rad Laboratories GmbH, Heidenmannstrasse 164, D-8000 München 45, Germany

Denton Vacuum, Cherry Hill, NJ, USA

Digital Equipment Co., Maynard, MA, USA

EDAX International Inc., 85 McKee Drive, Mahwah, NJ 07430, USA
  Agent: Philips Scientific Analytical Division, York Street, Cambridge CB1 2PX, UK
  Agent: Philips GmbH, EWI, Miramstrasse 87, 3500 Kassel, Germany

Electron Microscopy Sciences, 321 Morris Road, Box 251, Fort Washington, PA 19034, USA

SCI Science Services, Bad Strasse 7, 8000 München 70, Germany

Emscope Laboratories Ltd, Kingsnorth Technology Park, Wotton Road, Ashford, Kent TN23 2LW, UK
  Agent: Photometrics Inc., 4 Arrow Drive, Woburn, MA 01801, USA
  Agent: Helmut Schröder, Gauss Strasse 30/2, 7000 Stuttgart 1, Germany

Enzyme Systems Products USA, Dublin, MA, USA

Ernest F. Fullham Inc. 900 Albany Shaker Road, Latham NY 12110, USA
  Agent: Agar Scientific Ltd. 66a Cambridge Road, Stanstead, Essex CM24 8DA, UK
  Agent: Touzart et Matignon, 8 rue Eugene Henaff, Boite Postale 52, 94403 Vitry sur Seine Cedex, France

Gatan Ltd, 28–38 Desborough Street, High Wycombe, HP11 2NF, UK
Gatan Inc., 6678 Owens Drive, Pleasanton, CA 94566, USA
Gatan GmbH, Ingolstrasse 40, 8000 München 45, Germany
Japan Electron Optics Laboratory, Tokyo
Ladd Research Industries Inc., PO Box 1005, Burlington, Vermont, 05402, USA
Agent: B.P. 1 Conflans-Sur-Lanterne, 70800 Saint-Loup-Sur-Semouse, France
Leica UK Ltd, Davy Avenue, Knowlhill, Milton Keynes, MK5 8LB. Reichert Jung GmbH, Heidelberger Strasse 17–19, 6907c Nussloch, Germany
Oxford Instruments Ltd, Eynsham, Oxford OX8 1TL, UK
Agent: Hans-Georg Pankau (EM), Oxford Intruments GmbH, Kreuzberger Ring 38, Postfach 4509, D-6200 Wiesbaden, Germany
Agent: Glenn Kinnear (EM) Oxford Instruments North America, Alfred Circle, Bedford, MA D1730, USA
Polaron Equipment Ltd, 21 Greenhill Crescent, Holywell Industrial Estate, Watford, Hertfordshire, WD1 8XG, UK
Polaron Instruments Inc., 2293, Amber Drive, Line Lexington Industrial Park, Hatfield, PA 19440, USA
Agent: Mikrotechnik, Dr Hert GmbH, Kunigundenstrasse 66, 8000 München 40, Germany
Polysciences Ltd, 24 Low Farm Place, Moulton Park, Northampton NN3 1HY, UK
Polysciences Inc, 400 Valley Road, Warrington, PA 18976, USA
Polysciences GmbH, Ulmenhof 28, 5401 St Goar, Germany
Raster Technologies, Weston, MA, USA
Sloan Technology Corp., Santa Barbara, CA, USA
Taab Laboratories Ltd, 3, Minerva House, Calleva Industrial Park, Aldermaston, Berkshire RG7 4QW, UK
Technical Products International, St. Louis, MI, USA
Pelco International, P.O. Box 492477, Redding, CA 96049–2477, USA
Agent: Agar Scientific Ltd, 66a Cambridge Road, Stanstead, Essex, CM24 8DA, UK
Agent: Plano W. Plannet GmbH, Marburger Strasse 90, D-3550 Marburg 7, Germany

Index

Index

Index

Index